Fractional Operators with Constant and Variable Order with Application to Geo-Hydrology

Fractional Operators with Constant and Variable Order with Application to Geo-Hydrology

Abdon Atangana

Institute for Groundwater Studies,
Faculty of Natural and Agricultural Sciences,
University of Free State,
Bloemfontein

ACADEMIC PRESS
An imprint of Elsevier

Library of Congress Cataloging-in-Publication Data
A catalog record for this book is available from the Library of Congress

British Library Cataloguing-in-Publication Data
A catalogue record for this book is available from the British Library

ISBN: 978-0-12-809670-3

For information on all Academic Press publications
visit our website at https://www.elsevier.com/books-and-journals

Working together
to grow libraries in
developing countries

www.elsevier.com • www.bookaid.org

Publisher: Candice Janco
Senior Acquisitions Editor: Graham Nisbet
Editorial Project Manager: Susan Ikeda
Production Project Manager: Poulouse Joseph
Designer: Christian J. Bilbow

Typeset by VTeX

I dedicate this book to the memories of my Lord Jesus Christ who died on the cross for our sins. To the memory of my late mother Ngono Antoinette who died in 2010; my grandmother who passed away in 2016.

Contents

Preface

The Lord God so loved the world and he gave it only begotten subject applied mathematics, such that whosoever learn it should not be ignorant, but use it, to understand the nature and also predict some natural phenomena accurately. And the Lord God said "My people are destroyed for lack of knowledge." Well, applied mathematics is the gate and key of sciences, therefore negligence of applied mathematics injures the whole corpus of knowledge, since he who is ignorant of it cannot fully grasp other sciences or things of this world. As an example, how do we explain the movement of groundwater through the geological formation? This can only be achieved via mathematical equations.

Acknowledgment

The fear of the Lord is the foundation of wisdom, and the knowledge of the holy one is understanding. I will first and foremost like to express my deepest thanks to the Lord Jesus, the King of kings, the Lord of lords, who died on the cross of Golgotha to save mankind from sins. I would thank God for his protection, heath, and wisdom. In the process of putting this book together, I become aware of how true this gift of writing is for me. The Lord God gave me the power to believe in my passion and peruse my dreams. I could not have completed this second book without faith I have in God the almighty.

I am indebted to all my students:

- To Miss Jessica Spannenberg my Ph.D. student for her assistance in proofreading.
- To my Postdoctoral fellow Dr. Kolade Owolabi M. for his help in latex and proofreading.
- To my Turkish Friend Dr. Ilknur Koca Hocam for her help with latex.
- An excellent wife is the crown of her husband, who can find? For her worth is far above jewels. Thus whosoever find a good wife finds a good thing, and obtains favor of the Lord, that is why the Lord God, after overseeing the time-fractional differentiation of Abdon Atangana said, "It is time-locally but not time-fractionally good for Abdon Atangana to be alone, I will make him a helper fit for him." And the rib that the Lord God had taken from Abdon Atangana, he made into a beautiful, caring, loving, wise, intelligent, God-fearing woman (Ernestine Alabaraoye) and brought her to Abdon Atangana. Then Abdon Atangana said, "This at last is bone of my bones, a controller of my fractal and chaotic behaviors, flesh of my flesh, blood of my blood, love of my heard, my fractional derivative in Atangana–Baleanu sense, she will be called Ernestine Atangana, because she was taken out of me. Because of this, Abdon Atangana shall leave his father and mother and hold fast to his God-fearing wife and they shall become one flesh." From my soul I would like deeply to thank my wife for her love, good food, prayers, and constant support.
- Thanks to my two-year-old son Atangana Ndzengue Biliga Melchizedek who always helps me to wake up at night in order to complete this book.
- To my family: Eleme Catherine, Ngono Adelaide, Eyebe Izidore, Ngono Antoinette, and all my nephews and nieces for their prayers.
- To my fathers: Dr. Pierre Ndzengue, the ambassador of Cameroon to Japan, and Tara Noah Jean for their prayers and support.

CHAPTER 1

Aquifers and Their Properties

1.1 Introduction

An aquifer is a geological formation in which groundwater flows through with ease. Aquifers should therefore have both permeability and porosity. Examples of these geological formations which form aquifers include sandstone, conglomerate, fractured limestone, and unconsolidated sand and gravel formations. Another example of an aquifer system is a fractured volcanic rock formation such as columnar basalt. The rubble zone where volcanic flow exists is usually both porous and permeable, thus allowing for good aquifer systems [1,2]. Furthermore, geological formations such as granite and schist have low porosity, and so are usually classified as poor aquifers. However, once they become fractured, they can produce good aquifers [1,2]. Furthermore, for a well to be productive, the well itself should be drilled into the ground to penetrate the aquifer. If groundwater is abstracted from the well at a rate faster than the aquifer is recovered, there is a decline in the water table, sometimes to a point that the well dries. As pumping occurs, the water table normally declines, resulting in a cone of depression at the well. Moreover, groundwater flow generally follows the slope of the water table, thus in this case, groundwater flows in the direction of the well being pumped. Not all aquifers can be seen as groundwater reservoirs [1–3]. These reservoirs are, however, found underground but only in cavernous geological formations where the formations surrounding fractures or cracks have undergone dissolution, forming open channels which allow rapid water movement similar to that of a river. Furthermore, since groundwater migrates at a slow rate through pore spaces of aquifer material, the only living organisms, that could float as it would in an actual river, are bacteria or viruses which are minute enough to migrate through pore spaces. Movement of groundwater through an aquifer occurs as groundwater is forced through a pore space of geological formations. Hence, porosity essentially defines an aquifer. To add, porosity of certain aquifers also allows them to act as good filters generating natural purification [2,3]. Since effort is required for forcing water movement through small pores, there is a loss in groundwater energy as it flows. This eventually results in decreased hydraulic head in the direction of groundwater flow. On the other hand, when pores are large in size, there is increased permeability, less energy loss, and rapid groundwater movement. Subsequently, groundwater migration is rapid for aquifers with large pores, such as in the case of the lower Portneuf River aquifer or in cases where porosity is a result of fractures which are interconnected. It is also significantly rapid in fractured rock aquifers such as the basalts of the eastern Snake River Plain. Despite being good aquifers, they are vulnerable to spreading of contamination which is challenging and often impossible to prevent [3].

Fractional Operators with Constant and Variable Order
with Application to Geo-Hydrology
DOI: 10.1016/B978-0-12-809670-3.00001-1

1.2 Classification of Aquifers

Hydrogeology is known as the field investigating the flow of water through aquifers. Terms related to aquifers include aquitard, which is a formation associated with low permeability along an aquifer; and aquiclude or aquifuge, which are associated with an impermeable formation existing above or below an aquifer [4–7]. In the case where the impermeable formation exists above the aquifer, the pressure created could cause the aquifer system to become confined. Furthermore, there exist two types of aquifers, namely unconfined and confined aquifers. Additionally, there are also semi-confined aquifers. Unconfined aquifers are also known as water table aquifers or phreatic aquifers, and this is merely due to their upper boundary being the water table or phreatic surface rather than a confining layer. In most locations of the world, the shallower aquifer is defined unconfined. This signifies the absence of a confining layer such as in the case of an aquitard or aquiclude found between the aquifer itself and the surface above it [4–7]. There also exist what are known as "perched" aquifers. These are aquifers defined by the accumulation of water above a formation of low permeability, such as a clay layer. Perched aquifers are usually associated with a local area of groundwater found at an elevation greater than a regional aquifer. Also, they differ from unconfined aquifers in size as they are smaller. On the other hand, confined aquifers are aquifers having a confining layer such as clay above it. Confining layers can be associated with allowing protection to the aquifer from surface contaminants. Furthermore, if one cannot distinctively distinguish between unconfined and confined aquifers such as in the case where there is no knowledge on whether a confining layer exists or when in a complex formation such as a fractured rock formation, an aquifer test can be used for estimation of a storativity value which is essential for distinguishing between aquifer types. It is, however, noteworthy that interpretation for aquifer tests done in unconfined aquifers is to be done differently to aquifer tests conducted in confined aquifers. Values of storativity can be used for distinguishing between aquifer systems as they differ for the two types of aquifers. To expand, confined aquifers are associated with having significantly low storativity ranging from 0.01 to as low as 10^{-5}. This means confined aquifers store water by means of expansion of the aquifer matrix and compression of water, which are both in general significantly small in quantity. Alternatively, the storativity in unconfined aquifers is known as specific yield; and these values are larger than 0.01, which on the other hand can be inferred as 1% of a bulk volume. Unconfined aquifers release water from storage by means of drainage of the aquifer material's pores, and since it drains, large quantities of water are released. This can be to the point at which an aquifer's drainable porosity or minimum volumetric water content is reached [4–6]. Aquifers can also be subdivided into isotropic and anisotropic aquifers, whereby the former is aquifer layers having the same hydraulic conductivity (K) in all flow directions and the latter is aquifer layers having different K values. Anisotropic aquifers can be found when there is variation in both horizontal (K_h) and vertical (K_v) conductivity values. Moreover, anisotropic aquifers can include a semi-confined aquifer with one or more aquitards, even in the case of the layers each being isotropic because in the end the K_h and K_v values would still differ. This can also be associated with hydraulic transmissivity and hydraulic resistance. Furthermore, in the event of calculating flow to drains and wells penetrating aquifer systems, the anisotropy should be accounted for in case there may be fault associated with the design of the drainage system. As indicated, there also exist aquitards and aquicludes. Aquitards are merely formations restricting groundwater flow from one aquifer to the next, thus they generally are defined by clay or non-porous rock having low hydraulic conductivity. These formations become aquicludes once the formation is completely impermeable.

1.3 Example of Some Aquifers in the World

The following are a number of aquifers from around the world:

- **The Great Artesian Basin** is located in Australia and is argued to be the biggest aquifer in the world. It has an extent of more than 1.7 million km^2, and has a fundamental role of water supply for Queensland and isolated locations of South Australia.
- **The Guarani Aquifer** is located below the surface of Argentina, Brazil, Paraguay, and Uruguay. Its name originates from the Guarani people, and it is also one of the world's biggest aquifers. The aquifer provides an essential freshwater resource and covers an area of 1,200,000 km^2. Moreover, it has a volume of approximately 40,000 km^3, a thickness ranging between 50 m and 800 m, and its maximum depth is more or less 1800 m.
- **The Ogallala Aquifer** is another large of aquifer system of the world, whose location is within the United States. It lies beneath the surface of eight different states. Although one of the largest aquifers, some parts of the aquifer are undergoing rapid depletion due to expanding municipal and agricultural activity. The aquifer is comprised mainly of fossil water originating from the previous glaciation. The average annual groundwater recharge for the more arid areas of the aquifer receives approximately 10% of the annual abstractions.
- **The Edwards Aquifer** is found on the eastern parts of the Edwards Plateau in Texas, United States. It discharges at a rate of about 900,000 acre feet (1.1 km^3), and is therefore one of the world's most productive artesian aquifers, providing water to about two million people. To add, this aquifer is essential to many different unique and endangered species.
- **The Arkell Spring Grounds** is found in southwestern Ontario, Canada, is a highly productive bedrock aquifer, yielding several freshwater springs along the Eramosa River situated in the northeast of the village of Arkell. In 1903, the City of Guelph started using these springs for municipal use, and then during 1963 and 1967 four additional boreholes were formed. In addition, this source of water primarily goes to the City of Guelph.

- **The Laurentian River** is an old river located in southern Ontario, Canada. The river system formed prior to the recent ice ages, and glacial debris filled this ancient river valley. Furthermore, the water still migrates down this river valley, and then eventually underground. To add, the source of this water is assumed to be located near Georgian Bay which is about 200 kilometers (120 mi) away.
- **The Oak Ridges Moraine** is also found within Ontario, Canada, but more specifically in the Mixed-wood Plains of south central Ontario, between Caledon and Rice Lake, near Peterborough. It is an ecologically essential geological landform which covers an area of about 1900 km^2 (730 sq mi). Moraine is a highly significant landform which obtained its name from the rolling hills and river valleys which extend about 160 km (99 mi) from the Niagara Escarpment toward the east, to Rice Lake. This landform was created more or less 12,000 years ago through glaciers that were advancing and retreating. Additionally, Moraine is presently of utmost importance because it is essential to large-scale urban development.
- **The Biscayne Aquifer** is located beneath parts of South Florida and was given its name after Biscayne Bay. The areas under which this aquifer is located includes Broward County, Miami-Dade County, Monroe County, and Palm Beach County; and these cover a total area of approximately 4000 square miles (10,000 km^2). To add, it is a shallow aquifer existing within a layer of significantly permeable limestone.
- **The Snake River Aquifer** is a large groundwater system found beneath the Snake River Plain which is located in the southern part of Idaho, United States. This aquifer system's water is sourced from rain and melting snow which migrate onto the Plain from the Snake River, Big Lost River, Bruneau River, as well as additional watercourses found in southern Idaho. Furthermore, the aquifer extends from east to west at a distance of about 400 miles (640 km). It is an essential water source for agricultural irrigation activity occurring in the Plain. Moreover, it is an aquifer subdivided into two aquifers due to the Salmon Falls Creek, namely the Eastern Snake River Plain Aquifer and Western Snake River Plain Aquifer.
- **The Floridan aquifer** forms part of a major artesian aquifer which extends into Florida. The aquifer lies beneath all of Florida and also large portions of coastal Georgia and regions within coastal Alabama and South Carolina. In essence, it exists beneath the coastal areas of south east United States. In addition, the Floridan aquifer is comprised of carbonate lithology and is also one of the world's most productive aquifers.
- **The Mahomet Aquifer** is an aquifer of primary importance in the east central regions of Illinois. It is comprised of sand and gravel, and forms part of the buried Mahomet Bedrock Valley. The Mahomet Aquifer with a thickness ranging from 50 to 200 feet (15 to 60 m) lies beneath 15 counties. Additionally, the aquifer provides more than

100,000,000 US gallons (380,000 m^3) of groundwater per day for public water use, as well as industrial and irrigational use.

- **The Ogallala Aquifer** was given its name in 1898 by a geologist, N.H. Darton, due to being located near a town of Ogallala in Nebraska. It is one of the world's largest shallow unconfined aquifers underlying an area of about 174,000 sq mi (450,000 km^2) of the Great Plains of the United States. In essence, the aquifer lies beneath parts of eight different states: South Dakota, Wyoming, Nebraska, Kansas, Colorado, Oklahoma, Texas, and New Mexico.
- **The Permian Basin** is a sedimentary basin primarily located in the western parts of Texas and south eastern parts of New Mexico, in the United States. The aquifer extends from south of Midland and Odessa toward the west, and then into the southeast portion of the neighboring New Mexico state.
- **The San Diego Formation** is a geological formation comprised of coastal transitional marine and non-marine pebble, cobble conglomerate deposits, marine sandstone material, and the Pliocene marine fossils, from a bay that was previously deposited during the Middle to Late Pliocene of the Cenozoic era (2 to 3 million years ago). This geological formation covers an area from the southern end of Mount Soledad located in San Diego County to Rosarito Beach in northern Baja California. This thus includes Mexico and Tijuana as well as the southwest end of San Diego County, from San Ysidro to Pacific Beach.
- **The San Joaquin River** is the biggest river in central California, United States. It extends up to a length of 366 miles (589 km), whereby the source is in the high regions of Sierra Nevada. The river migrates through agricultural areas of the northern San Joaquin Valley to where it discharges in Suisun Bay, San Francisco Bay, and the Pacific Ocean. The San Joaquin River is essential to irrigation and also acts as a wildlife corridor. To add, the river is one of the significantly dammed and diverted rivers of California.
- **The Sankoty aquifer** is located in Illinois, United States. It supplies groundwater to northwest and central communities of Illinois. Furthermore, it is an unconsolidated deposit found in a bedrock valley which was previously associated with the ancestral Mississippi River.
- **The Amazon basin** is a basin which constitutes the region of South America undergoing drainage from the Amazon River along with its tributaries. The basin has an area of approximately 6,915,000 km^2 (2,670,000 sq mi), or about 40% of the continent of South America. Moreover, the Amazon basin exists within a few different countries of South America, and which are Bolivia, Brazil, Colombia, Ecuador, Guyana, Peru, Suriname, and Ven.
- **The Hamza River** is the unofficial name given to an aquifer which appears to flow slowly through Brazil. The river is about 6000 km (3700 mi) in length. This water body was established in 2011, and its unofficial name is associated with named Valiya Mannathal

Hamza, a scientist of Brazil's National Observator, who conducted four decades of research in the area.

- **The Canning Basin** is a basin found in Western Australia, covering an area of about (506,000 km^2), of which more or less 430,000 km^2 is located on land. The basin is known for having abundant oil and gas, and thus has undergone extensive research. Moreover, in June 2003, 250 wells were drilled and there was 78,000 km of seismic shot.

- **The Jandakot Mound, or Jandakot Groundwater Mound** is an unconfined aquifer located in south west of Western Australia. Perth being the capital of Western Australia has two major shallow aquifers near the city itself, which are the Gnangara Mound located north of the Swan River, and the Jandakot Mound, or Jandakot Groundwater Mound itself. The latter aquifer is the smaller of the two that provides groundwater to the city, accounting for 40% of Perth's potable water. Furthermore, it extends from the Swan River located in north, to the Serpentine River in south; and then from the Indian Ocean which is on the western edges, to the Darling Scarp and Southern River located in the east. This covers an area of more or less 760 km^2.

- **The Yarragadee Aquifer** is an important freshwater aquifer found in the south west regions of Western Australia. It exists primarily below the Swan Coastal Plain which is found west of the Darling Scarp. Furthermore, the Yarragadee Aquifer has a north–south range from near Geraldton to the southern coast. However, the formation south of Perth is divided into two, whereby the southern portion is called the South West Yarragadee Aquifer. Moreover, the aquifer is significantly deep, located hundreds of meters below surface, and has a thickness extending up to more or less two kilometers.

- **The Bas Saharan Basin** is an artesian aquifer covering a large part of the Algerian and Tunisian Sahara. It extends up to Morocco and Libya, thus encompassing the entire Grand Erg Oriental.

- **The Lotikipi Basin Aquifer** is a big aquifer system found in the northwest area of Kenya. The discovery of the Lotikipi Basin Aquifer was made in September 2013, and it is said to be nine times bigger in size compared to the other aquifers located in Kenya. It contains about 200 billion cubic meters of fresh water, covering an area of about 4164 km^2. To add, it has the capacity to provide water to the population as it has sufficient fresh water which can last up to 70 years, provided there is adequate management of the aquifer.

- **The Nubian Sandstone Aquifer System (NSAS)** is the largest known fossil water aquifer found in the world. It underlies the Sahara Desert's eastern edge and covers a large area as it exists within four different countries of northeastern Africa. The NSAS has an area size of about just over two million km^2, covering northwest regions of Sudan, northeastern areas of Chad, southeastern parts of Libya, and almost all of Egypt. Additionally, the aquifer contains about 150,000 km^3 groundwater.

- **Alnarpsströmmen** is an artesian aquifer located in the Swedish province of Skåne. It is a subterranean aquifer which has been used for wells since the 18th century. Moreover, since 1901, it has been a freshwater resource for Malmö; and today it is used by numerous additional towns. Since it is artesian, it often results in the formation of fountains. Furthermore, more or less 400 liters per second of groundwater migrates from Alnarpsströmmen to Öresund through quaternary moraine. In addition, the aquifer has a width of about 20 km, and the flow of groundwater is only 10 meters per year.

- **Schwyll Aquifer** was initially called the Great Spring of Glamorgan. However, recently Welsh Water used the Schwyll Spring close to Ewenny as the primary water source for the Bridgend region. The aquifer now acts as a backup resource, and therefore has several related source protection zones which the Environmental Agency governs. The spring's outflow exceeds all other freshwater springs found in Wales; and it is larger than that of Wookey Hole or Cheddar Gorge risings. Finally, the aquifer has an underground waterway located in carboniferous limestone formations of South Wales.

- **The Upper Rhine aquifer** is one of the biggest aquifers in Europe, found beneath the plain and contains about 50,000 km^3 (110,000 cu mi) of fresh water. The aquifer provides water for about 3 million people in both France and Germany, and it accounts for 75% of potable water and 50% of industrial water use. To add, since the 1970s, the aquifer underwent pollution due to addition of nitrates, pesticides, chloride, and VOC.

- **The Yarkon-Taninim Aquifer** is an aquifer comprised of limestone material which exists within both Israel and Palestine. It is found beneath the foothills of the central parts of the country. About 340 million cubic meters per year and more or less 20 million cubic meters per year of water from the aquifer is used by Israel and Palestine, respectively. The aquifer starts in the northwest of Israel, then migrates down the Mediterranean Sea, toward the south. Recharge occurs in the Judean Mountains and when it forms the Taninim Springs which are on the coast of the Mediterranean sea. Because the aquifer lies beneath the Dead Sea, it is vulnerable to saltwater contamination.

- **The Tarim Basin** is an endorheic basin located in northwest China. It covers an area approximately 906,500 km^2 (350,000 sq mi). To add, since it is located in China's Xinjiang area, it is often synonymously referred to the southern half of the province, or Nanjiang.

1.4 Some Identified Aquifer Properties

Properties of the aquifer are essentially depending upon the composition of the aquifer. The most important properties of the aquifer are porosity and specific yield which in turn give its capacity to release water in the pores and its ability to transmit the flow with ease. In groundwater studies, we have also the permeability, hydraulic conductivity, transmissivity, and storativity [8,9].

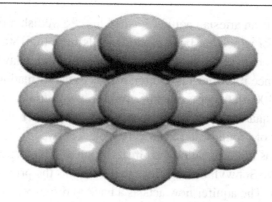

Figure 1.1: Cubic Packing.

Porosity: In soil mechanics and hydrology, porosity is defined as the ratio of volume of voids to the total volume of porous medium. Mathematically it can be expressed as

$$n = \frac{V_V}{V_T} = \frac{V_V}{V_S + V_V} = \frac{e}{1+e}, \quad e = \frac{V_S}{V_V}. \tag{1.1}$$

In this context e is the void ratio, n is porosity, V_V is the volume of void-space (air–water), V_S is the volume of solids, and V_T is the total or bulk volume of medium. The significance of the porosity is that it gives the idea of water storage capacity of the aquifer. Qualitatively, porosity less than 5% is considered to be small, between 5 and 20% as medium and the percentage exceeding 20% is considered large. The porosity depends on the cubic packing as presented in Fig. 1.1.

It is important to recall that the porosity does not depend on the diameter of grains. However, the size and shape of grains make a difference, see Fig. 1.2 below.

Porosity is a dimensionless quantity and can be reported either as a decimal fraction or as a percentage. Fig. 1.3 lists representative total porosity ranges for various geologic materials. In general, total porosity values for unconsolidated materials lie in the range of 0.25–0.7 (25 to 70%). Coarse-textured soil materials such as gravel and sand tend to have a lower total porosity than fine-textured soils such as silts and clays. The total porosity in soils is not a constant quantity because the soil, particularly clayey soil, alternately swells, shrinks, compacts, and cracks.

Specific yield: Porosity gives a measure of the water storage capability of soil but not all the water present in the soil pores is available for extraction by pumping for the use of humans or draining by gravity. The pores in the soil hold back sufficient quantity of water on account of forces like surface tension and molecular attraction. Hence the actual amount of water that

Greater porosity   Less porosity

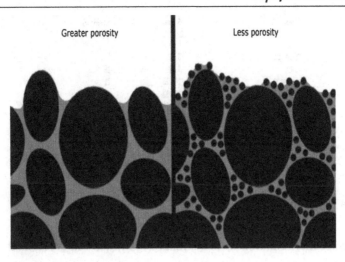

Figure 1.2: Heterogeneous particle sizes.

Formation	n (%)
Unconsolidated deposits	
Gravel	25 - 40
Sand	25 - 50
Silt	35 - 50
Clay	40 - 70
Rocks	
Fractured basalt	5 - 50
Karst limestone	5 - 50
Sandstone	5 - 30
Limestone, dolomite	0 - 20
Shale	0 – 10
Fractured crystalline rock	0 - 10
Dense crystalline rock	0 – 5

Figure 1.3: Range values of porosity.

can be extracted from the unit volume of aquifer by pumping of under the action of gravity is called *specific yield*. The fraction of water held back in the aquifer is known as *specific retention* [10]. Thus it can be said that porosity is the sum of specific yield and specific retention. Specific yield of soils differs from each other in the sense that some soil types have stronger molecular attraction with the water held in their pores than the others. It is found experimentally that cohesionless soils have higher specific yield than cohesive soils because the former have significantly less molecular attraction than the latter. Coarse-grained soils or rocks such as coarse sandstone can have specific yields that are closer to their actual porosity in the range 20 to 35%. The case of fine-grained materials is quite opposite to that range. More precisely, in mathematical formulations, the specific yield S_v is the ratio of the volume of water that drains from a saturated rock owing to the attraction of gravity to the total volume of the satu-

Formation	S_r(range)	S_r(average)
Clay	0 - 5	2
Sandy clay	3 - 12	7
Silt	3 - 19	18
Fine sand	10 - 28	21
Medium sand	15 - 32	26
Coarse sand	20 - 35	27
Gravelly sand	20 - 35	25
Fine gravel	21 - 35	25
Medium gravel	13 - 26	23
Coarse gravel	12 - 26	22
Limestone		14

Figure 1.4: Range values of porosity.

rated aquifer. The specific retention S_r is the rest of the water that is retained

$$n = S_v + S_r. \tag{1.2}$$

Fig. 1.4 shows the specific yield in percent.

Permeability: in fluid mechanics and the earth sciences permeability, commonly symbolized as κ, or k, is a measure of the ability of a porous material (often, a rock or an unconsolidated material) to allow fluids to pass through it. Permeability is the property of rocks that is an indication of the ability for fluids (gas or liquid) to flow through rocks. High permeability will allow fluids to move rapidly through rocks. Permeability is affected by the pressure in a rock. The unit of measure is called the Darcy, named after Henry Darcy (1803–1858) [10]. Sandstones may vary in permeability from less than one to over 50,000 millidarcys (md). Usually permeability is in the range of tens to hundreds of millidarcies. A rock with 25% porosity and a permeability of 1 md will not yield a significant flow of water. Such "tight" rocks are usually artificially stimulated (fractured or acidified) to create permeability and yield a flow [10]. The mathematical formula for permeability is:

$$k = v\frac{\mu\Delta x}{\Delta P}, \tag{1.3}$$

v being the superficial fluid flow velocity through the medium (that is to say, the average velocity calculated as if the fluid were the only phase present in the porous medium) (m/s);

k is the permeability of a medium given in m^2;

μ is the dynamic viscosity of the fluid given in (Pa s);

ΔP is the applied pressure difference given in (Pa);

Δx is the thickness of the bed of the porous medium given in (m).

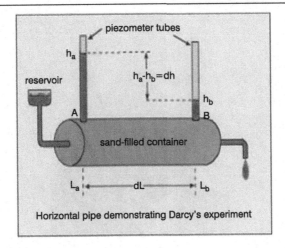

Horizontal pipe demonstrating Darcy's experiment

Figure 1.5: Darcy model.

Hydraulic conductivity: The hydraulic conductivity, symbolically represented as K, is a property of vascular plants, soils, and rocks, that describes the ease with which a fluid (usually water) can move through pore spaces or fractures. It depends on the intrinsic permeability of the material, the degree of saturation, and on the density and viscosity of the fluid. The mathematical formula is:

$$K = \frac{QL}{Ah}.$$

(1.4)

Here, Q is the discharge rate of the aquifer,

h is the difference of hydraulic head over the distance L,

A is the cross area of the specimen. The following figures (Fig. 1.5, Fig. 1.6) show the model used by Henry Darcy and a photo of Darcy himself who is considered the father of groundwater hydrology.

It is important to note that, when the horizontal and vertical hydraulic conductivity (K_{ho} and K_{ve}) of the soil layer differ considerably, the layer is said to be anisotropic with respect to hydraulic conductivity. An aquifer is called semi-confined when a saturated layer with a relatively small horizontal hydraulic conductivity, the semi-confining layer or aquitard overlies a layer with a relatively high horizontal hydraulic conductivity so that the flow of groundwater in the first layer is mainly vertical and in the second layer mainly horizontal [4–7]. The resistance of a semi-confining top layer of an aquifer can be determined from pumping tests. When calculating flow to drains or to a well field in an aquifer with the aim to control the water table, the anisotropy is to be taken into account, otherwise the result may be erroneous [4–7].

Figure 1.6: Sir Henry Darcy.

Because of their high porosity and permeability, sand and gravel aquifers have higher hydraulic conductivity than clay or unfractured granite aquifers. It would thus be easier to extract water from sand or gravel aquifers, for instance, using a pumping well because of their high transmissivity, compared to clay or unfractured bedrock aquifers. Hydraulic conductivity has units with dimensions of length per time, for instance m/s, ft/day and (gal/day)/ft^2, while transmissivity has units with dimensions of length squared per time.

Hydraulic conductivity (K) is one of the most complex and important properties of aquifers in hydrogeology, as its values found in nature [2–4,7,8]:

- range over many orders of magnitude, the distribution is often considered to be log-normal;
- vary a large amount through space, sometimes considered to be randomly spatially distributed, or stochastic in nature;
- are directional, in general K is a symmetric second-rank tensor; for example the vertical K values can be several orders of magnitude smaller than horizontal K values;
- are scale dependent (testing a cubic meter of aquifer will generally produce different results than a similar test on only a cubic-centimeter sample of the same aquifer);
- must be determined indirectly through field pumping tests, laboratory column flow tests, or inverse computer simulation, sometimes also from grain size analyses;
- are very dependent in a non-linear way on the water content, which makes solving the unsaturated flow equation difficult. In fact, the variably saturated K for a single material varies over a wider range than the saturated K values for all types of materials.

Transmissivity: The transmissivity is a measure of how much water can be transmitted horizontally, such as to a pumping well. The mathematical formula of transmissivity is:

$$T = bK \tag{1.5}$$

with b the depth of the aquifer.

Storativity: Storage properties are physical properties that characterize the capacity of an aquifer to release groundwater. These properties are storativity S, specific storage S_s, and specific yield S_y. The *storativity* or the storage coefficient is therefore the volume of water released from storage per unit decline in hydraulic head in the aquifer, per unit area of the aquifer. Storativity is a dimensionless quantity, and ranges between 0 and the effective porosity of the aquifer. In mathematical expression, the storativity is:

$$S = \frac{dV_w}{Adh} = S_s b + S_y, \tag{1.6}$$

where:

- V_w is the volume of water released from storage ($[L^3]$);
- h is the hydraulic head ($[L]$);
- S_s is the specific storage;
- S_y is the specific yield;
- b is the thickness of aquifer;
- A is the area ($[L^2]$).

The specific storage is the amount of water that a portion of an aquifer releases from storage, per unit mass or volume of aquifer, per unit change in hydraulic head, while remaining fully saturated. Mass specific storage is the mass of water that an aquifer releases from storage, per mass of aquifer, per unit decline in hydraulic head. The mathematical equation is provided as

$$(S_s)_m = \frac{1}{m_a} \frac{dm_w}{dh}, \tag{1.7}$$

where

- $(S_s)_m$ is the mass specific storage ($[L^{-1}]$);
- m_a is the mass of that portion of the aquifer from which the water is released ($[M]$);
- dm_w is the mass of water released from storage ($[M]$);
- dh is the decline in hydraulic head ($[L]$).

CHAPTER 2

Principle of Groundwater Flow

Groundwater flow occurs as water migrates through subsurface material such as soil and rock. It is stored within the pore spaces, fractures and cavities existing within subsurface material. Storage within these geological structures defines a groundwater system which can be divided into an unconfined and confined system. An unconfined system is associated with an exposed water surface, whereas a confined system does not have this association but rather an upper boundary causing it to be under significant pressure. Furthermore, water migrates across the Earth's surface, then enters the subsurface; and upon entering the subsurface, downward migration still occurs. The rate at which groundwater migrates downward depends on the nature of the geological material through which it moves as well as the amount of water already within the groundwater system. To add, as water migrates downward, it eventually reaches the water table of a groundwater system [3–7]. Furthermore, hydrologists have the capacity to predict and measure the flow and level of water, as well as the hydraulic gradient. Groundwater has a movement similar to that of surface water, whereby it flows in the direction of a hydraulic gradient. The two, however, contrast due to groundwater migration occurring at a much slower rate. For example, groundwater migration over a distance of a mile can take up to 10 years; and due to this, remediation of groundwater depletion and pollution becomes of interest [3–7].

2.1 Groundwater Within Geological Formations

The aforementioned groundwater system can be inferred as an aquifer system which is merely a unit rock of an unconsolidated deposit yielding ample amount of groundwater for use. The depth at which the water table exists indicates the point at which pore spaces and fractures are completely saturated. Furthermore, an aquifer system undergoes both natural recharge from the surface and natural discharge to the surface resulting in the formation of springs, seeps, oases, and wetlands. Groundwater discharge can also occur through wells by means of abstraction associated with agricultural, municipal, and industrial activities [3–7]. The study entailing the movement and distribution of groundwater is termed hydrogeology, also known as groundwater hydrology. Generally, groundwater is thought to flow through shallow aquifers; however, it can also be associated with soil moisture, permafrost, immobile water in significantly low permeable bedrock, and deep geothermal or oil formation water. It is put

Fractional Operators with Constant and Variable Order
with Application to Geo-Hydrology
DOI: 10.1016/B978-0-12-809670-3.00002-3

forward that groundwater allows lubrication which can potentially influence fault movement. Moreover, it is probable that a large amount of the subsurface holds some water which often may be mixed with other fluids. In addition, groundwater is not only confined to existence on Earth. This is said due to theories stipulating that some of the formations of the land forms found on Mars may be influenced by groundwater; and existence of liquid within the subsurface of Jupiter's moon, Europa [3–7].

2.1.1 Groundwater Cycle

Groundwater comprises approximately 20% of Earth's fresh water supply. This is about 0.61% of Earth's entire water content, including oceans and permanent ice. The global storage of groundwater more or less equals the total amount of freshwater stored in snow and ice packs, including both north and south poles. This makes groundwater a fundamental resource which can act as a buffer to shortages in surface water such as in cases of drought. Due to the slow movement of groundwater, an aquifer can act as a long-term 'reservoir' of the natural hydrological cycle as it has residence times ranging from days to millennia. This significantly contrasts to the residence times associated with the atmosphere and fresh surface water – minutes to years. Evidence for the long residence time of groundwater can be seen when referring to the Great Artesian Basin located in central and eastern Australia. The basin houses one of the world's largest confined aquifer systems, extending up to nearly 2 million km^2. Hydrogeologist have analyzed trace elements within water sourced from the deep formations and determined the water could possibly be more than 1 million years old [3–7]. After making comparisons in groundwater age, the age appeared to increase across the basin. To expand, along the Eastern Divide where recharge occurs, a younger age is found; and as groundwater flows toward the west across the continent, the age of groundwater increases. This in essence infers that in order to have moved over a distance of 1000 km from the recharge area in 1 million years, groundwater flowing through the Great Artesian Basin migrates at an average rate of approximately 1 meter per year. Furthermore, research proves that evaporation of groundwater plays a considerable role in the local hydrological cycle, particularly in arid areas. As a result, researchers in Saudi Arabia made propositions to recapture and recycle this evaporative moisture content for crop irrigation. A 50-centimeter-square reflective carpet, comprised of small neighboring plastic cones, was placed in a plant-free desert environment for a period of five months, in the absence of rainfall or irrigation. The ground vapor captured and condensed allowed life to be given to seeds buried below ground, with a green area of approximately 10% the carpet area. Moreover, it is assumed that a greater extent in green area would be found if seeds were put down prior placing the carpet [3–7]. Groundwater is also ecologically essential but this importance is often overlooked even within the field of freshwater biology and ecology. This is an inadequate approach because groundwater maintains lakes, rivers, wetlands, as well as subterranean ecosystems found in alluvial and karst aquifers [3–7]. There are, however, cases where ecosystems often do not require groundwater. For example, the existence of some terrestrial ecosystems such as those found in open deserts and similar arid regions rely solely on irregular rainfall and the moisture transfer to soil from the atmosphere. Even though there exist terrestrial ecosystems in similar environments where groundwater has no significant role, groundwater still plays a role in many of the world's major ecosystems. Furthermore, water flow occurs between groundwater and surface water. This is why most lakes, rivers, and wetlands are groundwater fed. On the other hand, it is also often found that these surface water bodies feed underlying groundwater systems. Groundwater feeds soil moisture by means of percolation. Subsequently, several terrestrial vegetation communities have direct dependence on either groundwater or percolated soil moisture above the groundwater system, for at least a certain period of each year. Systems either largely or completely dependent on groundwater are associated with the term, ecotones, of which examples include hyporheic zones which are the mixing zones of stream water and groundwater, as well as riparian zones.

2.1.2 Groundwater Overdraft

Groundwater is a fundamental and every so often an abundant resource. However, over-exploitation can result in significant issues in both society and the environment. The most apparent issue related to society is the concern associated with lowering of groundwater tables beyond the extent of existing wells. Consequently, wells should be drilled at greater depths for some parts of the Earth such as California, Texas, and India. For the Punjab area of India groundwater tables have declined by 10 meters since 1979. This is even more of a concern, considering the rate of groundwater depletion is increasing even more. Unfortunately, these are still not the only issues, as depleting groundwater systems become vulnerable to groundwater-related subsidence and saltwater intrusion.

2.1.3 Overview of Groundwater Within Subsurface

Certain issues affect groundwater use all over the world. Similarly to a case where river water is overused and causes pollution worldwide, aquifers are affected the same way [11–13]. The major difference is that aquifers cannot physically be seen. Furthermore, a significant issue for water management agencies is calculating the "sustainable yield" of an aquifer or river water because often the same water is calculated twice, once in the aquifer and then when there is a hydraulic connection between an aquifer and a river. Despite this issue being understood for many years, no change was made. This was partially due to the inadequate action taken by government agencies. This is seen in Australia where groundwater and surface water were managed by separate government agencies before the Council of Australian Government's water reform framework in the 1990s gave initiation to the statutory reforms. This was an approach influenced by opposition and inadequate communication. In general, water management agencies gave no consideration to the lag time associated with the vigorous response of groundwater to groundwater development. Moreover, only decades after scientific understanding were the two different waters managed together. To add, the effect of groundwater exploitation may be observed decades or centuries later [11–13]. In 1982, a study by Bredehoeft and colleagues involved modeling a situation where groundwater abstraction from an intermontane basin caused withdrawal of the entire annual recharge, leaving no resource for groundwater-dependent ecosystems. With no water, even for boreholes close to the vegetation of the ecosystem, 30 percent of the original demand could still be maintained by the aforementioned lag after 100 years. However, by the 500th year, the percentage of the original demand decreased causing the vegetation to die out. Although for decades science could be used to make these calculations, water management agencies have not considered effects observed outside the approximate time-frame of political elections 3 to 5 years. Marios Sophocleous claimed that when preparing groundwater planning, management agencies should define and use suitable time frames. This means calculations of groundwater abstraction permits are based on predicted effects occurring decades or even centuries forward. Furthermore, as water migrates through the landscape, soluble salts, very often sodium chloride, are collected. In these cases when water evaporates to the atmosphere, the salts remain behind. Another issue, very often seen in irrigation lands, is insufficient drainage of soil and surface groundwater systems, which can lead to groundwater tables extending to the surface in low-lying regions. A consequence of this is significant land degradation associated with soil salinity and water logging, combined with increased salt content in surface water [11–13]. As a result, local economies and environments are negatively impacted. Four noteworthy impacts arise, whereby the first involves flood mitigation schemes. Despite these schemes being developed for protection of infrastructure built on flood plains, it unintentionally decreases aquifer recharge associated with natural flooding. The second impact is a prolonged reduction in groundwater resources in widespread aquifers, as it may lead to land subsidence and often infrastructural damage. The third impact is a saline intrusion; and the fourth is drainage of sulfate soils often located in low-lying coastal plains, as this can lead to acidification and pollution of freshwater and estuarine streams. Furthermore, another issue of groundwater abstraction from over-allocated aquifers could possibly cause significant undesirable consequences to both terrestrial and aquatic ecosystems. This is often noticeable in some areas, but there are areas where it is not as noticeable due to extended periods over which the damage arises [11–13].

2.2 Concept of Groundwater Flow Motion

The mathematical equation governing groundwater flow describes the flow of groundwater through an aquifer. Transient flow of groundwater is described using the diffusion equation in a similar way to heat transfer where the flow of heat is described in a solid (heat conduction). On the other hand, a form of the Laplace equation which is a form of potential flow, is used to describe steady-state groundwater flow. Furthermore, the derivation of the groundwater flow equation is for a small representative elemental volume (REV) where a medium's properties are thought to be effectively constant. Moreover, a mass balance is done for groundwater flow in and out of the REV; and the flux in terms of hydraulic head can be expressed using Darcy's law which is a mathematical equation requiring slow groundwater flow.

2.2.1 Darcy's Law and Its Application

Darcy's law is a phenomenologically derived constitutive equation that describes the flow of a fluid through a porous medium. The law was formulated by Henry Darcy based on the results of experiments on the flow of water through beds of sand. It also forms the scientific basis of fluid permeability used in the earth sciences, particularly in hydrogeology. Although Darcy's law as an expression of conservation of momentum was determined experimentally by Darcy, it has since been derived from the Navier–Stokes equations via homogenization. It is analogous to the Fourier's law in the field of heat conduction, Ohm's law in the field of electrical networks, or Fick's law in diffusion theory. One application of Darcy's law is to water flow through an aquifer; Darcy's law along with the equation of conservation of mass is equivalent to the groundwater flow equation, one of the basic relationships of hydrogeology. Darcy's law is also used to describe oil, water, and gas flows through petroleum reservoirs. Darcy's law at constant elevation is a simple proportional relationship between the instantaneous discharge rate through a porous medium, the viscosity of the fluid, and the pressure drop over a given distance.

$$Q = \frac{-kA(p_b - p_a)}{\mu L} \tag{2.1}$$

The total discharge Q has units of volume per time, for instance m^3/s, is equal to the product of the intrinsic permeability of the medium, m^2, the cross-sectional area of the flow, A, with units of area, for instance m^2, and the total pressure drop $(p_b - p_a)$ (Pa), all divided by the viscosity, μ (Pa), and the length over which the pressure drop is taking place, L. The negative sign is needed because fluid flows from high pressure to low pressure. Fig. 2.1 illustrates the definitions and directions for Darcy's law. The elevation head must be taken into account if the inlet and outlet are at different elevations. If the change in pressure is negative (where

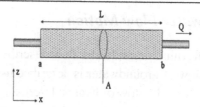

Figure 2.1: Diagram showing definitions and directions for Darcy's law.

$p_a > p_b$), then the flow will be in the positive 'x' direction. Dividing both sides of the equation by the area and using more general notation leads to:

$$q = \frac{-k}{\mu}\Delta p \qquad (2.2)$$

where q is the flux discharge per unit area, with units of length per time, m/s, Δp and is the pressure gradient vector (Pa/m). This value of flux, often referred to as the Darcy flux, is not the velocity which the fluid traveling through the pores is experiencing. The fluid velocity (v) is related to the Darcy flux (q) by the porosity (ϕ). The flux is divided by porosity to account for the fact that only a fraction of the total formation volume is available for flow. The fluid velocity would be the velocity a conservative tracer would experience if carried by the fluid through the formation.

$$v = \frac{q}{\phi} \qquad (2.3)$$

Darcy's law is a simple mathematical statement which neatly summarizes several familiar properties that groundwater flowing in aquifers exhibits, including:

- if there is no pressure gradient over a distance, no flow occurs, these are hydrostatic conditions;
- if there is a pressure gradient, flow will occur from high pressure towards low pressure opposite the direction of increasing gradient, hence the negative sign in Darcy's law;
- the greater the pressure gradient through the same formation material, the greater the discharge rate; and
- the discharge rate of fluid will often be different, through different formation materials, or even through the same material in different directions, even if the same pressure gradient exists in both cases.

Darcy's law is only valid for slow, viscous flow; fortunately, most groundwater flow cases fall in this category. Typically any flow with a Reynolds number less than unity is clearly laminar, and it would be valid to apply Darcy's law. Experimental tests have shown that flow regimes with Reynolds numbers up to 10 may still be Darcian, as in the case of groundwater flow. The Reynolds number (a dimensionless parameter) for porous media flow is typically expressed as

$$Re = \frac{\rho v d}{\mu} \tag{2.4}$$

where ρ is the density of water with units of mass per volume, v is the specific discharge – not the pore velocity – with units of length per time, d is a representative grain diameter for the porous media often taken as the 30% passing size from a grain size analysis using sieves with units of length, and μ is the viscosity of the fluid.

2.2.2 Derivation of Darcy's Law

At the stationary situation, creeping, incompressible flow, we have

$$\frac{d(\rho u_i)}{dt} = 0 \tag{2.5}$$

From the above equation, the well-known the Navier–Stokes equation is converted to the Stokes equation given below as

$$\mu \Delta^2 u_i + \rho g_i - \partial p = 0 \tag{2.6}$$

where μ is the viscosity, u_i is the velocity in the ith direction, g_i is the gravity component in the ith direction, and p is the pressure. Assuming the viscous resisting force is linear with the velocity, we may write

$$-(k_{ij})^{-1}\mu\phi u_j + \rho g_i - \partial_i p = 0, \tag{2.7}$$

where ϕ is the porosity, and k_{ij} is the second order permeability tensor. This gives the velocity in the n direction

$$k_{ni}(k_{ij})^{-1}u_j = \delta_{nj}u_n = -\frac{k_{ni}}{\phi\mu}(\partial_i p - \rho g_i), \tag{2.8}$$

which gives Darcy's law for the volumetric flux density in the nth direction,

$$q_n = -\frac{k_{ni}}{\mu}(\partial_i p - \rho g_i). \tag{2.9}$$

In isotropic porous media the off-diagonal elements in the permeability tensor are zero, $k_{ij} = 0$ for i different from j, and the diagonal elements are identical, $k = k_{ii}$, and the common form is obtained

$$q = -\frac{k}{\mu}(\Delta p - \rho g). \tag{2.10}$$

Darcy's law is valid for laminar flow through the soil. In fine grained soil, the dimensions of interstices are small and thus flow is laminar. Coarse-grained soils also behave similarly but in very coarse-grained soil, the flow is of turbulent nature [14]. Hence Darcy's law is not valid in such soils. For flow through commercial pipes, the flow is laminar when Reynolds number is less than 2000, but in some soils it has been found that flow is laminar when the value of Reynolds number is less than unity [14].

2.3 Theis Model of Groundwater Flow

Most commonly an aquifer test is conducted by pumping water from one well at a steady rate and for at least one day, while carefully measuring the water levels in the monitoring wells. When water is pumped from the pumping well the pressure in the aquifer that feeds that well declines. This decline in pressure will show up as drawdown which is the change in hydraulic head in an observation well. Drawdown decreases with radial distance from the pumping well and increases with the length of time that the pumping continues. The aquifer characteristics which are evaluated by most aquifer tests are:

- Hydraulic conductivity: The rate of flow of water through a unit cross-sectional area of an aquifer, at a unit hydraulic gradient. In US units the rate of flow is in gallons per day per square foot of cross-sectional area; in SI units hydraulic conductivity is usually quoted in m^3 per day per m^2. Units are frequently shortened to meters per day or equivalent.
- Specific storage or storativity: A measure of the amount of water a confined aquifer will give up for a certain change in head;
- Transmissivity: The rate at which water is transmitted through whole thickness and unit width of an aquifer under a unit hydraulic gradient. It is equal to the hydraulic conductivity times the thickness of an aquifer.

The Theis equation was created by Charles Vernon Theis working for the US Geological Survey in 1935 [15], from heat transfer literature (with the mathematical help of C.I. Lubin, for two-dimensional radial flow to a point source in an infinite, homogeneous aquifer). To derive this equation, the following assumptions were made:

- The flow in the aquifer is adequately described by Darcy's law (meaning $Re < 10$);
- homogeneous, isotropic, confined aquifer;
- the well is fully penetrating (open to the entire thickness (b) of aquifer);
- the well has zero radius (it is approximated as a vertical line) therefore no water can be stored in the well;
- the well has a constant pumping rate Q;
- the head loss over the well screen is negligible;
- aquifer is infinite in radial extent;

- horizontal (not sloping), flat, impermeable (non-leaky) top and bottom boundaries of aquifer;
- groundwater flow is horizontal;
- no other wells or long-term changes in regional water levels (all changes in potentiometric surface are the result of the pumping well alone).

Even though these assumptions are rarely all met, depending on the degree to which they are violated, for instance if the boundaries of the aquifer are well beyond the part of the aquifer which will be tested by the pumping test, the solution may still be useful.

2.3.1 Derivative of Theis Groundwater Flow Equation

To derive the groundwater flow equation, we make use of the principle of continuity equation of the flow, that is the difference between the rate inflow and the rate outflow from annular cylinder which is the equation of change of volume of water within the annular space. Therefore, we have the following relation:

$$Q_1 - Q_2 = \partial_t V, \tag{2.11}$$

where Q_1 is the rate inflow, Q_2 is the rate of outflow, and on the right-hand side is the rate of change of volume of water that accounts for the locality of the flow within the annular space. The slope of the hydraulic gradient at the inner surface is therefore provided as:

$$\partial_t V. \tag{2.12}$$

Let assume that h is the height of piezometric surface above the impervious stratum, then the slope of hydraulic gradient line at the outer surface is provided as:

$$J = \frac{\partial h(r,t)}{\partial r} + \frac{\partial^2 h(r,t)}{\partial r^2}. \tag{2.13}$$

Nonetheless, using Darcy's law, we obtain

$$Q = KJA, \tag{2.14}$$

with of course A the area of the flow. Therefore, in the case of the inner flow, we have the following expression:

$$Q_1 = K\left(\frac{\partial h(r,t)}{\partial r} + \frac{\partial^2 h(r,t)}{\partial r^2}\right)(2\pi(r+dr)b). \tag{2.15}$$

In the case of outflow, we have the following expression:

$$Q_2 = K \left(\frac{\partial h(r,t)}{\partial r} \right) (2\pi r b). \tag{2.16}$$

Here, b is the thickness of the confined aquifer. Now from the definition of storage coefficient (S), S is the volume of water released per unit surface area per unit change in head normal to the surface. Therefore, the change in volume is given as:

$$\delta V = 2\pi r S dr dh. \tag{2.17}$$

This leads us to the following equation

$$\partial_t V = 2\pi r dr \partial_t h(r,t). \tag{2.18}$$

Here t is the time since the beginning of pumping. However, putting together all the equations, we obtain the following:

$$K \left(\frac{\partial h(r,t)}{\partial r} + \frac{\partial^2 h(r,t)}{\partial r^2} \right) (2\pi (r+dr)b) - K \left(\frac{\partial h(r,t)}{\partial r} \right) (2\pi r b) = 2\pi r dr \partial_t h(r,t). \tag{2.19}$$

However, dividing both sides by and neglecting the higher order terms, we obtain

$$\frac{\partial h(r,t)}{\partial r} + \frac{\partial^2 h(r,t)}{\partial r^2} = \frac{S}{Kb} \partial_t h(r,t). \tag{2.20}$$

Thus in terms of transmissivity, we have

$$\frac{\partial h(r,t)}{\partial r} + \frac{\partial^2 h(r,t)}{\partial r^2} = \frac{S}{T} \partial_t h(r,t). \tag{2.21}$$

The above equation is the groundwater equation of the unsteady flow towards the well. In this equation, h is the hydraulic head, r is the radial distance from the well, S is the storage coefficient, T is the transmissivity, and t is the time since the beginning of pumping.

2.3.2 Derivation of Exact Solution

The most simple method used here is to derive the exact solution of the groundwater flow equation. This method is often used to some class of parabolic partial differential equation. This method uses the concept of reduction of dimension; in particular, the method uses the Boltzmann transformation. In this method, define for an arbitrary $t_0 < t$ the equation:

$$u_0 = \frac{S r^2}{4T(t - t_0)}. \tag{2.22}$$

Let us considered now the following function:

$$h(r, t) = \frac{c}{t - t_0} \exp[-u_0] \tag{2.23}$$

with c being any arbitrary constant. If we assume that r_b is the ratio of the borehole from which the groundwater is being taken out from the aquifer, thus the total volume of the water withdrawn from the aquifer is provided by:

$$Q_0 \Delta t_0 = \pi c T \tag{2.24}$$

Here

$$h(r, t) = \frac{c}{t - t_0} \exp[-u_0] \tag{2.25}$$

is the drawdown which will be experimental at a detachment, r, from the pumping well after the time space of Δt_0. Now assume that the above formula is continual m-times, meaning, water is being removed for a very small period of time, Δt_k, at a consecutive times, $t_{k+1} = t_k + \Delta t_k$, $(k = 0, 1, 2, 3, ...n)$. At this instance, since the new groundwater flow equation is linear, it follows that the total drawdown at any time $t > t_k$ will be given by

$$h(r, t) = \frac{1}{4\pi T} \sum_{k=0}^{n} \frac{\Delta t_k Q_k}{t - t_k} \exp[-u_0]. \tag{2.26}$$

Note that, in the above equation, the summation can be transformed into an integral if $\Delta t \longrightarrow 0$, then the summation symbol in the above equation is transformed to integral sign to obtain:

$$h(r, t) = \frac{1}{4\pi T} \int_{t_0}^{t} Q(x) \frac{\exp[-u_0]}{t - x} dx. \tag{2.27}$$

A particularly important solution which arises when the solution is considered at the origin zero and at the point where the discharge rate is not time independent then, Eq. (2.27) becomes

$$h(r, t) = \frac{Q}{4\pi T} \int_{u}^{\infty} \frac{\exp[-x]}{x} dx. \tag{2.28}$$

The above equation (2.28) is therefore the exact solution of the Theis groundwater flow equation within a confined aquifer under the conditions presented earlier. In terms of the well function, we have the following expression:

$$h(r, t) = \frac{Q}{4\pi T} W \left(u = \frac{Sr^2}{4\pi T} \right) \tag{2.29}$$

C.E. Jacob

Figure 2.2: Sir C.E. Jacob.

2.3.3 Cooper and Jacob Approximation of Theis's Solution of Groundwater Flow Equation

The Cooper and Jacob (1946) [16] solution or Jacob's modified non-equilibrium method is useful for determining the hydraulic properties (transmissivity and storativity) of non-leaky confined aquifers. Analysis involves matching a straight line given by the solution to draw-down data plotted as a function of the logarithm time since the start of pumping. Recent advances in aquifer test interpretation incorporate derivative analysis to improve the reliability of the Cooper and Jacob matching procedure. The following picture (Fig. 2.2) is that of C.E. Jacob. The approximate solution was obtained under the following conditions:

- aquifer has infinite areal extent;
- aquifer is homogeneous, isotropic, and of uniform thickness, item control well is fully penetrating;
- flow to control well is horizontal;
- aquifer is confined;
- flow is unsteady;
- water is released instantaneously from storage with decline of hydraulic head;
- diameter of pumping well is very small so that storage in the control well can be neglected, values of u are small that is to say r is small and t is large.

The Cooper and Jacob solution is an approximation of the Theis non-equilibrium method. The approximation was suggested to be used because of the difficulty of computing the well function; to achieve this, Cooper and Jacob used the concept of Taylor series of an infinitely differentiable function around a particular point, say a.

Definition 2.3.1. *The Taylor series of a real or complex-valued function $f(x)$ that is infinitely differentiable at a real or complex number a is the power series:*

$$f(x) = \sum_{n=0}^{\infty} \frac{f^{(n)}(a)}{n!}(x-a)^n \tag{2.30}$$

where $f^{(n)}(a)$ is the nth derivative of the function f at the point a, n! is the n-factorial.

From the above equation (2.30), we obtain the Taylor series of the well function as

$$W(u) = -0.5772 - \ln(u) + \sum_{n=0}^{\infty} \frac{(-1)^{n+1}u^n}{n.n!} \tag{2.31}$$

For small values of u, that is to say large values of t and small values of r, Cooper and Jacob found that the Theis well function may be approximated using only the first two terms as follows

$$W(u) = -0.5772 - \ln(u). \tag{2.32}$$

The critical value of u for the Cooper and Jacob approximation is alternately given as $u \leq$ 0.05 [18] and $u \leq 0.01$ [17–20]. A smaller value for the critical value of u leads to a more accurate approximation of the Theis well function.

Combining Eq. (2.29) and Eq. (2.32), Cooper and Jacob approximate drawdown in a non-leaky confined aquifer with the following linear equation:

$$h(r, t) = \frac{Q}{4\pi T}(-0.5772 - \ln(u)). \tag{2.33}$$

Converting to decimal logarithms, the Cooper and Jacob equation is as follows:

$$h(r, t) = \frac{2.303 Q}{4\pi T} \log\left(\frac{2.25Tt}{Sr^2}\right), \tag{2.34}$$

which is an equation for a line.

2.4 Groundwater Flow Within an Unconfined Aquifer

Abstraction from a well penetrating an unconfined aquifer results in actual dewatering of the aquifer. This leads to an overturned cone shaped volume known as the cone of depression or cone of influence. The aforementioned dewatering occurs as a result of gravity drainage to the lowest point of the cone's apex, which is at the well. Furthermore, once abstraction stops, the cone slowly starts to fill with water. The proportion of the amount of water draining from this cone which is under the influence of gravity to the amount of the cone is termed as the specific yield which is usually given in percentage or as a decimal fraction. Although drainage occurs under gravity, it is impossible for all water to drain from the aquifer's geological formation that was originally saturated within the cone. This is because some of the water clings onto the geological material due to a force of greater strength than gravity. This is associated with molecular attraction of water to the geological material's surface, or adhesion. Subsequently, specific yield equates the geological material's total volume of pore space, and this is given in percentage of the total volume of the geological material, minus the specific retention which is the volume of water kept or retained by the aforementioned molecular attraction or adhesion to the material. In almost all unconfined aquifers, specific yield ranged between 10 and 30%. This means that 10 to 30% of water held in an aquifer system can be released through abstraction or other forms of discharge. Aquifers which are coarse-grained have a specific yield greater than aquifer systems comprised of fine-grained material. To add, specific yield differs from maximum yield which is a property influenced by the size of an aquifer system. As indicated, when water is abstracted from a well, a cone of depression forms. The shape of the cone as well as its expansion rate across the top is dependent on transmissivity, storage of the aquifer, and the rate at which abstraction or pumping occurs. The initial water being abstracted is sourced from the pores in the immediate vicinity of the well. As abstraction continues, the cone expands in size until a recharge source is made available. In an unconfined aquifer, the cone of depression firstly expands at a rate ranging between less than 100 meters to often more than 1000 meters per day. Furthermore, as the cone expands more, water continues to be released from the formation within which the cone exists. In the case where the cone of depression reaches a stream of lake, induced infiltration occurs in a way that the demands of the well being pumped are provided for; and when these demands are met, the cone of depression stops expanding. The cone of depression's shape will change in the immediate area where direct recharge (a lake of stream) takes place. This change occurs by extending away from the source's axis. It is not always lakes or streams that groundwater within an unconfined aquifer are intercepted but sometimes water from springs which do not flow. In essence, the cone of depression will stop expanding until there is enough base area to meet the demand of the well undergoing abstraction, at significant recharge rates. Often the cone of depression is significantly large so that it goes beyond the initial groundwater divide of an aquifer and therefore causes water to be induced from drainage basins situated on both sides of the basin within which the well being pumped is found. The cone modifies itself when there is change in the influence of recharge and discharge within the aquifer. To expand, when there are periods of precipitation and aquifer recharge occurs, the cone of depression gets smaller in size, and the size to which it decreases depends on the volume of recharge the aquifer receives. Alternatively, when a period of drought occurs, the cone gets both deeper and larger in extent, withdrawing water from storage to meet the demand of a pumping well.

Furthermore, pumping tests are generally conducted to estimate hydraulic properties of unconfined aquifers. The response in drawdown over time after conducting a pumping test for an unconfined aquifer normally yields an S-shaped curve which shows three chronologically sequenced segments. To expand, after pumping occurs, there is a rapid increase in drawdown for the first segment. Thereafter, the second segment flattens out. This is followed by a gradual increase in drawdown for the third segment. All three segments were explained by Boulton [21,22] by referring to the release of water storage as two different mechanisms which are the instantaneous yield and the gravity yield. The former is a result of compression and expansion of the aquifer material and water, respectively; and this is quantified by means of the storage coefficient. On the other hand, gravity yield (gravity drainage) is seen to occur at a high rate. Subsequently, there is an exponential decrease which can be described in terms of delayed yield. This property is essentially described by both specific yield and the delay index, where the specific yield describes total gravity yield and the delayed index determines its time distribution. Moreover, Boulton's solution is associated with radial flow which can be described by means of transmissivity. Subsequently, Boulton's solution for drawdown in unconfined aquifer systems is comprised of three hydraulic properties, namely, specific yield, storage coefficient, and transmissivity. Additionally, there is also the fitting parameter which is the delay index. Storage coefficient is the primary storage parameter for confined aquifers or leaky confined aquifers. In contrast, it has a minor role in the case of unconfined aquifers, where it affected mainly the first segment of a time versus drawdown curve. Furthermore, the differential equation which governs unsteady-state flow towards a well fully penetrating an unconfined aquifer, in polar-coordinate notation is given as [22]:

$$T\left(\frac{\partial^2 s}{\partial r^2} + \frac{\partial s}{r\partial r}\right) = S\frac{\partial s}{\partial t} + DS_y \int_0^t \frac{\partial s}{\partial z} \exp\left(-D(t-z)\right) dz \tag{2.35}$$

where T is the transmissivity, S is the storage coefficient, D is the inverse of delay index, S_y is the specific yield, s is the drawdown, r is the radial distance from the center of the well, t is the time, and τ is the dummy variable. Initial conditions:

$$t = 0; \qquad s = 0, \qquad 0 \le r < \infty, \tag{2.36}$$

with boundary conditions:

$$0 \le t; \qquad s \to 0, \qquad r \to \infty \tag{2.37}$$

$$\lim_{r\to 0} r\frac{\partial s}{\partial r} = -\frac{Q}{2\pi T} \tag{2.38}$$

where Q is the pumping discharge in the well. Boulton (1963) obtained an analytical solution for Eq. (2.35) subject to initial and boundary conditions (2.36), (2.37) and (2.38) as:

$$s(r,t) = \frac{Q}{4\pi T} \int_0^\infty \frac{2}{x} [1 - \exp[-u_1](\cosh u_2 + M \sinh u_1)] J_0\left(\frac{rx}{\mu B}\right) dx \tag{2.39}$$

where

- $$u_1 = \frac{1}{2} Dt N(1 + x^2), \tag{2.40}$$

- $$u_2 = \frac{1}{2} DT S_q, \tag{2.41}$$

- $$N = \frac{S + S_y}{S}, \tag{2.42}$$

- $$\mu = \sqrt{\frac{N - 1}{N}}, \tag{2.43}$$

- $$B = \sqrt{\frac{T}{DS_y}}, \tag{2.44}$$

- $$M = \frac{N(1 - x^2)}{S_q}, \tag{2.45}$$

- $$S_q = \sqrt{N^2(1 + x^2)^2 - 4Nx^2}, \tag{2.46}$$

- $$J_0(x) = \sum_{j=0}^{\infty} \frac{(-1)^j}{\Gamma[j + 1]} \left(\frac{x}{2}\right)^{2j}, \tag{2.47}$$

whereas J_0 is the Bessel function of first kind of zero order, x is the variable of integration, and B is the drainage factor.

2.5 Groundwater Flow in a Deformable Aquifer

In 2014 [23] Atangana proposed a new groundwater flow equation within a deformable aquifer. He argued that, when investigating aquifer behavior, it is important to note that there exists a close relationship between the geometrical properties of the aquifer and the behavior of the solution [23]. The aim of his work was to solve the flow equation described by prolate spheroidal coordinates by means of perturbation and the Green's function method, where the spheroid is considered to be a perturbation of a sphere. He first transformed the spheroidal coordinates to spherical polar coordinates in the limit, as the shape factor tends to zero. The new groundwater flow equation was solved via an asymptotic parameter expansion and the Green's function method. This section presents the work done in his paper that was published in the Journal of Communication in Nonlinear Sciences and Numerical Simulation.

2.5.1 Perturbation From Prolate Coordinates to Spherical Coordinates

The transformation of the spheroidal coordinates to spherical polar coordinates in the limit as the shape factor tends to zero or as the prolate spheroid tends to a sphere and derivation of the groundwater equation for prolate spheroidal obstacle is presented here. The prolate spheroidal coordinates can be interconnected to the Cartesian coordinates as follows:

$$\begin{cases} x & = l \sinh \alpha \sin(\beta) \cos(\gamma) \\ y & = l \sinh(\alpha) \sin(\beta) \sin(\gamma) \\ z & = l \cosh(\alpha) \cos(\beta) \end{cases} \tag{2.48}$$

where $0 \le \alpha < \infty, 0 \le \beta < \pi$ and $0 \le \gamma < 2\pi$. It is worth noting that, by rotating an ellipse about the main z-axis, we get a prolate spheroid on one hand, and on the other hand, by rotating about the minor y-axis, we get an oblate spheroid. For small values of α the prolate spheroids are rod-shaped and so can be looked upon as estimation to a cable antenna. In the case of oblate spheroids the surface $\alpha = 0$ is a disc with equation:

$$y = 0, \qquad\qquad x^2 + z^2 = l^2 \tag{2.49}$$

From the coordinate system (2.48) we obtain the Laplacian as

$$\Delta^2 = \frac{1}{h_1 h^2 h^3} \left[\frac{\partial}{\partial \alpha} \left(\frac{h_1 h_3 \partial}{h_2 \partial \alpha} \right) + \frac{\partial}{\partial \beta} \left(\frac{h_2 h_3 \partial}{h_1 \partial \alpha} \right) + \frac{\partial}{\partial \alpha} \left(\frac{h_2 h_3 \partial}{h_1 \partial \gamma} \right) \right] \tag{2.50}$$

where h_1, h_2, and h_3 are metric coefficients given by the following formula:

$$h_i^2 = \left(\frac{\partial x}{\partial u_i} \right)^2 + \left(\frac{\partial y}{\partial u_i} \right)^2 + \left(\frac{\partial z}{\partial u_i} \right)^2 \; (i = 1, 2, 3) \tag{2.51}$$

such that

$$\begin{aligned} \Delta^2 \Phi &= \frac{1}{l^2 (\cosh^2(\alpha)) - \cosh^2(\beta))} \left(\frac{\partial^2 \Phi}{\partial \alpha^2} + \frac{\partial^2 \Phi}{\partial \beta^2} + \cosh(\alpha) \frac{\partial \Phi}{\partial \alpha} + \cosh(\beta) \frac{\partial \Phi}{\partial \beta} \right) \\ &+ \frac{1}{l^2 (\sinh(\alpha) \sin(\beta))^2} \frac{\partial \Phi}{\partial \gamma^2} \end{aligned} \tag{2.52}$$

On the other hand, the spherical polar coordinates are given as:

$$\begin{cases} x & = r\sin(\theta)\cos(\phi) \\ y & = r\sin(\theta)\sin(\phi) \\ z & = r\cos(\phi) \end{cases} \tag{2.53}$$

Here, $0 \leq r < \infty$, $0 \leq \theta < \pi$ and $0 \leq \phi < 2\pi$. The first aspect of this section is to obtain the prolate spheroid in spherical coordinates. The first guess here is that a Laplacian would be the sum of a sphere and a small term. To achieve this he first considers a relation between spheroidal coordinates and the polar spherical coordinates. The equation of an ellipse with center at the origin will then be:

$$\frac{x^2}{\alpha^2} + \frac{y^2}{\alpha^2\left(1 - \frac{l^2}{\alpha^2}\right)}. \tag{2.54}$$

One can notice that if $\frac{l}{a} \to 0$, then the ellipse is converted to a circle with ratio a. Thus, one will conclude that if $\frac{l}{a} \to 0$, then, the prolate spheroid is converted into a sphere, since a sphere is generated by a circle. For simplification, in this section, the sharpe factor will be $\tau = \frac{l}{a}$ where one half the inter-focal is l distance and a is the radius of the approximating sphere. Now comparing Eq. (2.48) and (2.53), we obtain:

$$\frac{y}{x} = \tan(\gamma) = \tan(\phi) \tag{2.55}$$

also

$$l\sinh(\alpha)\sin(\beta) = r\sin(\theta), \qquad l\cosh(\alpha)\cos(\beta) = r\cos(\theta) \tag{2.56}$$

The above implies $\phi = \gamma$. The relationship between (α, β, γ) and (r, θ, ϕ, τ) for small τ is:

$$\begin{cases} \alpha & = \alpha(r, \theta, \phi, \tau) \\ \beta & = \beta(r, \theta, \phi, \tau) \\ \gamma & = \gamma(r, \theta, \phi, \tau) \end{cases} \tag{2.57}$$

For $\alpha \gg 1$ looking at the dominant terms and with some asymptotic expansion tool, we obtain the following relations:

$$\exp(\alpha) = \frac{2r}{a\tau} - \frac{a}{2r}\cos(2\theta)\tau + O(\tau^3) \tag{2.58}$$

$$\beta = \theta + \left(\frac{a}{2r}\right)^2 \sin(2\theta)\tau^2 + O(\tau^2). \tag{2.59}$$

From Eqs. (2.58) and (2.59) we establish the following relations:

$$\frac{\partial r}{\partial \alpha} = r\left(1 - 2\left(\frac{a}{2r}\right)^2 \cos(2\theta)\tau^2\right) + O(\tau^4),$$

$$\frac{\partial r}{\partial \beta} = -\frac{a^2}{2r}\sin(2\theta)\tau^2 + O(\tau^4),$$

$$\frac{\partial \theta}{\partial \alpha} = \frac{1}{2}\left(\frac{a}{r}\right)^2 \sin(2\theta)\tau^2 + O(\tau^4),$$

$$\frac{\partial \theta}{\partial \alpha} = 1 - 2\left(\frac{a}{2r}\right)^2 \cos(2\theta)\tau^2 + O(\tau^4),$$

$$\frac{\partial^2 r}{\partial \alpha^2} = r\left(1 - \left(\frac{a}{r}\right)^2 \cos(2\theta)\tau^2\right) + O(\tau^2),$$

$$\frac{\partial^2 r}{\partial a^2} = -\left(\frac{a}{r}\right)^2 \cos(2\theta)\tau^2 + O(\tau^4),$$

$$\frac{\partial^2 \theta}{\partial \beta^2} = -\left(\frac{a}{r}\right)^2 \sin(2\theta)\tau^2 + O(\tau^4),$$

$$\frac{\partial^2 \theta}{\partial \alpha^2} = \left(\frac{a}{r}\right)^2 \sin(2\theta)\tau^2 + O(\tau^4),$$

$$\frac{\partial \Phi}{\partial \alpha} = \frac{\partial \Phi}{\partial \theta}\left(\frac{1}{2}\left(\frac{a}{r}\right)^2 \sin(2\theta)\tau^2 + O(\tau^4)\right) + \frac{\partial \Phi}{\partial r}\left(r\left(1 - 2\left(\frac{a}{2r}\right)^2 \cos(2\theta)\tau^2\right) + O(\tau^4)\right),$$

$$\frac{\partial \Phi}{\partial \beta} = \frac{\partial \Phi}{\partial \theta}\left(1 - 2\left(\frac{1}{2r}\right)^2 \cos(2\theta)\tau^2 + O(\tau^4)\right) + \frac{\partial \Phi}{\partial r}\left(-\frac{a^2}{2r}\sin(2\theta)\tau^2 + O(\tau^4)\right)$$

In general, we have the following

$$\frac{\partial^2 \Phi}{\partial u_i^2} = \frac{\partial^2 \Phi}{\partial r^2}\left(\frac{\partial r}{\partial u_i}\right)^2 + 2\frac{\partial^2 \Phi}{\partial r \partial \theta}\frac{\partial^2 \theta}{\partial u_i^2} + \frac{\partial^2 \Phi}{\partial \theta^2}\left(\frac{\partial \theta}{\partial u_i}\right)^2 + \frac{\partial \Phi \partial^2 r}{\partial r \partial u_i^2} + \frac{\partial \Phi \partial^2 \theta}{\partial r \partial u_i^2} \tag{2.60}$$

for $i = 1, 2$ and $u_1 = \alpha$, $u_2 = \beta$. Thus putting together all the above relations, we obtain the Laplacian for the prolate spheroidal:

$$\Delta^2 \Phi = \left(1 - 2\left(\frac{1}{2r}\right)^2 \cos(2\theta)\tau^2 + O(\tau^4)\right)^{-1}$$

$$\left\{\frac{\partial^2 \Phi}{\partial r^2} + \frac{2\partial \Phi}{r \partial r} + \frac{\partial^2 \Phi}{r^2 \partial \theta^2} + \frac{\cot(\theta)\partial \Phi}{r^2 \partial \theta} + \left(\frac{a}{r}\right)^2\left(-\cos(2\theta)\frac{\partial^2 \Phi}{\partial r^2} + \frac{\sin(2\theta)}{r}\frac{\partial^2 \Phi}{\partial r \partial \theta}\right) + \right.$$

$$\left.\left(\frac{a}{r}\right)^2\left(-\frac{3(r+2)}{2r^2}\cos(\theta)\frac{\partial \Phi}{\partial r} + \frac{1}{2r^2}(1 + 2\sin(2\theta) - 2\cot(\theta)\frac{\partial \Phi}{\partial \theta})\right)\tau^2 + O(\tau^4)\right\}$$

$$+ \frac{1}{r^2 \sin^2(\theta)}\frac{\partial^2 \Phi}{\partial \phi^2} + O(\tau^4). \tag{2.61}$$

Therefore employing Eq. (2.61), the new groundwater flowing within a deformable aquifer is proposed as follows:

$$
S_0 \partial_t \Phi = K \left(1 - 2 \left(\frac{1}{2r} \right)^2 \cos(2\theta)\tau^2 + O(\tau^4) \right)^{-1}
$$

$$
\{ \frac{\partial^2 \Phi}{\partial r^2} + \frac{2\partial \Phi}{r \partial r} + \frac{\partial^2 \Phi}{r^2 \partial \theta^2} + \frac{\cot(\theta)\partial \Phi}{r^2 \partial \theta} +
$$

$$
\left(\frac{a}{r} \right)^2 \left(-\cos(2\theta)\frac{\partial^2 \Phi}{\partial r^2} + \frac{\sin(2\theta)}{r}\frac{\partial^2 \Phi}{\partial r \partial \theta} \right) +
$$

$$
\left(\frac{a}{r} \right)^2 \left(-\frac{3(r+2)}{2r^2}\cos(\theta)\frac{\partial \Phi}{\partial r} + \frac{1}{2r^2}(1 + 2\sin(2\theta) - 2\cot(\theta)\frac{\partial \Phi}{\partial \theta}) \right) \tau^2 + O(\tau^4) \}
$$

$$
+ \frac{1}{r^2 \sin^2(\theta)}\frac{\partial^2 \Phi}{\partial \phi^2} + O(\tau^4). \tag{2.62}
$$

2.5.2 Derivation of Solution via Asymptotic Method

The model of water flowing within a deformable aquifer can be reduced to

$$
\left(\Delta_s^2 + k^2 \partial_t + \tau^2 L_\tau \right) \Phi_p(r,t) = 0, \qquad k^2 = \frac{S_0}{K}. \tag{2.63}
$$

$$
\Delta_s^2 = \frac{\partial^2}{\partial r^2} + \frac{2\partial}{r \partial r} + \frac{\partial^2}{r^2 \partial \theta^2} + \frac{\cot(\theta)\partial}{r^2 \partial \theta} + \frac{1}{r^2 \sin^2(\theta)}\frac{\partial^2}{\partial \phi^2} \tag{2.64}
$$

$$
L_{\tau^2} = \left(\frac{a}{r} \right)^2 \{ -\cos(2\theta)\frac{\partial^2}{\partial r^2} + \frac{\sin(2\theta)}{r}\frac{\partial^2}{\partial r \partial \theta} - \frac{3(r+2)}{2r^2}\cos(\theta)\frac{\partial}{\partial r} +
$$

$$
\frac{1}{2r^2}(1 + 2\sin(2\theta) - 2\cot(\theta)\frac{\partial}{\partial \theta}) \}\tau^2 + O(\tau^4) \tag{2.65}
$$

The new equation can be viewed as a sum of two contributions, the first describing the groundwater flow through porous media and the second part viewed as deformation of geological formation in which the flow takes place [23]. It is very obvious from (2.65) that we get back the standard groundwater flow equation for spherical system, as expected when we put $\tau = 0$. To solve this equation we employ asymptotic method technique as follows: the piezometric head of prolate spheroidal as solution of this new equation will be expressed as:

$$
\Phi_p = \sum_{n=0}^{\infty} \tau^{2n} \Phi_{2n}(r, \theta, \phi, t) \tag{2.66}
$$

Putting Eq. (2.66) into Eq. (2.65) produces:

$$(\Delta_s^2 + k^2\partial_t + \tau^2 L_\tau)\left(\sum_{n=0}^{\infty} \tau^{2n}\Phi_{2n}(r,\theta,\phi,t)\right) = 0. \tag{2.67}$$

We can now use the regular perturbation method on Eq. (2.67) so that the constant term gives

$$(\Delta_s^2 + k^2\partial_t)\Phi_0 = 0, \tag{2.68}$$

and obviously the solution of Eq. (2.68) is the piezometric head of the flow in spherical system. This solution will be derived later using the method of separation of variables. We next consider the term of order τ^2 in Eq. (2.67):

$$(\Delta_s^2 + k^2\partial_t)\Phi_2 = -L_{\tau^0}\Phi_0. \tag{2.69}$$

Here $L_{\tau^{2n}}$, $(n = 0, 1, 2, 3, ...)$ represent the expression that contribute for the deformation in Eq. (2.63) of up to order $2n$ that is terms of $O(\tau^{2n})$, so that in general we obtain Φ_{2n} from

$$(\Delta_s^2 + k^2\partial_t)\Phi_{2n} = -L_{\tau^{2j}}\Phi_{2(n-1-j)}. \tag{2.70}$$

To solve the above equation, we go to construct a suitable Green's function for this case in point. Let $G(R, \tau_1)$ be the Green's function to be constructed, where $R = |r - r_0|$ and $\tau_1 = |t - t_0|$. G is chosen so as to satisfy homogeneous boundary conditions corresponding to the boundary conditions satisfied by Φ_{2n}, $n > 0$. It is important to notice that the homogeneous solution of Eq. (2.70) is similar to the diffusion equation if one replaces Φ_{2n} by ϕ, therefore the Green function involved here is the Green's function for the diffusion equation. Since the aquifer is said to be infinite, the Green function for flow equation for infinite aquifer is given by

$$G(R, \tau_1) = \frac{4\pi}{k}\left(\frac{k}{2\sqrt{(\pi\tau_1)}}\right)^3 \exp\left(-\frac{k^2 r_x^2}{4\tau_1}\right). \tag{2.71}$$

The above equation satisfies an important integral property which is valid for $n = 3$.

$$\int G(R, \tau_1)dV = \frac{4\pi}{k^2}, \qquad \tau_1 > 0. \tag{2.72}$$

This equation is an expression of groundwater flow. At a time t_0 and at a position r_0, the piezometer is introduced in the borehole that taps the aquifer. The water that is pumped out from the aquifer through the borehole is migrating through the porous media, but in such a way that the total amount of water in the aquifer is reduced as time goes on if there is no recharge. Since Eq. (2.72) still holds, we can observe that:

$$G(R, \tau_1) \to \frac{4\pi}{k^2}, \qquad \tau_1 \to 0. \tag{2.73}$$

In addition, the Green's function used for this purpose is a solution to the following equation

$$\Delta^2 G(R, \tau_1|R_0, \tau_{10}) - k^2 \frac{G(R, \tau_1|R_0, \tau_{10})}{\partial t} = -4\pi \delta(r - r_0)\delta(t - t_0) \tag{2.74}$$

The general solution of equation can then be given as function of the Green function as

$$\Phi_{2n}(r, t) = -\int_0^t \int (L_{\tau^{2j}} \Phi_{2(n-1-j)}(r_0, t_0) G(r, t|r_0, t_0))dt_0 dV_0 +$$

$$\frac{1}{4\pi} \int_0^t \int (G grad_0 \phi - \phi grad_0 G)dt_0 dS_0 + \frac{a^2}{4\pi} \int \{\phi G\}_{|t_0=0} dV_0 \tag{2.75}$$

replacing Eq. (2.75) into Eq. (2.71) providing $\Phi_{2n}(r, t)$ in terms of the Green function expression, initial and boundary conditions. For the case in point the two last integrals in Eq. (2.75) can be neglected so that Eq. (2.71) together with the boundary condition is reduced to:

$$\Phi_{2n}(r, t) = -\frac{q}{4\pi t} \int_0^t \int L_{\tau^{2j}} \Phi_{2(n-1-j)}(r_0, t_0) \frac{4\pi}{k} \left(\frac{k}{2\sqrt{(\pi \tau_1)}}\right)^3 \exp\left(-\frac{k^2 r_x^2}{4\tau_1}\right) dt_0 dV_0. \tag{2.76}$$

It is worth noting that if the first term is defined then the remaining terms, $\Phi_{2n}, n > 0$, can be completely determined such that each term is determined by using the previous terms, and the series solutions are thus entirely determined. Finally, we approximate the solution Φ_p by the truncated series:

$$\Phi_p^N(r, \theta, \phi, t) \sum_{n=0}^{N-1} \tau^{2n} \Phi_{2n}(r, \theta, \phi, t) \tag{2.77}$$

$$\Phi_p = \lim_{N \to 0} \Phi_p^N$$

We shall now give the analytical solution for Eq. (2.68) which in fact is the first term of the above series. We shall remember that the mathematical formulation of this problem is given as:

$$\frac{\partial^2 \Phi_0}{\partial r^2} + \frac{2\partial \Phi_0}{r \partial r} + \frac{1}{r^2} \frac{\partial^2 \Phi_0}{\partial \theta^2} + \frac{\cot(\theta)}{r^2} \frac{\partial \Phi_0}{\partial \theta} + \frac{1}{r^2 \sin^2(\theta)} \frac{\partial^2 \Phi_0}{\partial \phi^2} = k^2 \frac{\partial \Phi}{\partial t}. \tag{2.78}$$

We introduced a dimensionless variable, a typical time scale $\frac{1}{a} = k^2$ by writing $t^* = at$. To reduce Eq. (2.78) with the "*" we drop it for convenience to obtain

$$\frac{\partial^2 \Phi_0}{\partial r^2} + \frac{2\partial \Phi_0}{r\partial r} + \frac{1}{r^2}\frac{\partial^2 \Phi_0}{\partial \theta^2} + \frac{\cot(\theta)}{r^2}\frac{\partial \Phi_0}{\partial \theta} + \frac{1}{r^2 \sin^2(\theta)}\frac{\partial^2 \Phi_0}{\partial \phi^2} = \frac{\partial \Phi}{\partial t}. \tag{2.79}$$

To reduce this equation more, we put $\eta = \cos\theta$ and $T = r^{0.5}\Phi_0$. It is then possible for us to transform the above equation as follows:

$$\frac{\partial^2 T}{\partial r^2} + \frac{\partial T}{r\partial r} - \frac{T}{4r^2} + \frac{1}{r^2}\frac{\partial}{\partial \eta}\left((1-\eta^2)\frac{\partial T}{\partial \eta}\right) + \frac{1}{r^2(1-\eta^2)}\frac{\partial^2 T}{\partial \phi^2} = \frac{\partial T}{\partial t}. \tag{2.80}$$

The above equation (2.80) can then be solved via the method of separation of variables, meaning the solution of the above equation which can be assumed to be in the following.

$$T(r, \eta, \phi, t) = T_1(t)T_2(r)T_3(\phi)T_4(\eta). \tag{2.81}$$

Now substituting Eq. (2.81) into Eq. (2.80) produces the following set of ordinary differential equations that can all be solved separately

$$\frac{dT_1}{dt} + \zeta^2 T_1 = 0 \tag{2.82}$$

$$\frac{d^2 T_2}{d\phi^2} + \sigma^2 T_2 = 0 \tag{2.83}$$

$$\frac{d^2 T_3}{dr^2} + \frac{dT_3}{rdr}\left(\sigma^2 - \left(v + \frac{1}{2}\right)^2 \frac{1}{r^2}\right)T_3 = 0 \tag{2.84}$$

$$\frac{d}{d\theta}\left((1-\theta^2)\frac{dT_4}{d\theta}\right) + \left(v(v+1) - \frac{\sigma^2}{1-v^2}\right)T_4 = 0. \tag{2.85}$$

Here σ, ζ, and v are naturally the parameters of separation. Eq. (2.82) has an obvious solution as

$$T_1(t) = \exp(-\zeta), \tag{2.86}$$

while Eq. (2.83) has as solution

$$T_2(t) = A\cos(\sigma\phi) + B\sin(\sigma\phi). \tag{2.87}$$

Since the axial angle ϕ for the case in point is continuous from 0 to 2π, coming back to 0 as the radius vector is moved to and fro in a complete revolution of the sphere, the function $T_2(\pi)$ must be equal to the function $T_2(0)$ and generally speaking $T_2(2\pi + \phi) = T_2(\pi)$. In order to achieve this, the separation constant σ must be an integer. Assuming that $v + 1/2$ is not zero or integer, Eq. (2.84) the well-known Bessel function of order $v + 1/2$ has for elementary solutions:

$$J_{v+1/2}(r\zeta), \qquad J_{-v-1/2}(r\zeta). \qquad (2.88)$$

Eq. (2.85) is the well-known associated Legendre's differential equation and has two elementary solutions, namely P_v^σ and Q_v^σ. In this case in point the elementary solutions to retain are if we convert v to an integer number. The reason behind converting v to an integer is that, if v is non-integer, then P_v^σ has a branch point at $\eta = -1$, but when v is integer, P_v^σ is a finite polynomial in η and, of course, is analytic at $\eta \mp 1$, which in practical point of view is suitable for the case in point.

$$\exp\left(-\sigma^2 t\right), \qquad A\cos(\sigma\phi) + B\sin(\sigma\phi), \qquad P_v^\sigma, \qquad J_{v+1/2}(r\zeta) \qquad (2.89)$$

So that Eq. (2.80) can be given as: where $A_{\sigma\eta,j}$, $B_{\sigma\eta,j}$, and $\eta_{\sigma\eta,j}$ have to be determined using the boundaries conditions. For instance, by choosing $\eta_{\sigma\eta,j}$ to be positive, which actually in this case is an eigenvalue, we shall have the following:

$$J_{v+\frac{1}{2}}\left(\eta_{vj}\right) = 0 \qquad (2.90)$$

Let us look now at the physical aspect of the problem in point. At the time t_0 when the piezometers are introduced into the borehole, we have the following:

$$T(r, \eta, \phi, 0) = \sum_{j=1}^{\infty}\sum_{a=0}^{v} J_{v+\frac{1}{2}}\left(\eta_{vj}r\right) P_v^\sigma(\eta) A_{\zeta\eta j}(A_{\eta\zeta j}\cos(\sigma\phi) + B_{\eta\zeta j}\sin(\sigma\phi)). \qquad (2.91)$$

Now employing successively the following operators on the above equation:

$$\int_0^t r J_{v+\frac{1}{2}}(\zeta_{vj}r)dr, \qquad \int_{-1}^1 P_{vj}^\sigma(\eta)d\eta, \qquad \int_0^{2\pi} \sin(\sigma\phi), \qquad \int_0^{2\pi} \cos(\sigma\phi), \qquad (2.92)$$

and in addition to this, using the orthogonality relations of $J_{v+\frac{1}{2}}(\zeta_{vj}r)$ and $P_v^\sigma(\zeta_{vj}r)$ respectively, we obtain

$$A_{vj\zeta}\cos(\sigma\phi) + B_{vj\zeta}\sin(\sigma\phi) = \frac{(2v+1)(v-\sigma)!}{\pi(v+\sigma)! J_{v+\frac{1}{2}}(\zeta_{vj})J_{v-\frac{1}{2}}(\zeta_{vj})}$$

$$\int_0^{2\pi}\int_{-1}^1\int_1^1 r'^{\frac{3}{2}} J_{v+\frac{1}{2}}(\zeta_{vj}r') P_v^\eta(\eta')\cos(\sigma(\phi - \phi'))$$

$$\times T(r', \phi', \eta')dr'd\phi'd\eta'. \qquad (2.93)$$

Now replacing the above expression equation (2.89) into Eq. (2.93) and replacing T by its value we obtain

$$
\Phi_0(r, \eta, \phi, t) = \sum_{v=0}^{\infty} \sum_{j=1}^{\infty} \sum_{a=0}^{v} J_{v+\frac{1}{2}}(\eta_{vj} r) P_v^{\sigma}(\eta) A_{\zeta \eta j} \frac{(2v+1)(v-\sigma)!}{\pi (v+\sigma)! J_{v+\frac{1}{2}}(\zeta_{vj}) J_{v-\frac{1}{2}}(\zeta_{vj})}
$$

$$
\left(\int_0^{2\pi} \int_{-1}^{1} \int_1^{1} r'^{\frac{3}{2}} J_{v+\frac{1}{2}}(\zeta_{vj} r') P_v^{\eta}(\eta') \cos(\sigma (\phi - \phi')) T(r', \phi', \eta') dr' d\phi' d\eta' \right)
$$

$$
\times \exp \left(-\pi^2 \left(\frac{T}{S} \right)^2 \right). \tag{2.94}
$$

Now that we obtained the first component, the remaining terms can be determined using the recursive formula below, $\Phi_0(r, \eta, \phi, t)$:

$$
\Phi_{2n}(r, t) = \frac{Q}{4\pi T \sqrt{(r)}} \int_0^t dt_0 \int dV_0 L_{\tau_1^2}^{2j} \Phi_{2(n-1-j)}(r_0, t_0) \frac{4\pi}{k} \left(\frac{k}{2\sqrt{(\tau \pi)}} \right)^3 \exp \left(-\frac{k^2 R_x^2}{4\pi} \right). \tag{2.95}
$$

For instance,

$$
\Phi_2(r, t) = \frac{Q}{4\pi T \sqrt{(r)}} \int_0^t dt_0 \int dV_0 L_{\tau_1^2}^{2j} \Phi_0(r_0, t_0) \frac{4\pi}{k} \left(\frac{k}{2\sqrt{(\tau \pi)}} \right)^3 \exp \left(-\frac{k^2 R_x^2}{4\pi} \right), \tag{2.96}
$$

where Φ_0 is the expression in Eq. (2.93) and L_{τ^2} is the operator that accounts for the deformation.

2.6 Parallel Flow Model

Fractured formations are fractured rocks cut by an interconnected system of fractures resulting in blocks surrounded by a fracture network. It is worth noting that there is a difference between fractured formations with impervious blocks, for instance granite, and fractured formations with previous blocks (also known as matrix), this last one is called double porosity, for instance limestone and sandstone. The mathematical equation describing the flow of water via an aquifer can be found in [24,25]. We use the simple analytical solution describing the relationship between the apertures and the discharge rate. It is assumed that the upward flow of water along the faulty cement annuli can be approximated by the well-known cubic law (parallel plate model for fractures). We can represent a fracture as a planar void with two flat parallel surfaces. The hydraulic conductivity of this fracture is defined as follows:

$$
K_f = (2b)^2 \frac{\rho g}{12\mu}, \tag{2.97}
$$

where $2b$ is the fracture aperture, ρ is the density of water, g is acceleration due to gravity, and μ is the viscosity of water. The mean groundwater velocity through the fracture V_m can be calculated as the product of the fracture hydraulic conductivity and the hydraulic gradient:

$$V_m = K_f \frac{\delta_i}{\delta_z},$$

(2.98)

where $\frac{\delta_i}{\delta_z}$ is the hydraulic gradient. The transmissivity of an individual fracture is then

$$T_f = (2b)^3 \frac{\rho g}{12\mu},$$

(2.99)

and the flux along the fracture is

$$Q = T_f \frac{\delta_i}{\delta_z}$$

(2.100)

where Q is the flow in m^3/day per m width. The validity of the cubic law for laminar flow of fluids through open fractures consisting of parallel planar plates has been established over a wide range of conditions with apertures ranging down to a minimum of 0.2 μm. Artificially induced tension fractures and the laboratory setup used radial as well as straight flow geometries. Apertures ranged from 250 down to 4 μm, which was the minimum size that could be attained under a normal stress of 20 MPa. The cubic law was found to be valid whether the fracture surfaces were held open or were being closed under stress, and the results are not dependent on rock type. Permeability was uniquely defined by fracture aperture and was independent of the stress history used in these investigations. The apertures in this study are considered uncertain because it is very difficult even in the field or real world problem to measure accurately the apertures.

2.7 Uncertainties Analysis of Aquifer Parameters for Groundwater Flow Model

In statistics and quantitative research methodology, a data sample is a set of data collected or selected from a statistical population by a defined procedure. The elements of sample are known as sample points, sampling units or observations. Typically, the population is very large, making a census or a complete enumeration of all the values in the population impractical or impossible. The sample usually represents a subset of manageable size. Samples are collected and statistics are calculated from the samples so that one can make inferences or extrapolations from the sample to the population. The data sample may be drawn from a population without replacement, in which case it is a subset of a population; or with replacement, in which case it is a multi-subset. It is incumbent on the researcher to clearly define the target population. There are no strict rules to follow, and the researcher must rely on logic and judgment. The population is defined in keeping with the objectives of the study. Sometimes, the entire population will be sufficiently small, and the researcher can include the entire population in the study. This type of research is called a census study because data is gathered on every member of the population. Usually, the population is too large for the researcher to attempt to survey all of its members. A small, but carefully chosen sample can be used to represent the population. The sample reflects the characteristics of the population from which it is drawn.

Sampling methods are classified as either probability or non-probability. In probability samples, each member of the population has a known non-zero probability of being selected. Probability methods include random sampling, systematic sampling, and stratified sampling. In non-probability sampling, members are selected from the population in some non-random manner. These include convenience sampling, judgment sampling, quota sampling, and snowball sampling. The advantage of probability sampling is that sampling error can be calculated. Sampling error is the degree to which a sample might differ from the population. When inferring to the population, results are reported plus or minus the sampling error. In non-probability sampling, the degree to which the sample differs from the population remains unknown.

- **Random sampling** is the purest form of probability sampling. Each member of the population has an equal and known chance of being selected. When there are very large populations, it is often difficult or impossible to identify every member of the population, so the pool of available subjects becomes biased.
- **Systematic sampling** is often used instead of random sampling. It is also called an Nth name selection technique. After the required sample size has been calculated, every Nth record is selected from a list of population members. As long as the list does not contain any hidden order, this sampling method is as good as the random sampling method. Its only advantage over the random sampling technique is simplicity. Systematic sampling is frequently used to select a specified number of records from a computer file.
- **Stratified sampling** is a commonly used probability method that is superior to random sampling because it reduces sampling error. A stratum is a subset of the population that share at least one common characteristic. Examples of strata might be males and females, or managers and non-managers. The researcher first identifies the relevant strata and their actual representation in the population. Random sampling is then used to select a sufficient number of subjects from each stratum. "Sufficient" refers to a sample size large enough for us to be reasonably confident that the stratum represents the population. Strat-

ified sampling is often used when one or more of the strata in the population have a low incidence relative to the other strata.

- **Convenience sampling** is used in exploratory research where the researcher is interested in getting an inexpensive approximation of the truth. As the name implies, the samples are selected because they are convenient. This non-probability method is often used during preliminary research efforts to get a gross estimate of the results, without incurring the cost or time required to select a random sample.
- **Judgment sampling** is a common non-probability method. The researcher selects the sample based on judgment. This is usually and extension of convenience sampling. For example, a researcher may decide to draw the entire sample from one "representative" city, even though the population includes all cities. When using this method, the researcher must be confident that the chosen sample is truly representative of the entire population.
- **Quota sampling** is the non-probability equivalent of stratified sampling. Like stratified sampling, the researcher first identifies the strata and their proportions as they are represented in the population. Then convenience or judgment sampling is used to select the required number of subjects from each stratum. This differs from stratified sampling, where the strata are filled by random sampling.
- **Snowball sampling** is a special non-probability method used when the desired sample characteristic is rare. It may be extremely difficult or cost prohibitive to locate respondents in these situations. Snowball sampling relies on referrals from initial subjects to generate additional subjects. While this technique can dramatically lower search costs, it comes at the expense of introducing bias because the technique itself reduces the likelihood that the sample will represent a good cross section from the population.

Parameter uncertainty can be defined as uncertainty that arises in selecting values for parameters in the various models. There are many parameters in this assessment that are uncertain. First, there are insufficient data about the site climatic, geological, and hydrological conditions. As a result, such parameters as sorption coefficients, moisture content, river flow rate, river depth and width, hydraulic gradient in the aquifer, and erosion rate are taken from the general literature. Some parameters used need to be specified more accurately, for example, evaporation or distance between the disposal facility and the river or between the disposal facility and residences. On the other hand, the sensitivity analysis aims at quantifying the individual contribution from each parameter's uncertainty to the uncertainty of outputs. Correlations between parameters may also be inferred from sensitivity analysis. It is a frequent routine and is recommended to perform the uncertainty and sensitivity analysis in [26–28]. Here we shall present some approaches that can be used to evaluate parameters uncertainties in groundwater studies.

2.7.1 Samples Generation

The Latin Hypercube Sampling (LHS) is a type of stratified Monte Carlo (MC). The sampling region is partitioned into a specific manner by dividing the range of each component of x. We will only consider the case where the components of x are independent or can be transformed into an independent base. Moreover, the sample generation for correlated components with Gaussian distribution can be easily achieved. As originally described, in the following manner, LHS operates to generate a sample size N from the x variables $x_1, x_2, x_3, \ldots\ldots\ldots x_n$. The range of each variable is partitioned into N non-overlapping intervals on the basis of equal probability size $1/N$. One value from each interval is selected at random with respect to the probability density in the interval. The N values thus obtained for x_1 are paired in a random manner with the N values of x_2. These N pairs are combined in a random manner with the N values of x_3 to form Nn-triplets, and so on, until a set of Nn-tuples is formed. This set of Nn-tuples is the Latin hypercube sample. Thus, for given values of N and n, there exist $(N!)^{n-1}$ possible interval combinations for an LHS. A 10-run LHS for three normalized variables (range [0, 1]) with the uniform probability density function (p.d.f.) is listed below. In this case, the equal probability spaced values are 0, ..., 0.8, 1.

- **Efficiency of LHSMC** Considers the case that x denotes an n-vectors random variable with p.d.f. $f_x(x)$ for $x \in S$. Let h denote an objective function given by

$$h = q(x) \tag{2.101}$$

Consider now the following class of estimators:

$$T = \frac{1}{N} \sum_{i=1}^{N} g(H_i) \tag{2.102}$$

where $g(.)$ is an arbitrary known function and $H_i = q(x_i)$. If $g(h) = h$, that is, if h is a fixed point for g, then T represents an estimator of $[h]$. If $g(h) = H_i$, one obtains the rth sample moment. By choosing $g(h) = u(c - h)u(.)$ as a step function, one achieves the empirical distribution function of h at the point c. Now consider the following theorems.

Theorem 2.7.1. $E[T] = E[g(h)]$.

Since both methods have limitations and strengths, we propose a new approach that combines both methods; the new approach will be called Monte Carlo Hypercube Sampling Method (MCHSM).

- **Monte Carlo hypercube sampling method** It was demonstrated that the hypercube sample method was more efficient and less time consuming than the Monte Carlo simulation.

However, this Monte Carlo simulation still presents some worth. We propose a methodology that combines both the Monte Carlo simulation and the Latin hypercube sampling
as follows. Assume that the uncertain parameter is β and it ranges within $[a, b]$, then the
first step in this method consists of generating the sampling via the Monte Carlo sampling
within $[a, b]$. The next step is to reduce the number of sampling by calculating the mean,
the variance, and the standard deviation of the generated sample. These statistical parameters can then further be used to construct a distribution function, for instance, the normal
distribution. With constructed distribution in hand, one can further apply the hypercube
sample method to generate the final samples.

2.7.2 Evaluation of Uncertainties by Mean of Statistic Formulas

Iman and Conover [29] applied the LHS approach to cumulative distribution function
(c.d.f.) estimation of the three computer models: (1) environmental radionuclide movement,
(2) multi-component aerosol dynamics, and (3) salt dissolution in bedded salt formations.
They reported a good agreement of c.d.f. estimations. In this section, the application of Monte
Carlo Latin hypercube sampling to groundwater flow solution will be discussed. In agreement
with the real world problem, we assume that unknown parameters in the Theis groundwater
flowing within a confined aquifer are boundaries, that is we consider the solution to be a function of transmissivity T, storativity S, hydraulic conductivity K, and the discharge rate Q, so
that $f = c(\delta_i)$, $\delta_i \in [a_i, b_i]$ with $i = 1, 2, 3, 4$. Then, according to the Monte Carlo Latin hypercube technique, we first generate sample via the Monte Carlo sampling of for instance the
transmissivity.

Parameters uncertainties evaluation of groundwater flow equation

- **Cumulation function**: The distribution for the transmissivity, hydraulic conductivity, and
  discharge rate is presented as a cumulative distribution function (CDF) or as a complementary cumulative distribution function (CCDF), which is simply one minus the CDF.
  Hence, in our case the cumulative distribution function can be approximated as follows:

$$prob(f(\alpha) > F(\alpha)) = \sum_{i=1}^{N} \delta_{F(\alpha)}(f_i)\frac{1}{N}, \tag{2.103}$$

where

$$\delta_{F(\alpha)}(f_i) = \left\{ \begin{array}{ll} 1 & \text{if } f_i > F(\alpha) \\ 0 & \text{if } f_i < F(\alpha) \end{array} \right\}, \tag{2.104}$$

$prob(f(\alpha) > F(\alpha))$ is the probability that a value larger than $F(\alpha)$ will occur. The distribution function approximated above provides the most complete representation of the

uncertainty in the transmissivity, hydraulic conductivity, or discharge rate that is derived from the distributions.

- **Variance of the Sample**: the form of estimator of the variance is given as

$$Var(f) = \frac{1}{N-1} \sum_{i=1}^{N} (f_i - \overline{f})^2 \tag{2.105}$$

The goodness of an unbiased estimator can be measured by its variance. The variance approximated here provides a summary of this distribution but with the inevitable loss of resolution that occurs when the information is contained in 20 numbers [30].

- **Repeatability Uncertainty**: It is important noting that repeatability uncertainty is equal to the standard deviation of the sample data [31]. In the case under investigation, the mathematical expression is given as follows:

$$S(f) = \sqrt{\frac{1}{N-1} \sum_{i=1}^{N} (f_i - \overline{f})^2}. \tag{2.106}$$

- **Develop the Error Model**: An error model is an algebraic expression that defines the total error in the value of a quantity in terms of all relevant measurement process or component errors. The error model for the quantity $f(\alpha)$, that can be transmissivity or discharge rate, can be calculated with the formula

$$\epsilon_{f(\alpha)} = \sum_{i=1}^{4} \epsilon_{\alpha_i} f_{\alpha_i}, \tag{2.107}$$

where $\epsilon_{f(\alpha)}$ is the error of a given aquifer parameter and ϵ_{α_i} is the error in measurement of the given aquifer parameter, f_{α_i} are the first-order sensitivity coefficients that determine the relative contribution of the errors in the given parameter to the total error in $f(\alpha)$. For this purpose, we can chose the following definition of error:

$$\epsilon_{\alpha_i} = \frac{maximunvalue - minimunvalue}{100maximun}, i = 1, 2, 3, 4. \tag{2.108}$$

- **Uncertainty in Quantities**: The uncertainty in a quantity or variable is the square root of the variable's mean square error or variance. In mathematical terms, this is expressed as follows:

$$u_{\Phi(\alpha,r,t)} = \left(\sum_{i=1}^{4} \epsilon_{\alpha_i}^2 \Phi_{\alpha_i}^2 (r, t) \right)^{\frac{1}{2}}, \tag{2.109}$$

providing that the correlation coefficients for the error in the aquifer parameters are equal to zero.

- **Skewness and Kurtosis Tests**: Descriptive statistics, such as skewness and kurtosis, can provide relevant information about the normality of the data sample. Skewness is a measure of how symmetric the data distribution is about its mean. Kurtosis is a measure of the "peakedness" of the distribution [27–29]. In mathematical terms for the case under investigation, these are expressed as follows: since $f_j(\alpha)(j = 1, 2, ..N)$ are our sampled functions from a sample of size N, with mean $\overline{f}$ and standard deviation $var(f)$, then the sample coefficient of skewness c_3 and coefficient of kurtosis c_4 are given by [29–31]:

$$c_3 = \frac{1}{N-1}\frac{\sum_{i=1}^{N}\left(f_i - \overline{f}\right)^3}{var[(f^3)]}. \tag{2.110}$$

The following formula shows the response of the analytical expression of the sample coefficient of skewness:

$$c_4 = \frac{1}{N-1}\frac{\sum_{i=1}^{N}\left(f_i - \overline{f}\right)^4}{var[(f^4)]}. \tag{2.111}$$

The above study is very important in groundwater study because, to have a clear knowledge of aquifer parameters, several measurements of each parameter must be done, and once these parameters are known, they can be exposed to aleatory uncertainty analysis. One can conclude this section by noting that, as the increasing of human and also perhaps the climate pressures on groundwater resources all over the globe, accurate and reliable predictions of groundwater flow and pollutant transport are essential for sustainable groundwater management practices. Typically, nevertheless, the geological structure not adequately known and point measurements of subsurface properties or groundwater heads are sparse and prone to error. Consequently, incomplete or biased process representation, errors in the specification of initial and boundary conditions, as well as errors in the model parameters, render the predictions of groundwater dynamics and pollutant transport uncertain. Therefore, uncertainty of either parameters or other uncertainties assessments in groundwater flow and pollution modeling applications must for a better prediction typically associate all sources of uncertainty to errors in parameters and inputs; there it is worth noting that neglecting these sources of uncertainty, for instance, may introduce errors in the conceptualization of the dynamical system. Confining the set of acceptable dynamical system representations to a single mathematical formulation or model ends up in under-dispersive and prone-to-bias predictions which are observed nowadays in the field of geo-hydrology especially for those dealing with consulting in non-developed and developing countries. In this section, we tried to present some of a general and flexible approach that combines generalized likelihood uncertainty estimation and Bayesian model averaging to assess uncertainty in model predictions that arise from errors in model structure, inputs, and parameters for the groundwater model. Before starting our analysis, a set of plausible parameters of these groundwater models could be selected, and the joint prior input and parameter space could be sampled to form potential simulators of the system. For each parameter, the likelihood measures of acceptable simulators, assigned to them based on their ability to reproduce observed system behavior, could be integrated over the joint input and parameter space to obtain the integrated model likelihood. The latter could be used to weight the predictions of the respective model in the ensemble predictions.

CHAPTER 3

Groundwater Pollution

More than 50% of the world's population relies on groundwater for drinking water as well as irrigation purposes. However, groundwater is vulnerable to pollutants. Contamination of groundwater may occur when anthropogenic products such as gasoline, oil, road salts and chemicals enter a groundwater system, causing the groundwater to be unsuitable and harmful for human use. Material existing above groundwater can enter underlying soil and eventually reach the aquifer. This can be seen in cases where pesticides and fertilizers, as well as road salt, toxic materials from mine sites, and used motor oil migrate or seep through the subsurface and eventually to the aquifer over a period of time. To add, contamination of groundwater can also be a result of untreated waste from septic tanks, toxic chemicals coming from underground storage units, and leaky landfills. Consuming these contaminated waters may lead to serious health issues. To expand, contamination generated from septic tanks is associated with diseases such as hepatitis and dysentery; and toxins leached from wells into aquifer are associated with poisoning. Additionally, contaminated groundwater becomes a danger to wildlife species, and polluted water may lead to long-term health issues such as cancer.

3.1 History of Groundwater Pollution: Love Canal Disaster

The Love Canal is an area within Niagara Falls, New York. It is found in the LaSalle region of the city. It extends up to 36 blocks in the far southeastern edge of the city, along 99th street and Read Avenue. Two water bodies are associated with the northern and southern boundaries of the area, where Bergholtz Creek is to the north and Niagara River is about one-quarter mile (400 m) to the south. In the mid-1970s the Love Canal became of both national and international interest after reports in the press indicating the location of Love Canal was initially used for burying 22,000 tons of toxic waste by Hooker Chemical Company that is now known as Occidental Petroleum Corporation. In 1953, Hooker Chemical unwillingly sold the site of the Love Canal to the Niagara Falls School Board for $1, with a deed specifying the existence of contaminants and the liability limitation clause regarding contamination [31]. After a period of time after taking control of the land, the School Board proceeded with development. However, this development incorporated construction which eventually broke down containment structures in several different ways, causing previously trapped chemicals to migrate out. This downfall in containment structures together with heavy rainfall led to release and spreading of chemical waste. Consequently, this further led to a public health emergency and an ur-ban planning issue. This entire issue became a test case for liability clauses, and eventually Hooker Chemical were regarded as "negligent" in their disposing of waste but not neglect-ful in selling the site. Furthermore, the site was found and investigated from 1976 through the excavation in 1978 by a local newspaper – the Niagara Falls Gazette. Ten years later, David Axelrod from the New York State Health Department Commissioner, stated that Love Canal would for a long time be remembered as a "national symbol of a failure to exercise a sense of concern for future generations" [32]. The Love Canal incident was particularly important due to the inhabitants overflowing into the wastes [31,32].

3.1.1 Love Canal Disaster

During the time of the waste sites closure, Niagara Falls experienced rapid economic expan-sion and rapid population growth to a point where it exceeded growing by 33 percent in a space of 20 years ranging from 1940 to 1960 from 78,020 to 102,394 [31,32]. Land for devel-opment of new schools was required by the Niagara Falls City School District. Subsequently, the school district made an attempt to buy the land which was used to bury toxic waste from Hooker Chemical. The attempt was disallowed due to the associated safety issues. Despite this, the school district remained persistent [31,32]. After a period of time, Hooker Chemical was faced with parts of property being issued with condemnation and expropriation, and so they settled with selling the property but only provided the school district bought the entire property for one dollar. As a means of ensuring the school district knew what the potential consequences and risks associated with signing the deed and then owning the property were, Hooker Chemical, using their own expenditure conducted borehole tests on site with mem-bers of the school district as observers. Moreover, Hooker Chemical made sure the school district understood how unsuitable the site was for what the school district itself had planned. Regardless of that, the school district had no interest in making adjustments to their plans. Furthermore, on April 28, 1953, the agreement of selling this property to the school district was made, and in addition Hooker Chemical included a seventeen-line warning regarding the threats and risks of building on the property. In doing so, Hooker Chemical were assured they were no longer going to be affected by the legal issues associated with the consequences linked to the aforementioned threats and risks that could arise in the future [31,32]. "Prior to the delivery of this instrument of conveyance, the grantee herein has been advised by the grantor that the premises described above have been filled, in whole or in part, to the present grade level thereof with waste products resulting from the manufacturing of chemicals by the grantor at its plant in the City of Niagara Falls, New York, and the grantee assumes all risk and liability incident to the use thereof. It is therefore understood and agreed that, as a part of the consideration for this conveyance and as a condition thereof, no claim, suit, action or de-mand of any nature whatsoever shall ever be made by the grantee, its successors or assigns, against the grantor, its successors or assigns, for injury to a person or persons, including death resulting there from, or loss of or damage to property caused by, in connection with or by reason of the presence of said industrial wastes. It is further agreed as a condition hereof that each subsequent conveyance of the aforesaid lands shall be made subject to the foregoing provisions and conditions."

3.1.2 Construction of the 93rd Street School and the 99th Street School

Regardless of this clause, the school district proceeded with construction of the "99th Street School" where it was initially intended for location. In January 1954, the school's architect informed that education committee that during excavation, workers found two dump-sites containing 55-US-gallon (210 l; 46 imp gal) drums of chemical wastes. Additionally, the architect said it would be an inadequate approach to build on the property because it was not known what wastes existed below ground, and so there could have been potential damage to the concrete foundation [31–35]. Accordingly, the school district moved the site for development, about eighty to eighty-five feet further north [31–35]. Moreover, relocation was also given to the kindergarten playground, as it was directly above a chemical dump. In 1955, the development was complete, allowing 400 children placement in the school. This occurred along with openings of a number of other schools which were developed for other students. Within the same year, a twenty-five-foot area fell apart, exposing harmful chemical drums. These drums filled with water after occurrences of rain events. The children saw this an opportunity to be playful in large puddles of water [31–35]. Still in 1955, a second school opened six blocks away. Later in 1957, the City of Niagara Falls developed sewers systems for single-family and low-income residents. This was developed on lands neighboring the landfill site. Furthermore, the remaining property was sold off by the school district, and so homes were to be developed by private housing developers as well as the Niagara Falls Housing Authority, whose plan was to build the Griffon Manor housing project. During construction of gravel sewer beds, construction teams reached the protective clay seal which penetrated the canal walls [31–35]. The local government took parts of the protective clay layer for use as fill dirt for the 93rd Street School. Moreover, they pierced holes into the clay walls for development of water lines and the LaSalle Expressway. This made it possible for the toxic waste to seep out when rainwater was no longer restricted from entering the partly removed clay layer. This eventually led to movement of the toxic waste particles through the generated pierced holes of the clay walls [32–35]. Essentially, this promoted the built material to migrate from the canal. Furthermore, the property on which homes were built was not within the agreement between the school district and Hooker Chemical, and therefore none the home residents knew the canal's history [32–35]. No monitoring or evaluation was given to the waste stored beneath the ground. To add, the clay layer covering the canal which was intended to be impermeable began to break

[32–35]. On the other hand, construction of the LaSalle Expressway constricted groundwater flow to the Niagara River, and as a result in 1962 when there was a wet winter and spring, the expressway caused the breached canal to overflow with water. In addition, people started reporting about oil and colored liquid puddles on their properties and within their basements.

3.1.3 Health Problems and Site Cleanup of Love Canal

In 1976, David Pollak and David Russell, two reporters for the Niagara Falls Gazette, tested a number of sump pumps near to Love Canal, whereby the findings showed evidence of toxic chemicals. These findings were kept on the down low, and then in early 1978, Michael Brown, who was also a reporter, investigated the potential health effects by means of carrying an informal door-to-door survey. He found birth defects and several anomalies, of which enlarged feet, heads, hands, and legs were included. Thereafter, he gave advice to local residents to initiate a protest group which was under the leadership of Karen Schroeder, whose daughter had these birth defects. The New York State Health Department went with the same pattern and found an irregular frequency of miscarriages. Consequently, the dumpsite was proclaimed a phenomenal state crisis on August 2, 1978. Furthermore, Michael Brown who wrote numerous articles regarding the dump, tested the groundwater and later found the dump to be three times bigger than initially suspected, with conceivable implications past the first evacuation zone. Additionally, he found evidence of highly toxic dioxins. Lois Gibbs, a neighborhood mother had a son Michael Gibbs who started school in September 1977 and developed epilepsy about three months thereafter. He started suffering from asthma and a urinary tract infection, as well as a decrease in white blood cells. This was all related to his exposure to the leaked chemical waste. Later Lois Gibbs found out that the neighborhood in which she resided had buried chemical waste beneath it [32–35]. On August 2, 1978, she called a decision to head the Love Canal Homeowners' Association, and started rallying home owners [32–35]. In the following years, Gibbs directed effort in investigating the neighborhood's worries about the residents' health. Gibbs along with other residents made frequent complaints about unusual smells and "substances" found on their properties. Gibbs' neighborhood had a high rate of unexplained ailments, miscarriages, and mental illnesses [32–35]. Moreover, the basements of some homes were covered in thick black substances and vegetation started dying out. To add, on some home owner properties, only shrubby grasses was found to grow [32–35]. Despite asking city officials to investigate the area, the issue was not solved. Moreover, Michael O'Laughlin, mayor of Niagara Falls, disgracefully indicated that nothing was wrong in Love Canal. Based on the United States Environmental Protection Agency (EPA) in 1979, residents showed a "disturbingly high rate of miscarriages ... Love Canal can now be added to a growing list of environmental disasters involving toxic, ranging from industrial workers stricken by nervous disorders and cancers to the discovery of toxic materials in the milk of nursing mothers." There was a case where one Love Canal family had two of their four children having birth defects. To expand, a young girl was conceived deaf with a cleft palate, an additional row of teeth and a bit retarded; and a boy was with a deficit in his eye [32–35]. Initially, scientific studies were not very convincing in proving the chemicals were the cause of the residents' illnesses. Moreover, even though eleven known or assumed cancer causing chemicals were identified, in which a predominant chemical was benzene, scientists remained at odds with this issue. Dioxin (polychlorinated dibenzodioxin), a very dangerous substance, was also found in the water. Dioxin is measured in parts per trillion, and for the Love Canal, water samples appeared to have as much as 53 parts per billion [32–35]. Geologists were appointed for determining whether the chemicals found in residential areas were caused by underground swales. It was concluded that chemicals could seep into residents' basements as well as evaporate into the atmosphere of homes. Later in 1979, the EPA gave an announcement regarding the blood test results of residents in Love Canal. The results gave indication of a high presence of white blood cells which was a sign of leukemia [35–37] and damaged chromosomes. In actual fact, 33% of the residents were associated with damaged chromosomes, and this was significant because generally a 1% of the population have chromosomal damage [35–37]. In other studies which were conducted, no harm was found [32–37]. Furthermore, in 1991, the United States National Research Council (NRC) carried out a survey on the Love Canal's health studies. They found that the major issue concerned groundwater and not drinking water because groundwater migrated into basements and then was exposed through the atmosphere and soil [35–37]. It was found that many of the studies made reports that residents exposed to this were associated with significantly lower weights in newborn babies and birth defects [35–37]. There was evidence that the harmful and unwanted effects decreased when there was no exposure [35–37]. The NRC additionally found a study indicating exposed children were associated with "excess of seizures, learning problems, hyperactivity, eye irritation, skin rashes, abdominal pain, and incontinence" and stunted growth [35,36,27]. Furthermore, research was done on voles which are small rodents, in which it was found that their mortality significantly increased in comparison to controls (mean life expectancy in exposed animals "23.6 and 29.2 days, respectively, compared to 48.8 days" for control animals) [35–37]. To add, there is also an ongoing health study regarding the canal residents, by the New York State [35–38]. Still in the same year, the Albert Elia Building Co., Inc., now known as Sevenson Environmental Services, Inc., have appointed principal contractor to safe re-burying of the harmful waste at the Love Canal. In the end, the government called for relocation of more than 800 families and gave reimbursement for their homes. The United States Congress passed the Comprehensive Environmental Response, Compensation, and Liability Act (CERCLA), or the Superfund Act, due to the Superfund Act containing a "retroactive liability" provision. Furthermore, although the cleanup of the waste was governed by U.S laws in 1994 when the disposal was made, Occidental was held liable. John Curtin, a Federal District Judge, stated Hooker Chemical/Occidental was negligent but not reckless in their approach to handle the waste and sell the property to the Niagara Falls School District [35–38]. In addition, Curtin's decision contained an in-depth history of occasions paving the way to the Love Canal catastrophe. Finally, the EPA sued Occidental Petroleum who in 1995 agreed to pay $129 million in restitution [35–38]. Finally, during the years which followed, settlement of the residents' lawsuits was achieved.

3.2 Source of Pollution

We present in this section some sources of groundwater pollution.

- **Storage Tanks:** May contain gasoline, oil, chemicals, or other types of liquids and they can either be above or below ground. There are estimated to be over 10 million storage tanks buried in the United States and over time the tanks can corrode, crack, and develop leaks. If the contaminants leak out and get into the groundwater, serious contamination can occur.

- **Septic Systems:** On-site wastewater disposal systems used by homes, offices or other purpose arrangements that are not connected to a city sewer system. Septic systems are designed to slowly drain away human waste underground at a slow, harmless rate. An improperly designed, located, constructed, or maintained septic system can leak bacteria, viruses, household chemicals, and other contaminants into the groundwater causing serious problems.

- **Uncontrolled Hazardous Waste:** In many developed countries around the world, more precisely in the United State of America today, there are thought to be over 20,000 known abandoned and uncontrolled hazardous waste sites and the numbers grow every year. Hazardous waste sites can lead to groundwater contamination if there are barrels or other containers laying around that are full of hazardous materials. If there is a leak, these contaminants can eventually make their way down through the soil and into the groundwater.

- **Landfills:** Landfills are the places where our garbage is taken to be buried. Landfills are supposed to have a protective bottom layer to prevent contaminants from getting into the water. However, if there is no such layer or it is cracked, contaminants from the landfill (car battery acid, paint, household cleaners, etc.) can make their way down into the groundwater.

- **Chemicals and Road Salts:** The widespread use of chemicals and road salts is another source of potential groundwater contamination. Chemicals include products used on lawns and farm fields to kill weeds and insects and to fertilize plants, and other products used in homes and businesses. When it rains, these chemicals can seep into the ground and eventually into the water. Road salts are used in the wintertime to melt ice on roads to keep cars from sliding around. When the ice melts, the salt gets washed off the roads and eventually ends up in the water.

- **Atmospheric Contaminants:** Since groundwater is part of the hydrologic cycle, contaminants in other parts of the cycle, such as the atmosphere or bodies of surface water, can eventually be transferred into our groundwater supplies.

3.3 Type of Pollution

Contaminants found in groundwater cover a broad range of physical, inorganic chemical, organic chemical, bacteriological, and radioactive parameters. Principally, many of the same pollutants that play a role in surface water pollution may also be found in polluted groundwater, although their respective importance may differ. The following types can therefore be classified:

- **Pathogens** contained in human or animal feces can lead to groundwater pollution when they are given the opportunity to reach the groundwater, making it unsafe for drinking. Of the four pathogen types that are present in feces (bacteria, viruses, protozoa, and helminths or helminth eggs), the first three can be commonly found in polluted groundwater, whereas the relatively large helminth eggs are usually filtered out by the soil matrix. Groundwater that is contaminated with pathogens can lead to fatal fecal-oral transmission of diseases, for instance, cholera, diarrhoea [39]. If the local hydrogeological conditions which can vary within a space of a few square kilometers are ignored, pit latrines can cause significant public health risks via contaminated groundwater.
- **Volatile organic compounds.** Volatile organic compounds (VOCs) are a dangerous contaminant of groundwater. They are generally introduced to the environment through careless industrial practices. Many of these compounds were not known to be harmful until the late 1960s and it was some time before regular testing of groundwater identified these substances in drinking water sources.
- **Nitrate.** The issue of nitrate pollution in groundwater from pit latrines, which has led to numerous cases of "blue baby syndrome" in children, notably in rural countries such as Romania and Bulgaria [40]. Nitrate levels above 10 mg/L (10 ppm) in groundwater can cause "blue baby syndrome" (acquired methemoglobinemia) [40,41]. Nitrate can also enter the groundwater via excessive use of fertilizers, including manure. This is because only a fraction of the nitrogen-based fertilizers are converted to produce and other plant matter. The remainder accumulates in the soil or gets lost as runoff [41,42]. High application rates of nitrogen-containing fertilizers combined with the high water-solubility of nitrate lead to increased runoff into surface water as well as leaching into groundwater, thereby causing groundwater pollution [42–44]. The excessive use of nitrogen-containing fertilizers (be they synthetic or natural) is particularly damaging, as much of the nitrogen that is not taken up by plants is transformed into nitrate which is easily leached [43,44]. The nutrients, especially nitrates, in fertilizers can cause problems of natural habitats and

of human health if they are washed off soil into watercourses or leached through soil into groundwater.

- **Arsenic.** In the Ganges Plain of northern India and Bangladesh severe contamination of groundwater by naturally occurring arsenic affects 25% of water wells in the shallower of two regional aquifers. The pollution occurs because aquifer sediments contain organic matter that generates anaerobic conditions in the aquifer. These conditions result in the microbial dissolution of iron oxides in the sediment and, thus, the release of the arsenic, normally strongly bound to iron oxides, into the water. As a consequence, arsenic-rich groundwater is often iron-rich, although secondary processes often obscure the association of dissolved arsenic and dissolved iron.
- **Fluoride.** In areas that have naturally occurring high levels of fluoride in groundwater which is used for drinking water, both dental and skeletal fluorosis can be prevalent and severe [45].
- **Organic.** Organic pollutants can also be found in groundwater, such as insecticides and herbicides, a range of organohalides and other chemical compounds, petroleum hydrocarbons, various chemical compounds found in personal hygiene and cosmetic products, drug pollution involving pharmaceutical drugs and their metabolites. Inorganic pollutants might include ammonia, nitrate, phosphate, heavy metals, or radionuclides.

3.4 Heath Problems Caused by Groundwater Pollution

Waterborne diseases are caused by pathogenic microorganisms that most commonly are transmitted in contaminated fresh water. Infection commonly results during bathing, washing, drinking, in the preparation of food, or the consumption of food thus infected. Various forms of waterborne diarrheal diseases probably are the most prominent examples, and affect mainly children in developing countries; according to the World Health Organization, such diseases account for an estimated 4.1% of the total daily global burden of disease, and cause about 1.8 million human deaths annually. The World Health Organization estimates that 88% of that burden is attributable to unsafe water supply, sanitation, and hygiene [46]. The term *waterborne disease* is reserved largely for infections that predominantly are transmitted through contact with or consumption of infected water. Trivially, many infections may be transmitted by microbes or parasites that accidentally, possibly as a result of exceptional circumstances, have entered the water, but the fact that there might be an occasional freak infection need not mean that it is useful to categorize the resulting disease as *waterborne*. Nor is it common practice to refer to diseases such as malaria as *waterborne* just because mosquitoes have aquatic phases in their life cycles, or because treating the water they inhabit happens to be an effective strategy in control of the mosquitoes that are the vectors.

Microorganisms causing diseases that characteristically are waterborne prominently include protozoa and bacteria, many of which are intestinal parasites, or invade the tissues or circulatory system through walls of the digestive tract. Various other waterborne diseases are caused by viruses. In spite of philosophical difficulties associated with defining viruses as *organisms*, it is practical and convenient to regard them as microorganisms in this connection [47]. Yet other important classes of waterborne diseases are caused by metazoan parasites. Typical examples include certain Nematoda, that is to say roundworms. As an example of waterborne Nematode infections, one important waterborne nematodal disease is Dracunculiasis. It is acquired by swallowing water in which certain copepoda occur that act as vectors for the Nematoda. Anyone swallowing a copepod that happens to be infected with Nematode larvae in the genus Dracunculus becomes liable to infection. The larvae cause guinea worm disease [46–48]. Another class of waterborne metazoan pathogens are certain members of the Schistosomatidae, a family of blood flukes. They usually infect victims that make skin contact with the water [46,47]. Blood flukes are pathogens that cause Schistosomiasis of various forms, more or less seriously affecting hundreds of millions of people worldwide [47,48]. Long before modern studies had established the germ theory of disease, or any advanced understanding of the nature of water as a vehicle for transmitting disease, traditional beliefs had cautioned against the consumption of water, rather favoring processed beverages such as beer, wine, and tea. For example, in the camel caravans that crossed Central Asia along the Silk Road, the explorer Owen Lattimore noted, "The reason we drank so much tea was because of the bad water. Water alone, unboiled, is never drunk". There is a superstition that it causes blisters on the feet [49]. Waterborne diseases can have a significant impact on the economy, locally as well as internationally. People who are infected by a waterborne disease are usually confronted with related costs and not seldom with a huge financial burden. This is especially the case in less developed countries. The financial losses are mostly caused by, for instance, costs for medical treatment and medication, costs for transport, special food, and by the loss of manpower. Many families must even sell their land to pay for treatment in a proper hospital. On average, a family spends about 10% of the monthly households income per person infected [50]. In general, we have the following types of infections:

- Algal Infections, with disease: Desmodesmus Infection
- Viral infections, with diseases: SARS (Severe Acute Respiratory Syndrome), Hepatitis A, Poliomyelitis (Polio), Polyomavirus infection.
- Bacterial infections, with disease: Vibrio Illness, Typhoid fever, Salmonellosis, Otitis Externa (swimmer's ear), Leptospirosis, Legionellosis (two distinct forms: Legionnaires' disease and Pontiac fever), Dysentery, M. marinum infection, E. coli Infection, Cholera, Campylobacteriosis, Botulism.
- Parasitic infections, with disease: Enterobiasis, Ascariasis, Coenurosis, Echinococcosis (Hydatid disease), Hymenolepiasis (Dwarf Tapeworm Infection), Fasciolopsiasis, Taeniasis, Dracunculiasis (Guinea Worm Disease), Schistosomiasis (immersion).

- Protozoal infections, with disease: Microsporidiosis, Giardiasis (fecal-oral) (hand-to-mouth), Cyclosporiasis, Cryptosporidiosis (oral), Amoebiasis (hand-to-mouth).

3.5 Convection Dispersion Model

The convection–diffusion equation is a combination of the diffusion and convection (advection) equations, and describes physical phenomena where particles, energy, or other physical quantities are transferred inside a physical system due to two processes: diffusion and convection. Depending on context, the same equation can be called the convection–diffusion equation, drift-diffusion equation, or (generic) scalar transport equation.

3.5.1 Derivation of the Mathematical Model

The convection–diffusion equation can be derived in a straightforward way from the continuity equation, which states that the rate of change for a scalar quantity in a differential control volume is given by flow and diffusion into and out of that part of the system along with any generation or consumption inside the control volume:

$$\frac{\partial c}{\partial t} + \nabla . \vec{j} = R, \tag{3.1}$$

where $\vec{j}$ is the total flux and R is a net volumetric source for c. There are two sources of flux in this situation. First, diffusive flux arises due to diffusion. This is typically approximated by Fick's first law:

$$\vec{j}_{diffusion} = -D\nabla c. \tag{3.2}$$

The above equation means the flux of the diffusing material relative to the bulk motion in any part of the system is proportional to the local concentration gradient. Second, when there is overall convection or flow, there is an associated flux called adjective flux:

$$\vec{j}_{advection} = \vec{v} c. \tag{3.3}$$

The total flux (in a stationary coordinate system) is given by the sum of these two:

$$\vec{j} = \vec{j}_{advection} + \vec{j}_{diffusion} = \vec{v} c - D\nabla c. \tag{3.4}$$

Plugging into the continuity equation, we obtain:

$$\frac{\partial c}{\partial t} + \Delta . (\vec{v} c - D\nabla c) = R \tag{3.5}$$

which is the model used to predict the movement of pollution within the geological formation called aquifers. In this equation the parameters involved are:

- c is the variable of interest (species concentration for mass transfer, temperature for heat transfer),
- D is the diffusivity (also called diffusion coefficient), such as mass diffusivity for particle motion or thermal diffusivity for heat transport,
- $\vec{v}$ is the average velocity with which the quantity is moving. For example, in advection, c might be the concentration of salt in a river, and then $\vec{v}$ would be the velocity of the water flow. As another example, c might be the concentration of small bubbles in a calm lake, and then $\vec{v}$ would be the average velocity of bubbles rising towards the surface by buoyancy (see below). For multiphase flows and flows in porous media, $\vec{v}$ is the (hypothetical) superficial velocity.
- R describes "sources" or "sinks" of the quantity c. For example, for a chemical species, $R > 0$ means that a chemical reaction is creating more of the species, and $R < 0$ means that a chemical reaction is destroying the species. For heat transport, $R > 0$ might occur if thermal energy is being generated by friction.
- ∇ represents gradient and $\nabla\cdot$ represents divergence.

Understanding the terms involved. The right-hand side of the equation is the sum of three contributions.

- The first, $\nabla \cdot (D\nabla c)$, describes diffusion. Imagine that c is the concentration of a chemical. When concentration is low somewhere compared to the surrounding areas (e.g. a local minimum of concentration), the substance will diffuse in from the surroundings, so the concentration will increase. Conversely, if concentration is high compared to the surroundings (e.g. a local maximum of concentration), then the substance will diffuse out and the concentration will decrease. The net diffusion is proportional to the Laplacian (or second derivative) of concentration.
- The second contribution, $-\nabla \cdot (\vec{v}c)$, describes convection (or advection). Imagine standing on the bank of a river, measuring the water's salinity (amount of salt) each second. Upstream, somebody dumps a bucket of salt into the river. A while later, you would see the salinity suddenly rise, then fall, as the zone of salty water passes by. Thus, the concentration at a given location can change because of the flow.
- The final contribution, R, describes the creation or destruction of the quantity. For example, if c is the concentration of a molecule, then R describes how the molecule can be created or destroyed by chemical reactions. R may be a function of c and of other parameters. Often there are several quantities, each with its own convection–diffusion equation, where the destruction of one quantity entails the creation of another. For example, when methane burns, it involves not only the destruction of methane and oxygen but also the creation of carbon dioxide and water vapor. Therefore, while each of these chemicals has its own convection–diffusion equation, they are "coupled together" and must be solved as a system of simultaneous differential equations.

3.5.2 Derivation of Exact Solution

In this section, we shall consider a more complex advection dispersion equation (ADE). Here, the analytical solution of the one-dimensional ADE for linear pulse time dependent boundary condition is derived using Laplace transform. The ADE is considered with constant parameters where the decay is also taken into account. The ADE with these properties is as follows:

$$\frac{\partial C}{\partial t} = -v\frac{\partial C}{\partial x} + D\frac{\partial^2 C}{\partial x^2} - kC, 0 < x, t < \infty. \tag{3.6}$$

In which, C is the solute concentration, v is the constant flow velocity, D is the constant diffusion coefficient, k the coefficient of first order reaction, while t and x represent the variables of time and space respectively. Eq. (3.6) is solved with respect to the following initial and boundary conditions:

$$C(x,0) = C_0, \qquad C(0,t) = (at+b)[u(t-t_1) - u(t-t_2)], \qquad \frac{\partial C(\infty,t)}{\partial x}. \tag{3.7}$$

Here, C_0 is the initial concentration, a and b are the parameters of the linear pulse boundary condition at $x = 0$, t_1 and t_2 are the beginning and ending times of the source activation, respectively, while $u(t-t_i)$ is the shifted Heaviside function, defined to be zero for $t < t_i$ and 1 for $t \geq t_i$. Applying the Laplace transform to Eq. (3.6) and its boundary conditions yields the corresponding problem in the Laplace domain that is:

$$s\overline{C} - C(x,0) = -v\frac{d\overline{C}}{dx} + D\frac{d^2\overline{C}}{dx^2} - k\overline{C}, \qquad \overline{C} = L(C). \tag{3.8}$$

L is the Laplace transform operator, s representing the Laplace transform variable. Eq. (3.8) is a linear inhomogeneous ordinary differential equation that is of the following general solution:

$$\overline{C(x,s)} = c_1 \exp\left(\frac{vx}{2D}\sqrt{\frac{v^2x^2}{4D^2} + \frac{kx^2}{D} + \frac{sx^2}{D}}\right) + c_1 \exp\left(\frac{vx}{2D}\sqrt{\frac{v^2x^2}{4D^2} + \frac{kx^2}{D} + \frac{sx^2}{D}}\right)$$
$$+ \frac{C_0}{s+k} \tag{3.9}$$

where c_1 and c_2 are arbitrary constants that can be specified using initial and boundaries conditions. For example, we chose

$$c_1 = 0 \tag{3.10}$$

$$c_2 = (at_1+b)\frac{\exp(-t_1 s)}{s} + a\frac{\exp(-t_1 s)}{s^2} - (at_2+b)\frac{\exp(-t_2 s)}{s} - a\frac{\exp(-t_2 s)}{s^2} - \frac{C_0}{s+k}, \tag{3.11}$$

however, replacing c_1 and c_2 by their value and applying the inverse Laplace transform in (3.9).

3.6 Groundwater Remediation: Techniques and Actions

Groundwater is the main source of drinking water as well as agricultural and industrial usage. Unfortunately, groundwater quality has been degraded due to improper waste disposal practices and accidental spillage of hazardous chemicals. Therefore, it is critical that the groundwater contamination be prevented and the contaminated groundwater at numerous sites worldwide be re-mediated in order to protect public health and the environment.

3.6.1 Remediation Technique

A systematic approach for the assessment and remediation of contaminated sites is necessary in order to facilitate the remediation process and avoid undue delays. The most important aspects of the approach include site characterization, risk assessment, and selection of an effective remedial action [51]. Innovative integration of various tasks can often lead to a faster, cost-effective remedial program. Site characterization is often the first step in a contaminated site remediation strategy. It consists of the collection and assessment of data representing contaminant type and distribution at a site under investigation. The results of a site characterization form the basis for decisions concerning the requirements of remedial action. Additionally, the results serve as a guide for design, implementation, and monitoring of the remedial system. Each site is unique; therefore, site characterization must be tailored to meet site-specific requirements. An inadequate site characterization may lead to the collection of unnecessary or misleading data, technical misjudgment affecting the cost and duration of possible remedial action, or extensive contamination problems resulting from inadequate or inappropriate remedial action. Site characterization is often an expensive and lengthy process; therefore, it is advantageous to follow an effective characterization strategy to optimize efficiency and cost.

An effective site characterization includes the collection of data pertaining to site geology, including site stratigraphy and important geologic formations; site hydrogeology, including major water-bearing formations and their hydraulic properties; and site contamination, including type, concentration, and distribution. Additionally, surface conditions both at and around the site must be taken into consideration. Because little information regarding a particular site is often known at the beginning of an investigation, it is often advantageous to follow a phased approach for the site characterization. A phased approach may also minimize financial impact by improving the planning of the investigation and ensuring the collection of relevant data. Phase I consists of the definition of investigation purpose and the performance of a preliminary site assessment. A preliminary assessment provides the geographical location, background information, regional hydrogeologic information, and potential sources of contamination pertaining to the site. The preliminary site assessment consists of two tasks, a literature review and a site visit. Based on the results of the Phase I activities, the purpose and scope of the Phase II exploratory site investigation need to be developed. If contamination was detected at the site during the course of the preliminary investigation, the exploratory site investigation must be used to confirm such findings as well as obtain further data necessary for the design of a detailed site investigation program. A detailed work plan should be prepared for the site investigations describing the scope of related field and laboratory testing. The work plan should provide details about sampling and testing procedures, sampling locations, and frequency, a quality assurance/quality control (QA/QC) plan, a health and safety (S and H) plan, a work schedule, and a cost assessment. Phase III includes a detailed site investigation in order to define the site geology and hydrogeology as well as the contamination profile. The data obtained from the detailed investigation must be adequate to properly assess the risk posed at the site as well as to allow for effective designs of possible remedial systems. As with the exploratory investigations, a detailed work plan including field and laboratory testing programs as well as QA/QC and S and H plans should be outlined. Depending on the size, accessibility, and proposed future purpose of the site, this investigation may last anywhere from a few weeks to a few years. Because of the time and the effort required, this phase of the investigation is very costly. If data collected after the first three phases is determined to be inadequate, Phase IV should be developed and implemented to gain additional information. Additional phases of site characterization must be performed until all pertinent data has been collected.

Depending on the logistics of the project, site characterization may require regulatory compliance and/or approval at different stages of the investigation. Thus, it is important to review the applicable regulations during the preliminary site assessment (Phase I). Meetings with regulatory officials may also be beneficial to insure that investigation procedures and results conform to regulatory standards. This proactive approach may prevent delays in obtaining the required regulatory permits and/or approvals. Innovative site characterization techniques are increasingly being used to collect relevant data in an efficient and cost-effective manner. Recent advances in cone penetrometer and sensor technology have enabled contaminated sites to be rapidly characterized using vehicle-mounted direct push probes. Probes are available for directly measuring contaminant concentrations in situ, in addition to measuring standard stratigraphic data, to provide flexible, real-time analysis. The probes can also be reconfigured to expedite the collection of soil, groundwater, and soil gas samples for subsequent laboratory analysis. Noninvasive, geophysical techniques such as ground-penetrating radar, cross-well radar, electrical resistance tomography, vertical induction profiling, and high resolution seismic reflection produce computer-generated images of subsurface geological conditions and are qualitative at best. Other approaches such as chemical tracers are used to identify and quantify contaminated zones, based on their affinity for a particular contaminant and the measured change in tracer concentration between wells employing a combination of conservative and partitioning tracers.

Risk assessment

Once site contamination has been confirmed through the course of a thorough site characterization, a risk assessment is performed. A risk assessment is a systematic evaluation used to determine the potential risk posed by the detected contamination to human health and the environment under present and possible future conditions. If the risk assessment reveals that an unacceptable risk exists due to the contamination, a remedial strategy is developed to assess the problem. If corrective action is deemed necessary, the risk assessment will assist in the development of remedial strategies and goals necessary to reduce the potential risks posed at the site. The USEPA and the American Society for Testing and Materials (ASTM) have developed comprehensive risk assessment procedures. The USEPA procedure was originally developed by the United States Academy of Sciences in 1983. It was adopted with modifications by the USEPA for use in Superfund feasibility studies and RCRA corrective measure studies [52]. This procedure provides a general, comprehensive approach for performing risk assessments at contaminated sites. It consists of four steps:

- Hazard identification.
- Exposure assessment.
- Toxicity assessment.
- Risk characterization.

3.6.2 Remediation Action

When the results of a risk assessment reveal that a site does not pose risks to human health or the environment, no remedial action is required. In some cases, however, monitoring of a site may be required to validate the results of the risk assessment. Corrective action is required when risks posed by the site are deemed unacceptable. When action is required, remedial strategy must be developed to insure that the intended remedial method complies with all technological, economic, and regulatory considerations. The costs and benefits of various remedial alternatives are often weighed by comparing the flexibility, compatibility, speed, and cost of each method. A remedial method must be flexible in its application to ensure that it is adaptable to site-specific soil and groundwater characteristics. The selected method must be able to address site contamination while offering compatibility with the geology and hydrogeology of the site. Generally, remediation methods are divided into two categories: in situ remediation methods and ex situ remediation methods. In situ methods treat contaminated groundwater in-place, eliminating the need to extract groundwater. In situ methods are advantageous because they often provide economic treatment, little site disruption, and increased safety due to lessened risk of accidental contamination exposure to both on-site workers and the general public within the vicinity of the remedial project. Successful implementation of in situ methods, however, requires a thorough understanding of subsurface conditions. Ex situ methods are used to treat extracted groundwater. Surface treatment may be performed either on-site or off-site, depending on site-specific conditions. Ex situ treatment methods are attractive because consideration does not need to be given to subsurface conditions. Ex situ treatment also offers easier control and monitoring during remedial activity implementation.

Some techniques for groundwater remediation

Groundwater remediation techniques span biological, chemical, and physical treatment technologies. Most ground water treatment techniques utilize a combination of technologies. Some of the biological treatment techniques include bio-augmentation, bioventing, biosparging, bioslurping, and phyto-remediation. Some chemical treatment techniques include ozone and oxygen gas injection, chemical precipitation, membrane separation, ion exchange, carbon absorption, aqueous chemical oxidation, and surfactant enhanced recovery. Some chemical techniques may be implemented using nano-materials. Physical treatment techniques include, but are not limited to, pump and treat, air sparging, and dual phase extraction. **Biological treatment technologies.** Within this category, we have the following well-known techniques:

- Bio-augmentation: If a treatability study shows no degradation (or an extended lab period before significant degradation is achieved) in contamination contained in the groundwater, then inoculation with strains known to be capable of degrading the contaminants may be helpful. This process increases the reactive enzyme concentration within the bioremediation system and subsequently may increase contaminant degradation rates over the non-augmented rates, at least initially after inoculation [53].
- Bioventing is an in situ remediation technology that uses microorganisms to biodegrade organic constituents in the groundwater system. Bioventing enhances the activity of indigenous bacteria and archaea and stimulates the natural in situ biodegradation of hydrocarbons by inducing air or oxygen flow into the unsaturated zone and, if necessary, by adding nutrients [54]. During bioventing, oxygen may be supplied through direct air injection into residual contamination in soil. Bioventing primarily assists in the degradation of adsorbed fuel residuals, but also assists in the degradation of volatile organic compounds (VOCs) as vapors move slowly through biologically active soil [55].
- Biosparging is an in situ remediation technology that uses indigenous microorganisms to biodegrade organic constituents in the saturated zone. In biosparging, air (or oxygen) and nutrients (if needed) are injected into the saturated zone to increase the biological activity of the indigenous microorganisms. Biosparging can be used to reduce concentrations of petroleum constituents that are dissolved in groundwater, adsorbed to soil below the water table, and within the capillary fringe.

- Bioslurping combines elements of bioventing and vacuum-enhanced pumping of free-product that is lighter than water (light non-aqueous phase liquid or LNAPL) to recover free-product from the groundwater and soil, and to bio-remediate soils. The bioslurper system uses a "slurp" tube that extends into the free-product layer. Much like a straw in a glass draws liquid, the pump draws liquid (including free-product) and soil gas up the tube in the same process stream. Pumping lifts LNAPLs, such as oil, off the top of the water table and from the capillary fringe meaning an area just above the saturated zone, where water is held in place by capillary forces. The LNAPL is brought to the surface, where it is separated from water and air. The biological processes in the term "bioslurping" refer to aerobic biological degradation of the hydrocarbons when air is introduced into the unsaturated zone.
- In the phyto-remediation process certain plants and trees are planted, whose roots absorb contaminants from groundwater over time, and are harvested and destroyed. This process can be carried out in areas where the roots can tap the groundwater. Few examples of plants that are used in this process are Chinese Ladder fern Pteris vittata, also known as the brake fern, is a highly efficient accumulator of arsenic. Genetically altered cottonwood trees are good absorbers of mercury and transgenic Indian mustard plants soak up selenium well.

Chemical treatment technologies

- Chemical precipitation is commonly used in wastewater treatment to remove hardness and heavy metals. In general, the process involves addition of agent to an aqueous waste stream in a stirred reaction vessel, either batchwise or with steady flow. Most metals can be converted to insoluble compounds by chemical reactions between the agent and the dissolved metal ions. The insoluble compounds (precipitates) are removed by settling and/or filtering.
- Ion exchange for groundwater remediation is virtually always carried out by passing the water downward under pressure through a fixed bed of granular medium (either cation exchange media and anion exchange media) or spherical beads. Cations are displaced by certain cations from the solutions and ions are displaced by certain anions from the solution. Ion exchange media most often used for remediation are zeolites (both natural and synthetic) and synthetic resins.
- Carbon absorption: The most common activated carbon used for remediation is derived from bituminous coal. Activated carbon absorbs volatile organic compounds from groundwater by chemically binding them to the carbon atoms.
- Chemical oxidation: In this process, called In Situ Chemical Oxidation or ISCO, chemical oxidants are delivered in the subsurface to destroy (converted to water and carbon dioxide or to nontoxic substances) the organic molecules. The oxidants are introduced as either liquids or gasses. Oxidants include air or oxygen, ozone, and certain liquid chemicals

such as hydrogen peroxide, permanganate and persulfate. Ozone and oxygen gas can be generated on site from air and electricity and directly injected into soil and groundwater contamination. The process has the potential to oxidize and/or enhance naturally occurring aerobic degradation. Chemical oxidation has proven to be an effective technique for dense non-aqueous phase liquid or DNAPL when it is present.

- Surfactant enhanced recovery increases the mobility and solubility of the contaminants absorbed to the saturated soil matrix or present as dense non-aqueous phase liquid. Surfactant-enhanced recovery injects surfactants (surface-active agents that are primary ingredient in soap and detergent) into contaminated groundwater. A typical system uses an extraction pump to remove groundwater downstream from the injection point. The extracted groundwater is treated above-ground to separate the injected surfactants from the contaminants and groundwater. Once the surfactants have separated from the groundwater they are reused. The surfactants used are non-toxic, food-grade, and biodegradable. Surfactant enhanced recovery is used most often when the groundwater is contaminated by dense non-aqueous phase liquids (DNAPLs). These dense compounds, such as trichloroethylene (TCE), sink in groundwater because they have a higher density than water. They then act as a continuous source for contaminant plumes that can stretch for miles within an aquifer. These compounds may biodegrade very slowly. They are commonly found in the vicinity of the original spill or leak where capillary forces have trapped them.

Physical treatment technologies

- Pump and treat is one of the most widely used groundwater remediation technologies. In this process groundwater is pumped to the surface and is coupled with either biological or chemical treatments to remove the impurities.
- Air sparging is the process of blowing air directly into the groundwater. As the bubbles rise, the contaminants are removed from the groundwater by physical contact with the air and are carried up into the unsaturated zone. As the contaminants move into the soil, a soil vapor extraction system is usually used to remove vapors.
- Dual-phase vacuum extraction
  Dual-phase vacuum extraction (DPVE), also known as multi-phase extraction, is a technology that uses a high-vacuum system to remove both contaminated groundwater and soil vapor. In DPVE systems a high-vacuum extraction well is installed with its screened section in the zone of contaminated soils and groundwater. Fluid/vapor extraction systems depress the water table and water flows faster to the extraction well. DPVE removes contaminants from above and below the water table. As the water table around the well is lowered by pumping, unsaturated soil is exposed. This area, called the capillary fringe, is often highly contaminated, as it holds undissolved chemicals, chemicals that are lighter

than water, and vapors that have escaped from the dissolved groundwater below. Contaminants in the newly exposed zone can be removed by vapor extraction. Once above ground, the extracted vapors and liquid-phase organic and groundwater are separated and treated. Use of dual-phase vacuum extraction with these technologies can shorten the cleanup time at a site, because the capillary fringe is often the most contaminated area.

- Monitoring-wells are often drilled for the purpose of collecting groundwater samples for analysis. These wells, which are usually six inches or fewer in diameter, can also be used to remove hydrocarbons from the contaminant plume within a groundwater aquifer by using a belt style oil skimmer. Belt oil skimmers, which are simple in design, are commonly used to remove oil and other floating hydrocarbon contaminants from industrial water systems. A monitoring-well oil skimmer remedies various oils, ranging from light fuel oils such as petrol, light diesel or kerosene to heavy products such as No. 6 oil, creosote, and coal tar. It consists of a continuously moving belt that runs on a pulley system driven by an electric motor. The belt material has a strong affinity for hydrocarbon liquids and for shedding water. The belt, which can have a vertical drop of 100+ feet, is lowered into the monitoring well past the LNAPL/water interface. As the belt moves through this interface it picks up liquid hydrocarbon contaminant, which is removed and collected at ground level as the belt passes through a wiper mechanism. To the extent that DNAPL hydrocarbons settle at the bottom of a monitoring well, and the lower pulley of the belt skimmer reaches them, these contaminants can also be removed by a monitoring-well oil skimmer. Typically, belt skimmers remove very little water with the contaminant, so simple weir type separators can be used to collect any remaining hydrocarbon liquid, which often makes the water suitable for its return to the aquifer. Because the small electric motor uses little electricity, it can be powered from solar panels or a wind turbine, making the system self-sufficient and eliminating the cost of running electricity to a remote location.

3.7 Sensibility Analysis of Model Parameters

Sensitivity analysis is the study of how the uncertainty in the output of a mathematical model or system, numerical or otherwise, can be apportioned to different sources of uncertainty in its inputs [53,54]. A related practice is uncertainty analysis, which has a greater focus on uncertainty quantification and propagation of uncertainty. Ideally, uncertainty and sensitivity analyses should be run in tandem. The process of recalculating outcomes under alternative assumptions to determine the impact of variable under sensitivity analysis can be useful for a range of purposes [55], including:

- Testing the robustness of the results of a model or system in the presence of uncertainty.
- Increased understanding of the relationships between input and output variables in a system or model.

- Uncertainty reduction: identifying model inputs that cause significant uncertainty in the output and should therefore be the focus of attention if the robustness is to be increased perhaps by further research [53–56].
- Searching for errors in the model by encountering unexpected relationships between inputs and outputs [53–56].
- Model simplification fixing model inputs that have no effect on the output, or identifying and removing redundant parts of the model structure.
- Enhancing communication from modelers to decision makers, for instance by making recommendations more credible, understandable, compelling, or persuasive [53–56].
- Finding regions in the space of input factors for which the model output is either maximum or minimum or meets some optimum criterion; see optimization and Monte Carlo filtering [53–56].
- In case of calibrating models with large number of parameters, a primary sensitivity test can ease the calibration stage by focusing on the sensitive parameters. Not knowing the sensitivity of parameters can result in time being uselessly spent on non-sensitive ones [53–56].

Taking an example from groundwater studies, in a groundwater flow or pollution model there are always variables that are uncertain. Advection coefficients, velocity of the plume, retardation factor, transmissivity, storativity, hydraulic conductivity and other variables may not be known with great precision. Sensitivity analysis answers the question, *If these deviate from expectations, what will the effect be on the business, model, system, or whatever is being analyzed, and which variables are causing the largest deviations?*. A mathematical model is defined by a series of equations, input variables, and parameters aimed at characterizing some process under investigation. Some other examples beside groundwater flow and pollution problems might be a climate model, an economic model, or a finite element model in engineering. Increasingly, such models are highly complex, and as a result their input/output relationships may be poorly understood. In such cases, the model can be viewed as a black box, meaning the output is an opaque function of its inputs. Good modeling practice requires that the modeler provides an evaluation of the confidence in the model. This requires, first, a quantification of the uncertainty in any model results uncertainty analysis, and second, an evaluation of how much each input is contributing to the output uncertainty. Sensitivity analysis addresses the second of these issues although uncertainty analysis is usually a necessary precursor, performing the role of ordering by importance the strength and relevance of the inputs in determining the variation in the output [53,54,56].

3.7.1 Some Commonly Used Methods for Sensitivity Analysis

There are a large number of approaches to performing a sensitivity analysis, many of which have been developed to address one or more of the constraints discussed above [53]. They are also distinguished by the type of sensitivity measure, be it based on, for instance, variance decomposition, partial derivatives, or elementary effects. In general, however, most procedures adhere to the following outline:

a Quantify the uncertainty in each input for example ranges, probability distributions. Note that this can be difficult and many methods exist to elicit uncertainty distributions from subjective data [57].

b Identify the model output to be analyzed (the target of interest should ideally have a direct relation to the problem tackled by the model).

c Run the model a number of times using some design of experiments [58], dictated by the method of choice and the input uncertainty.

d Using the resulting model outputs, calculate the sensitivity measures of interest.

In some cases this procedure will be repeated; for example, in high-dimensional problems, where the user has to screen out unimportant variables before performing a full sensitivity analysis, we present some approaches used for sensibility analysis starting with the best one. **One-at-a-time** approach is that of changing one-factor-at-a-time (OFAT or OAT), to see what effect this produces on the output [59–61]. OAT customarily involves the following steps:

- Moving one input variable, keeping others at their baseline nominal values.
- Returning the variable to its nominal value, then repeating for each of the other inputs in the same way.

Sensitivity may then be measured by monitoring changes in the output, for example by partial derivatives or linear regression. This appears a logical approach as any change observed in the output will unambiguously be due to the single variable changed. Furthermore, by changing one variable at a time, one can keep all other variables fixed to their central or baseline values. This increases the comparability of the results (all effects are computed with reference to the same central point in space) and minimizes the chances of computer programme crashes, more likely when several input factors are changed simultaneously. OAT is frequently preferred by modelers because of practical reasons. In case of model failure under OAT analysis the modelers immediately know which is the input factor responsible for the failure [59–61].

- **Screening method.** Screening is a particular instance of a sampling-based method. The objective here is rather to identify which input variables are contributing significantly to the output uncertainty in high-dimensionality models, rather than exactly quantifying sensitivity, that is to say, in terms of variance. Screening tends to have a relatively low computational cost when compared to other approaches, and can be used in a preliminary analysis to weed out uninfluential variables before applying a more informative analysis to the remaining set [62,63]. One of the most commonly used screening methods is the elementary effect method.

- **Scatter plots method.** A simple but useful tool is to plot scatter plots of the output variable against individual input variables, after randomly sampling the model over its input distributions. The advantage of this approach is that it can also deal with "given data," implying a set of arbitrarily-placed data points, and gives a direct visual indication of sensitivity. Quantitative measures can also be drawn, for example, by measuring the correlation between Y and X_i, or even by estimating variance-based measures by nonlinear regression [64].

- **Regression analysis,** in the context of sensitivity analysis, involves fitting a linear regression to the model response and using standardized regression coefficients as direct measures of sensitivity. The regression is required to be linear with respect to the data like a hyperplane, hence with no quadratic terms, and so on, as regressors because otherwise it is difficult to interpret the standardized coefficients. This method is therefore most suitable when the model response is in fact linear; linearity can be confirmed, for instance, if the coefficient of determination is large. The advantages of regression analysis are that it is simple and has a low computational cost [65,66].

- **Local methods.** Local methods involve taking the partial derivative of the output Y with respect to an input factor X_i:

$$\left| \frac{\partial Y}{\partial X_i} \right|_{\mathbf{x}^0},$$
(3.12)

where the subscript X^0 indicates that the derivative is taken at some fixed point in the space of the input, hence the local in the name of the class. Adjoint modeling and Automated Differentiation are methods in this class [65,66].

- **Variance-based method.** Variance-based methods [67–69] are a class of probabilistic approaches which quantify the input and output uncertainties as probability distributions, and decompose the output variance into parts attributable to input variables and combinations of variables. The sensitivity of the output to an input variable is therefore measured by the amount of variance in the output caused by that input. These can be expressed as conditional expectations; that is to say, considering a model $Y = f(X)$ for $X = X1, X2, ...Xk$, a measure of sensitivity of the ith variable X_i is given as

$$\text{Var}_{X_i}\left(E_{\mathbf{X}_{\sim i}}\left(Y \mid X_i \right) \right)$$
(3.13)

where "*Var*" and "*E*" denote the variance and expected value operators respectively, and X_i denotes the set of all input variables except X_i. This expression essentially measures the contribution X_i alone to the uncertainty (variance) in Y (averaged over variations in other variables), and is known as the first-order sensitivity index or main effect index. Importantly, it does not measure the uncertainty caused by interactions with other variables [67–69]. A further measure, known as the total effect index, gives the total variance in Y caused by X_i and its interactions with any of the other input variables. Both quantities are typically standardized by dividing by $Var(Y)$ [67–69].

These methods allow full exploration of the input space, accounting for interactions, and nonlinear responses. For these reasons they are widely used when it is feasible to calculate them. Typically this calculation involves the use of Monte Carlo methods, but since this can involve many thousands of model runs, other methods such as emulators can be used to reduce computational expense when necessary. Note that full variance decomposition is only meaningful when the input factors are independent from one another [67–69].

3.7.2 Limitations of Sensibility Analysis Methods

One-at-a-time. Despite its simplicity, however, this approach does not fully explore the input space, since it does not take into account the simultaneous variation of input variables. This means that the OAT approach cannot detect the presence of interactions between input variables. *Local method.* Similar to OAT/OFAT, local methods do not attempt to fully explore the input space, since they examine small perturbations, typically one variable at a time. *Variance based-method.* It is designed to test against any and all alternatives to the null hypothesis and thus may be suboptimal for testing against a specific hypothesis. It is optimal when losses are proportional to the square of the differences among the unknown population means, but may not be optimal otherwise. For example, when losses are proportional to the absolute values of the differences among the unknown population means, expected losses would be minimized via a test that makes use of the absolute values of the differences among the sample means. It is designed for use when the observations are drawn from a normal distribution and though it is remarkably robust, it may not yield exact p-values when the observations come from distributions that are heavier in the tails than the normal. Even in cases when the analysis of variance yields almost exact p-values, it may be less powerful than the corresponding permutation test when the observations are drawn from non-normal distributions under the alternative. *Scatter plots method.* One major limitation of scatter-plots is that they are most effective with small numbers of dimensions, as increasing the dimensionality results in decreasing the screen space provided for each projection. Strategies for addressing this limitation include using three dimensions per plot or providing panning or zooming mechanisms. Other limitations include being generally restricted to orthogonal views and difficulties in discovering relationships which span more than two dimensions. Advantages of scatter-plots include ease of interpretation and relative insensitivity to the size of the data set.

3.8 Problems of Transboundary Aquifers

Water draws people together because water is life. However, when many people, animals, and industries are competing over limited water, things can get tense. Transboundary aquifers are sources of groundwater that defy our political boundaries and often lead to intense conversation about what should be done in order to give everyone a fair share. In the past decades, researchers, policy makers, and citizens have been actively working together under international guidelines to make major improvements to helping solve transboundary water issues. We can now take a broad look around the world to see what is working and what is not.

The Internationally Shared Aquifer Resource Management (ISARM) Initiative has recently published a methodological guide outlining best practices. The worldwide ISARM Initiative is a UNESCO and International Association of Hydrogeologists (IAH) led multi-agency effort aimed at improving the understanding of scientific, socio-economic, legal, institutional, and environmental issues related to the management of transboundary aquifers. The guidebook, Towards the Concerted Management of Transboundary Aquifer Systems, uses both case studies and analysis in order to identify the features of successful water management programs around the world. The guide comes in three parts:

- the need for a more comprehensive approach based on Integrated Water Resources Management (IWRM) principles,
- a range of technical, legal, organizational, economic, training, and cooperation tools that can help improve the knowledge and management of resources, and
- a progressive, multi-pronged approach for implementing the concerted, equitable, and sustainable management of transboundary aquifer systems, as well as potential mechanisms for creating and sustainability operating appropriate institutional structures to manage these shared groundwater resources.

CHAPTER 4

Limitations of Groundwater Models With Local Derivative

In both hydrogeology research and applied studies, computer-based models using the concept of local derivative are frequently employed to examine fundamental questions about rates and directions of groundwater flow and associated transport of groundwater pollution. Presumably, modeling allows us to examine in a quantitative fashion both theoretical and applied questions in hydrogeology. For instance, we may wish to understand the role of heterogeneity in hydraulic conductivity on groundwater and lake interaction, or movement of a zone of contaminated groundwater in a municipality that uses groundwater for municipal water supply. Modeling of groundwater systems can give us answers to these and other problems. It will also give answers with high precision in some cases. However, using the concept of local derivative, the precision offered by these groundwater models will not in any way imply that it provides an accurate description of the nature and its complexities. Even if we understand the physical and chemical processes of groundwater flow and associated transport perfectly, our understanding of geology and recharge of any region is always too vague to allow for accurate assessment of real-world problems. The ability of mathematical models of groundwater based upon the local derivative to accurately predict the behavior of groundwater flow and pollution in the real-world situations is poor. At best groundwater models based on local derivative, despite their high degree of precision, are qualitative predictors of future behavior. A major cause of the lack of accuracy is the severe discrepancy between the scale of measurement necessary to understand aquifer parameters for accurate modeling and the scale of measurement generally made under the constraints of limited time and limited budgets. Another problem is the introduction of aquifer complexity into the mathematical equations using the local derivative. Undoubtedly any model created employing local derivative is an extreme idealization of natural and is non-unique. A non-unique model is one that inherently possesses error.

4.1 Limitations of Groundwater Flow Model

Groundwater flow models are necessarily simplified mathematical representations of complex natural systems. Because of this, there are limits to the accuracy with which groundwater systems can be simulated. These limitations must be known when using models and interpreting model results. There are many sources of error and uncertainty in models. Model error commonly stems from practical limitations of grid spacing, time discretization, parameter

Fractional Operators with Constant and Variable Order
with Application to Geo-Hydrology
DOI: 10.1016/B978-0-12-809670-3.00004-7

structure, insufficient calibration data, and the effects of processes not simulated by the model. These factors, along with unavoidable error in observations, result in uncertainty in model predictions.

Actual groundwater flow systems are much more complex than the conceptual models can typically represent. The accuracy of predictive simulations is thus difficult to assess, so it's wise to assume a fair amount of uncertainty when using models to make predictions. In particular, the subsurface has complex distributions of materials with transient groundwater fluxes. No matter how much effort is spent drilling, sampling, and testing the subsurface, only a small fraction of it is ever sampled or tested hydraulically. The available data will provide only an incomplete picture of the actual subsurface system. Because of the inherent difficulty of characterizing subsurface regions, substantial uncertainty is always introduced in the conceptual model created. Typically, the complex distribution of subsurface materials is represented in the conceptual model as regions with locally homogeneous and isotropic hydraulic conductivity and anisotropic representation are also common. The parameters assigned to these regions are chosen to represent the large-scale average (or effective) hydraulic behaviors of flow and transport. Complex transient fluxes like recharge or pumping rates are represented in the conceptual model as either steady-state average values or as transient rates that change in some simplified manner. In a well-constructed mathematical model, most of the uncertainty in the results stems from discrepancies between the real system and the conceptual system. Most mathematical models provide a fairly accurate simulation of the conceptual system. Therein lies the danger. Accurate simulation of the conceptual system is often taken to mean accurate simulation of the real system. With sophisticated model-generated graphics, there is a tendency to forget the unavoidable uncertainties in representing the real system with a simpler conceptual model. With the groundwater flow equation, Darcy's law is a simple mathematical statement which neatly summarizes several familiar properties that groundwater flowing in aquifers exhibits, including:

- if there is no pressure gradient over a distance, no flow occurs (these are hydrostatic conditions);
- if there is a pressure gradient, flow will occur from high pressure towards low pressure (opposite the direction of increasing gradient – hence the negative sign in Darcy's law);
- the greater the pressure gradient (through the same formation material), the greater the discharge rate; and
- the discharge rate of fluid will often be different through different formation materials (or even through the same material, in different directions) even if the same pressure gradient exists in both cases.

The ∇p used in the Darcy law is the pressure gradient vector, which does not really replicate the real-world situation as the local derivative cannot describe the change of pressure in different scales in the real aquifer. Darcy's law is valid for laminar flow through the soil. In fine grained soil, the dimensions of interstices are small and thus flow is laminar. Coarse-grained soils also behave similarly but in very coarse-grained soil, the flow is of turbulent nature. Hence Darcy's law is not valid in such soils. For flow through commercial pipes, the flow is laminar when Reynolds number is less than 2000, but in some soils it has been found that flow is laminar when the value of Reynolds number is less than unity. Almost all aquifers found in the real world are not homogeneous, but the soil properties differ from one point to another. The derivative with integer order cannot portray such real-world problem. Theis groundwater flow model was borrowed from heat equation. The Theis solution is based on the following assumptions:

- The flow in the aquifer is adequately described by Darcy's law, that is to say $Re < 10$. Homogeneous, isotropic, confined aquifer;
- the well is fully penetrating (open to the entire thickness (b) of aquifer);
- the well has zero radius (it is approximated as a vertical line) – therefore no water can be stored in the well;
- the well has a constant pumping rate Q;
- the head loss over the well screen is negligible;
- aquifer is infinite in radial extent;
- horizontal (not sloping), flat, impermeable (non-leaky) top and bottom boundaries of aquifer;
- groundwater flow is horizontal;
- no other wells or long-term changes; in regional water levels all changes in potentiometric surface are the result of the pumping well alone.

These assumptions are rarely all met, especially when the rates are described using the local derivative. Another major limitation of the groundwater flow equations using the concept of derivative with integer number is the ability to remember the history of the flow of water within a given aquifer as the local derivative cannot describe the memory effect.

4.2 Limitations of Groundwater Convection Dispersion Model

Characterizing the collective behavior of particle transport on the Earth surface is a key ingredient in describing landscape evolution. We seek equations that capture essential features of transport of an ensemble of particles on hillslopes, valleys, river channels, or river networks, such as mass conservation, super-diffusive spreading in flow fields with large velocity variation, or retardation due to particle trapping. An important class of problems in the Earth surface sciences involves describing the collective behavior of particles in transport. Familiar examples include transport of solute and contaminant particles in surface and subsurface water flows, the behavior of soil particles and associated soil constituents undergoing biome-chanical transport and mixing by bioturbation, and the transport of sediment particles and sediment-borne substances in turbulent shear flows, whether involving shallow flows over soils, deeper river flows, or ocean currents. These and many other examples share three essential features. First is the behavior of a well defined ensemble of particles. These particles may be considered "tracers," whose total mass is conserved or otherwise accounted for if radioactive decay, physical transformations, or chemical reactions are involved. Second, these tracers typically alternate between states of motion and rest over many time scales, and indeed, most tracers of interest in Earth surface systems are at rest much of the time. Third, when in transport, some tracers move faster, and some move slower, than the average motion due to spatiotemporal variations in the mechanisms inducing their motion. Tracer motions thus may be considered as consisting of quasi-random walks with rest periods. Natural geological deposits with highly contrasting permeability may form mobile and relatively immobile zones, where the potential mass exchange between mobile and immobile zones results in a wide time distribution for solute *trapping*. The transport process, groundwater, is, by its very nature, always in contact with the matrix of an aquifer. There is thus a possibility that the solutes may interact with the rock matrix and one another. A true mathematical model for groundwater pollution must therefore be able to account for interactions between the dissolved solids and matrix of the aquifer. It will thus be advantageous to look at the nature of the interactions between dissolved solids and a porous medium that may be expected in groundwater pollution. Experimental evidence indicates that when a dissolved solid comes in contact with the matrix of a porous medium it may

(a) pass through the medium with no apparent effect,
(b) be absorbed by the porous matrix, and
(c) react with the porous matrix and other substances dissolved in the fluid.

The dissolved solids encountered in porous flow are, for this reason, often classified as conservative, non-conservative, and reactive tracers. This behavior implies that the quantity of dissolved solids in a porous medium depends not only on the flow pattern but also on the nature of the porous matrix and the solution. These situations (a), (b), and (c) cannot be characterized efficiently by the time-local model, including the time advection dispersion equation. If the high-permeable material tends to form preferential flow paths, such as the interconnected paleochannels observed in alluvial depositional systems, then the solute transport may show a heavy leading edge, which cannot be described by the convection dispersion equation. Development of partial differential equations such as the convection–dispersion equation (ADE) begins with assumptions about the random behavior of a single particle: possible velocities it may experience in a flow field and the length of time it may be immobilized. When assumptions underlying the ADE are relaxed, other kinds of derivatives are needed to be applied on time or space terms. These derivatives must be nonlocal in order to describe transport affected by hydraulic conditions at a distance. Another model of convection dispersion within a geological formation arises when velocity variations are heavy tailed and describe particle motion that accounts for variation in the flow field over the entire system. But when the geological formation, under which this pollution moves, is changing in time and space, the ADE cannot be used.

An important physical problem was considered in the work by [70]; in their work there was considered the possibility that the hop length pdf declines with distance in such a way that the variance is undefined, for example, a power law distribution. In this case, long displacement distances, albeit relatively rare, are nonetheless more numerous than with, say, an exponential distribution, so that hop length density possesses a so-called *heavy tail*. For example, during downstream transport, sediment tracers dispersed by turbulent eddies and by momentary excursions into parts of the mean flow that are either faster or slower than the average may occasionally experience *super-dispersive* events during excursions into sites with unusually high flow velocities [71]. In these situations where the hop length density is heavily tailed, the scaling of dispersion is non-Fickian, that is, $\sigma = (Dt)^{\frac{1}{\alpha}}$, where $1 < \alpha < 2$. Because σ therefore grows at a rate faster than *normal* Fickian dispersion, this behavior is referred to as superdiffusive. Moreover, this behavior may be characterized as being *nonlocal* in that during a small interval, tracers released from position x mostly move to nearby positions, but also involve an *unusually large* number of motions to positions far from x. Conversely, during the same interval, tracers arriving at position x mostly originate from nearby positions, but also involve a significant number or motions originating far from x. This means that the behavior at a particular position does not necessarily depend only on local (nearby) conditions, but rather may also depend on conditions upstream or downstream. For example, because the local rate of change in tracer concentration consists of the divergence of the flux of tracers, then if the flux is related to the bed stress, say, in the case of sediment transport, the possibility exists that changes in the local concentration depend on stress conditions, and therefore on system configuration, *far* upstream. As described below, the behavior of the tracer concentration $C(x, t)$ in this nonlocal case may not be described by the advection-dispersion equation.

CHAPTER 5

Fractional Operators and Their Applications

5.1 Introduction to the Concept of Fractional Calculus

In applied mathematics and mathematical analysis, fractional derivative is a derivative of any arbitrary order, real or complex. Even though the term "fractional" is a misnomer, it has been widely accepted for such a derivative for a long time. The concept of a fractional derivative was coined by the famous mathematician Leibnitz in 1695 in his letter to L'Hôpital. In recent years, the Fractional Calculus (FC) draws increasing attention due to its applications in many fields. The history of the theory goes back to 17th century, when in 1695 the derivative of order $\frac{1}{2}$ was described by Leibnitz in his letter to L'Hospital [72–74]. Since then, the new theory turned out to be very attractive to mathematicians as well as physicists, biologists, engineers, and economists. The first application of fractional calculus was due to Abel in his solution to the Tautocrone problem [75]. It also has applications in biophysics, quantum mechanics, wave theory, polymers, continuum mechanics, Lie theory, field theory, spectroscopy and in group theory, among other applications [76–78]. In [79], Samko et al. provide an encyclopedic treatment of the subject. Various types of fractional derivatives were studied: Riemann–Liouville, Caputo, Hadamard, Erdelyi–Kober, Grunwald–Letnikov, Marchaud, and Riesz are just a few to name. Atangana and Aydin also provided an encyclopedic treatment of the subject including the difference between these derivatives with fractional order, their possible application in real-world problems, their advantages and disadvantages [80]. In fractional calculus, the fractional derivatives are defined via fractional integrals. According to the literature, the Riemann–Liouville fractional derivative (RLFD), hence the Riemann–Liouville fractional integral, plays a major role in FC [76–80]. The Caputo fractional derivative has also been defined via a modified Riemann–Liouville fractional integral [81]. Butzer et al. investigate properties of the Hadamard fractional integral and the derivative in [76–80]. In [82], they also obtained the Mellin transforms of the Hadamard fractional integral and differential operators and in [83], Pooseh et al. obtained expansion formulas of the Hadamard operators in terms of integer order derivatives. Many other interesting properties of those operators and others are summarized in [79] and the references therein. In [84], the author introduced a new fractional integral, which generalizes the Riemann–Liouville and the Hadamard integrals into a single form. Further properties such as expansion formulas, variational calculus applications, control theoretical applications, convexity and integral inequalities and Hermite–Hadamard type inequalities of this new operator and similar operators can be found in [79]. Recently a new derivative with fractional order was introduced by Caputo and Fabrizio [85] and applied in some interesting real-world problems [86–89]. Another modified version was proposed by Doungmo and Atangana and was called A.D. derivative [90,91].

5.2 Riemann–Liouville Type

The classical form of fractional calculus is given by the Riemann–Liouville integral, which is essentially what has been described above. The theory for periodic functions (therefore including the 'boundary condition' of repeating after a period) is the Weyl integral. It is defined on Fourier series, and requires the constant Fourier coefficient to vanish, thus, it applies to functions on the unit circle whose integrals tend to 0. A fairly natural question to ask is whether there exists a linear operator H, or half-derivative, such that:

$$H^2 f(x) = Df(x) = \frac{d}{dx} f(x) = f'(x). \tag{5.1}$$

It turns out that there is such an operator, and indeed for any $a > 0$, there exists an operator P such that

$$(P^a f)(x) = f'(x), \tag{5.2}$$

or to put it another way, the definition of $d^n y/dx^n$ can be extended to all real values of n.

Let $f(x)$ be a function defined for $x > 0$. Form the definite integral from 0 to x. Call this

$$(Jf)(x) = \int_0^x f(t)\, dt. \tag{5.3}$$

Repeating this process gives:

$$(J^2 f)(x) = \int_0^x (Jf)(t)dt = \int_0^x \left(\int_0^t f(s)\, ds \right) dt, \tag{5.4}$$

and this can be extended arbitrarily.

The Cauchy formula for repeated integration, namely

$$(J^n f)(x) = \frac{1}{(n-1)!} \int_0^x (x-t)^{n-1} f(t)\, dt, \tag{5.5}$$

leads in a straightforward way to a generalization for real n.

Using the gamma function to remove the discrete nature of the factorial function gives us a natural candidate for fractional applications of the integral operator.

$$(J^\alpha f)(x) = \frac{1}{\Gamma(\alpha)} \int_0^x (x-t)^{\alpha-1} f(t)\, dt \qquad (5.6)$$

This is in fact a well-defined operator.

It is straightforward to show that the J operator satisfies

$$(J^\alpha)(J^\beta f)(x) = (J^\beta)(J^\alpha f)(x) = (J^{\alpha+\beta} f)(x) = \frac{1}{\Gamma(\alpha+\beta)} \int_0^x (x-t)^{\alpha+\beta-1} f(t)\, dt. \quad (5.7)$$

This question can also be answered by employing the Laplace transform and the convolution theorem. Noting that

$$\mathcal{L}\{Jf\}(s) = \mathcal{L}\left\{ \int_0^t f(\tau)\, d\tau \right\}(s) = \frac{1}{s}(\mathcal{L}\{f\})(s) \qquad (5.8)$$

and

$$\mathcal{L}\left\{ J^2 f \right\} = \frac{1}{s}(\mathcal{L}\{Jf\})(s) = \frac{1}{s^2}(\mathcal{L}\{f\})(s) \qquad (5.9)$$

and so on, we assert

$$J^\alpha f = \mathcal{L}^{-1}\left\{ s^{-\alpha}(\mathcal{L}\{f\})(s) \right\}. \qquad (5.10)$$

For instance,

$$J^\alpha \left(t^k \right) = \mathcal{L}^{-1}\left\{ \frac{\Gamma(k+1)}{s^{\alpha+k+1}} \right\} = \frac{\Gamma(k+1)}{\Gamma(\alpha+k+1)} t^{\alpha+k} \qquad (5.11)$$

as expected. Indeed, given the convolution rule

$$\mathcal{L}\{f * g\} = (\mathcal{L}\{f\})(\mathcal{L}\{g\}) \qquad (5.12)$$

and short handing $p(x) = x^{\alpha-1}$ for clarity, we find that

$$(J^\alpha f)(t) = \frac{1}{\Gamma(\alpha)} \mathcal{L}^{-1}\{(\mathcal{L}\{p\})(\mathcal{L}\{f\})\} \qquad (5.13)$$

$$= \frac{1}{\Gamma(\alpha)}(p * f) \qquad (5.14)$$

$$= \frac{1}{\Gamma(\alpha)} \int_0^t p(t-\tau) f(\tau) \, d\tau \tag{5.15}$$

$$= \frac{1}{\Gamma(\alpha)} \int_0^t (t-\tau)^{\alpha-1} f(\tau) \, d\tau, \tag{5.16}$$

which is what Cauchy gave us above.

Laplace transforms "work" on relatively few functions, but they are often useful for solving fractional differential equations

$$_aD_t^{-\alpha} f(t) = {_aI_t^{\alpha}} f(t) = \frac{1}{\Gamma(\alpha)} \int_a^t (t-\tau)^{\alpha-1} f(\tau) d\tau. \tag{5.17}$$

5.2.1 Some Useful Properties

We present the following properties of the Riemann–Liouville fractional integral which are very useful. Fix a bounded open interval (a, b). The operator I^{α} associates to each integrable function f on (a, b) the function $I^{\alpha} f$ on (a, b) which is also integrable by Fubini's theorem. Thus I^{α} defines a linear operator on $L^1(a, b)$:

$$I^{\alpha} : L^1(a, b) \to L^1(a, b). \tag{5.18}$$

Fubini's theorem also shows that this operator is continuous with respect to the Banach space structure on L^1, and that the following inequality holds:

$$\|I^{\alpha} f\|_1 \le \frac{|b-a|^{\text{re}(\alpha)}}{\text{re}(\alpha)|\Gamma(\alpha)|} \|f\|_1. \tag{5.19}$$

Here $\| \cdot \|_1$ denotes the norm on $L^1(a, b)$.

More generally, by Hölder's inequality, it follows that if $f \in L^p(a, b)$, then $I^{\alpha} f \in L^p(a, b)$ as well, and the analogous inequality holds

$$\|I^{\alpha} f\|_p \le \frac{|b-a|^{\text{re}(\alpha)/p}}{\text{re}(\alpha)|\Gamma(\alpha)|} \|f\|_p \tag{5.20}$$

where $\| \cdot \|_p$ is the L^p norm on the interval (a, b). Thus I^{α} defines a bounded linear operator from $L^p(a, b)$ to itself. Furthermore, $I^{\alpha} f$ tends to f in the L^p sense as $\alpha \to 0$ along the real axis. That is,

$$\lim_{\substack{\alpha \to 0 \\ \alpha > 0}} \|I^{\alpha} f - f\|_p = 0 \tag{5.21}$$

for all $p \leq 1$. Moreover, by estimating the maximal function of I, one can show that the limit $I^{\alpha} f \to f$ holds pointwise almost everywhere.

The operator I^{α} is well-defined on the set of locally integrable functions on the whole real line R. It defines a bounded transformation on any of the Banach spaces of functions of exponential type $X^{\alpha} = L_1(e - \sigma|t|dt)$, consisting of locally integrable functions for which the norm

$$\|f\| = \int_{-\infty}^{\infty} |f(t)|e^{-\sigma|t|} \, dt \tag{5.22}$$

is finite. For f in X^{σ}, the Laplace transform of $I^{\alpha} f$ takes the particularly simple form

$$(\mathcal{L}I^{\alpha} f)(s) = s^{-\alpha} F(s) \tag{5.23}$$

for $re(s) > \sigma$. Here $F(s)$ denotes the Laplace transform of f, and this property expresses that I^{α} is a Fourier multiplier.

Definition 5.2.1. *The Mellin transform of a function $f(x)$, denoted by $f^*(x)$, is defined by*

$$f^*(s) = mf(x), s = \int_0^{\infty} x^{s-1} f(x) dx, x > 0. \tag{5.24}$$

The inverse Mellin transform is given by the contour integral

$$f(x) = m^{-1} f^*(s), x = \frac{1}{2\pi i} \int_{\alpha-i\infty}^{\alpha+i\infty} f^*(s) x^{-s} ds, \ i = \sqrt{-1}, \tag{5.25}$$

where α is a real number.

Theorem 5.2.2. *The following result holds true.*

$$m \left({}_0 J_x^{\alpha} f \right)(s) = \frac{\Gamma(1 - \alpha - s)}{\Gamma(1 - \alpha)} f^*(s + \alpha), \tag{5.26}$$

where $\Re(\alpha) > 0$ and $\Re(\alpha + s) < 1$.

Riemann–Liouville fractional derivative: The corresponding derivative is calculated using Lagrange's rule for differential operators. Computing nth order derivative over the integral of order $(n - \alpha)$, the α order derivative is obtained. It is important to remark that n is the nearest integer bigger than α – that is to say, $n = \lceil \alpha \rceil$.

$$_a D_t^{\alpha} f(t) = \frac{d^n}{dt^n} {}_a D_t^{-(n-\alpha)} f(t) = \frac{d^n}{dt^n} {}_a I_t^{n-\alpha} f(t) \tag{5.27}$$

The Laplace transform of the Riemann–Liouville fractional derivative is given as:

$$L_a D_t^\alpha f(t)(p) = s^\alpha F(s) - \sum_{k=0}^{n-1} s^k \left(D^{\alpha-k-1} f(t) \right)_{t=0}.$$ (5.28)

The Fourier transform of Riemann–Liouville fractional derivative is given as:

$$F_a D_t^\alpha f(t), w = (-iw)^\alpha G(w).$$ (5.29)

Theorem 5.2.3. *If $n \in \mathcal{N}$ and $\lim_{t\to\infty} t^{s-1} f^v(t) = 0$, $v = 0, 1, \ldots n$, then*

$$m\left(f^n(t), (s)\right) = (-1)^n \frac{\Gamma(s)}{\Gamma(s-n)} m(f(t); s-n).$$ (5.30)

5.3 Caputo Type

In this section we present the Caputo derivative with fractional order that was proposed by Italian Caputo in 1967. This form of fractional derivative is given as:

$$^C D_0^\alpha (f(t)) = \int_0^t \frac{f^n(y)}{(t-y)^{\alpha+1-n}} dy, n - 1 < \alpha < n \in \mathcal{N}.$$ (5.31)

Proposition 5.3.1. *Let $n - 1 < \alpha < n$, $n \in \mathcal{N}$, $\alpha \in \mathcal{R}$ and $f(t)$ be a function such that $_0^C D_t^\alpha (f(t))$ exists. Then the following for the Caputo fractional derivative hold:*

$$\lim_{\alpha \to n} {}^C D_t^\alpha (f(t)) = f^n(t),$$ (5.32)

$$\lim_{\alpha \to n-1} {}^C D_t^\alpha (f(t)) = f^{(n-1)}(t) - f^{(n-1)}(0).$$ (5.33)

Proof. The proof is achieved by mean of integration by parts as follows:

$$
\begin{aligned}
^C D_0^\alpha (f(t)) &= \int_0^t \frac{f^n(y)}{(t-y)^{\alpha+1-n}} dy \\
&= \frac{1}{n-\alpha+1} \left(-f^n(\tau) \frac{(t-\tau)^{n-\alpha}}{n-\alpha} \Big\|_{\tau=0}^t + \int_0^t f^{n-1}(\tau) \frac{(t-\tau)^{n-\alpha}}{n-\alpha} \right) \\
&= \frac{1}{n-\alpha+1} \left(f^n(0) t^{(n-\alpha)} + \int_0^t f^{(n+1)}(\tau)(t-\tau)^{n-\alpha} d\tau \right).
\end{aligned}
$$ (5.34)

Nevertheless, taking the limit for $\alpha \to n$ and $\alpha \to n-1$, respectively, we obtain

$$\lim_{\alpha \to n} {}^C D_t^\alpha (f(t)) = f^n(t),$$

$$\lim_{\alpha \to n-1} {}^C D_t^\alpha (f(t)) = f^{(n-1)}(t) - f^{(n-1)}(0).$$ (5.35)

$\square$

Theorem 5.3.2. *Assuming that the Laplace transform $F(s)$ of the function f exists, then the Laplace transform of Caputo fractional derivative of $f(t)$ is given by:*

$$L\left(^C D_t^\alpha(f(t))\right)(s) = s^\alpha F(s) - \sum_{k=0}^{n-1} s^{n-\alpha-1} f^k(0). \tag{5.36}$$

Theorem 5.3.3. *Let $t > 0$, $\alpha \in \mathcal{R}$, $n - 1 < \alpha < n \in \mathcal{N}$; if in addition the function $f(t)$ is n-times differentiable, then, the following relation between the Riemann–Liouville and Caputo derivatives of fractional order holds*

$$^C D_t^\alpha(f(t)) = D_t^\alpha(f(t)) - \sum_{k=0}^{n-1} \frac{t^{k-\alpha}}{\Gamma(k+1-\alpha)}. \tag{5.37}$$

Proof. Since the function is n-times differentiable, using the well-known Taylor series expansion around the point 0, we obtain

$$f(t) = \sum_{k=0}^{n-1} \frac{t^k}{k!} f^{(k)}(0) + R_{n-1}, \tag{5.38}$$

where

$$R_{n-1} = \frac{1}{\Gamma(n)} \int_0^t f^{(n)}(\tau)(t-\tau)^{n-1} d\tau. \tag{5.39}$$

However, employing the linearity property of the Riemann–Liouville fractional derivative and also the Riemann–Liouville fractional derivative of power function, then:

$$\begin{aligned}
D^\alpha f(t) &= D^\alpha \left(\sum_{k=0}^{n-1} \frac{t^k}{k!} f^{(k)}(0) + R_{n-1} \right) \\
&= \sum_{k=0}^{n-1} \frac{D^\alpha(t^k)}{k!} f^{(k)}(0) + D^\alpha R_{n-1} \\
&= \sum_{k=0}^{n-1} \frac{t^{k-\alpha}}{\Gamma(k+1)} \frac{\Gamma(k+1)}{\Gamma(k-\alpha+1)} f^{(k)}(0) + J^{n-\alpha} f^{(n)}(t) \\
&= \sum_{k=0}^{n-1} \frac{t^{k-\alpha}}{\Gamma(k+1)} \frac{\Gamma(k+1)}{\Gamma(k-\alpha+1)} f^{(k)}(0) + {}^C D_t^\alpha(f(t)). \tag{5.40}
\end{aligned}$$

Thus,

$$D^\alpha f(t) = \sum_{k=0}^{n-1} \frac{t^{k-\alpha}}{\Gamma(k-\alpha+1)} f^{(k)}(0) + {}^C D_t^\alpha(f(t)). \tag{5.41}$$

This completes the proof.   □

Corollary 5.3.4. *Leibniz rule for Caputo derivative: Let $t > 0$, $\alpha \in \mathcal{R}$, $n - 1 < \alpha < n \in \mathcal{R}$. If both functions $f(t)$ and $g(t)$ are continuous together with their derivatives in $[0, t]$ then the following relation is valid:*

$$^C D_t^\alpha(f(t)g(t)) = \sum_{k=0}^{\infty} C_\alpha^k \left(D^{\alpha-n} f(t)\right) g^{(k)}(t) - \sum_{k=0}^{n-1} \frac{t^{k-\alpha}}{\Gamma(k+1-\alpha)} \left((f(t)g(t))^k(0)\right). \tag{5.42}$$

5.4 Beta-Type

In the literature of fractional derivatives, most of evolution equations are expressed in the form

$$D_t^\alpha u(t) \;=\; Au(t), \quad 0 < \alpha < 1, \ t > 0, \tag{5.43}$$
$$u(0) \;=\; f, \tag{5.44}$$

where A is a linear closed operator densely defined in a Banach space H (endowed with the norm $\|\cdot\|$) and D_t^α is one of the most popular fractional derivatives, namely the Riemann–Liouville and the Caputo derivatives, respectively, defined as [95]:

$$D_x^\alpha(f(x)) = \frac{1}{\Gamma(1-\alpha)} \frac{d}{dx} \int_0^x (x-t)^{-\alpha} f(t)\, dt \tag{5.45}$$

and

$$D_x^\alpha(f(x)) = \frac{1}{\Gamma(1-\alpha)} \int_0^x (x-t)^{-\alpha} \frac{d}{dt} f(t) dt. \tag{5.46}$$

Models of type (5.43) have been intensively studied and investigated in a large number of works [93,94,96,98,97,99,100].

This raises the question of the right choice for the fractional derivative one uses to analyze a phenomenon that evolves. Thus, a new type of definition for a derivative was proposed and developed by Atangana and Goufo [92]. This proposed version representing a derivative with a new parameter, named β-derivative is seen as the extension of the conventional first order derivative. It is defined as

$$
{}_{0}^{A}D_{t}^{\beta}u(t) = \begin{cases} \lim\limits_{\varepsilon \to 0} \dfrac{u\left(t+\varepsilon\left(t+\frac{1}{\Gamma(\beta)}\right)^{1-\beta}\right) - u(t)}{\varepsilon} & \text{for all } t \geq 0, \ 0 < \beta \leq 1, \\[4mm] u(t) & \text{for all } t \geq 0, \ \beta = 0, \end{cases}
\tag{5.47}
$$

where u is a function such that $u : [0, \infty) \to \mathbb{R}$ and Γ is the gamma function

$$
\Gamma(\zeta) = \int_{0}^{\infty} t^{\zeta-1}e^{-1}dt.
$$

If the above limit exists then u is said to be β-differentiable. Note that for $\beta = 1$, we have ${}_{0}^{A}D_{t}^{\beta}u(t) = \frac{d}{dt}u(t)$. Moreover, unlike other fractional derivatives, the β-derivative of a function can be locally defined at a certain point, the same way as first order derivative.

It is important to mention the definition of the β-derivative for any $\beta \geq 0$, that reads as

$$
{}_{0}^{A}D_{t}^{\beta}u(t) = \begin{cases} \lim\limits_{\varepsilon \to 0} \dfrac{u\left(t+\varepsilon\left(t+\frac{1}{\Gamma(\beta)}\right)^{\lceil\beta\rceil-\beta}\right) - u(t)}{\varepsilon} & \text{for all } t \geq 0, \ n-1 < \beta \leq n, \\[4mm] u(t) & \text{for all } t \geq 0, \ \beta = 0, \end{cases}
\tag{5.48}
$$

with $n \in \mathbb{N}$ and $\lceil \beta \rceil$ the smallest integer greater or equal to β. We present, in this section, some useful theorems and properties of the derivative with new parameter.

Theorem 5.4.1. *Assume that a given function, say $f : [a, \infty] \to \mathbb{R}$, is beta-differentiable at a point say $x_0 \geq a$, $\beta \in [0, 1]$, then f is continuous at x_0.*

Proof. Assume that f is β-differentiable, then

$$
{}_{0}^{A}D_{t}^{\beta}(f(t_0)) = \lim\limits_{\varepsilon \to 0} \dfrac{f\left(t_0 + \varepsilon\left(t_0 + \frac{1}{\Gamma(\beta)}\right)^{1-\beta}\right) - f(t_0)}{\varepsilon}
\tag{5.49}
$$

exists. However,

$$
\lim\limits_{\varepsilon \to 0} f\left(t_0 + \varepsilon\left(t_0 + \frac{1}{\Gamma(\beta)}\right)^{1-\beta}\right) - f(t_0) = \lim\limits_{\varepsilon \to 0} \dfrac{f\left(t_0 + \varepsilon\left(t_0 + \frac{1}{\Gamma(\beta)}\right)^{1-\beta}\right) - f(t_0)}{\varepsilon}\varepsilon
\tag{5.50}
$$

$$= {}^{A}_{0}D^{\beta}_{t}(f(t_0)).0 \tag{5.51}$$
$$= 0, \tag{5.52}$$

therefore,

$$\lim_{\varepsilon \to 0} f\left(t_0 + \varepsilon \left(t_0 + \frac{1}{\Gamma(\beta)}\right)^{1-\beta}\right) = \left(\lim_{\varepsilon \to 0} f\left(t_0 + \varepsilon \left(t_0 + \frac{1}{\Gamma(\beta)}\right)^{1-\beta}\right) - f(t_0)\right)$$
$$+ f(t_0) \tag{5.53}$$
$$= 0 + f(t_0) \tag{5.54}$$
$$= f(t_0). \tag{5.55}$$

Nevertheless, if we assume that $t = t_0 + \varepsilon \left(t_0 + \frac{1}{\Gamma(\beta)}\right)^{1-\beta}$ so that

$$\varepsilon = \frac{(t - t_0)}{\left(t_0 + \frac{1}{\Gamma(\beta)}\right)^{1-\beta}}, \tag{5.56}$$

then

$$\lim_{\frac{(t-t_0)}{\left(t_0 + \frac{1}{\Gamma(\beta)}\right)^{1-\beta}} \to 0} f(t) = f(t_0), \tag{5.57}$$

since $\left(t_0 + \frac{1}{\Gamma(\beta)}\right)^{1-\beta} \neq 0$. Then Eq. (5.57) can be rewritten as:

$$\lim_{t - t_0 \to 0} f(t) = f(t_0), \quad \text{or,} \quad \lim_{t \to t_0} f(t) = f(t_0). \tag{5.58}$$

This completes the proof. □

Theorem 5.4.2. *Assume that a given function, say $f : [a, \infty] \to \mathbb{R}$, is locally differentiable, then f is also beta-differentiable.*

Proof. If f is differentiable, then the following limit exists:

$$\lim_{h \to 0} \frac{f(t + h) - f(t)}{h}. \tag{5.59}$$

This implies the existence of the following limit too:

$$\lim_{h \to 0} \frac{f(t + h) - f(t)}{h} \left(t + \frac{1}{\Gamma(\beta)}\right)^{1-\beta}. \tag{5.60}$$

Nevertheless, letting $\varepsilon = h \left(t + \frac{1}{\Gamma(\beta)} \right)^{\beta - 1}$ that will imply $h = \varepsilon \left(t + \frac{1}{\Gamma(\beta)} \right)^{1-\beta}$, thus Eq. (5.60) can be reformulated as

$$\lim_{\varepsilon \to 0} \frac{f \left(t + \varepsilon \left(t + \frac{1}{\Gamma(\beta)} \right)^{1-\beta} \right) - f(t)}{\varepsilon} = {}_0^A D_t^\beta (f(t)). \tag{5.61}$$

This completes the proof. □

Theorem 5.4.3. *Formal statement of the mean value theorem for variable order derivative: Let $f : [a,b] \to \mathbb{R}$ be a continuous function on the closed interval $[a,b]$, and beta-differentiable and differentiable on the open interval (a,b), where $a < b$. Then there exists some c in (a,b) such that*

$$
{}_0^A D_t^\beta (f(c)) = h(\beta, c, a, b) \frac{f(b) - f(a)}{b - a}. \tag{5.62}
$$

Proof. According to the basic idea of the local derivative, the slope that joins the points $(a, f(a))$ and $(b, f(b))$ is given by the expression $\frac{f(b)-f(a)}{b-a}$ [61]. This is in other words the chord of the graph of the function f, which in physical or geometrical interpretation gives the $f'(x)$ slope of the tangent to the curve at the point $(t, f(t))$. Let us define

$$J(t) = f(t) - d \left(t + \frac{1}{\Gamma(\beta)} \right)^{1-\beta} t, \tag{5.63}$$

where d and c are constant. Thus f is continuous on $[a,b]$ and differentiable on (a,b), therefore J also is. However, choose d such that the Rolle theorem can be satisfied including

$$J(a) = J(b) \equiv f(a) - d \left(a + \frac{1}{\Gamma(\beta)} \right)^{1-\beta} a = f(b) - d \left(b + \frac{1}{\Gamma(\beta)} \right)^{1-\beta} b. \tag{5.64}$$

Rearranging, we obtain the following

$$d = \frac{f(b) - f(a)}{b \left(b + \frac{1}{\Gamma(\beta)} \right)^{1-\beta} - a \left(a + \frac{1}{\Gamma(\beta)} \right)^{1-\beta}}. \tag{5.65}$$

Thus by Rolle theorem since J is differentiable and $J(a) = J(b)$, we can find a constant $c \in (a,b)$ such that $J'(c) = 0$. However,

$$J'(x) = f'(x) - d \left(t + \frac{1}{\Gamma(\beta)} \right)^{1-\beta} - d(1 - \beta) \left(t + \frac{1}{\Gamma(\beta)} \right)^{-\beta}, \tag{5.66}$$

thus, $J'(c) = 0$ implies

$$d = \frac{f'(c)}{\left(c + \frac{1}{\Gamma(\beta)}\right)^{1-\beta} + \left(c + \frac{1}{\Gamma(\beta)}\right)^{-\beta}}. \tag{5.67}$$

Comparing Eqs. (5.65) and (5.67), we obtain:

$$\frac{f'(c)}{\left(c + \frac{1}{\Gamma(\beta)}\right)^{1-\beta} + \left(c + \frac{1}{\Gamma(\beta)}\right)^{-\beta}} = \frac{f(b) - f(a)}{b\left(b + \frac{1}{\Gamma(\beta)}\right)^{1-\beta} - a\left(a + \frac{1}{\Gamma(\beta)}\right)^{1-\beta}}. \tag{5.68}$$

Now multiply both sides by

$$\left(c + \frac{1}{\Gamma(\beta)}\right)^{1-\beta}\left[\left(c + \frac{1}{\Gamma(\beta)}\right)^{1-\beta} + \left(c + \frac{1}{\Gamma(\beta)}\right)^{-\beta}\right], \tag{5.69}$$

where

$$\left(c + \frac{1}{\Gamma(\beta)}\right)^{1-\beta} = \frac{f(b) - f(a)}{b\left(b + \frac{1}{\Gamma(\beta)}\right)^{1-\beta} - a\left(a + \frac{1}{\Gamma(\beta)}\right)^{1-\beta}}. \tag{5.70}$$

Now with f differentiable, we have that

$$f'(c) = \lim_{h \to 0} \frac{f(t + h) - f(t)}{h}. \tag{5.71}$$

Now we consider a very small change of variable as follows: $h = \left(b + \frac{1}{\Gamma(\beta)}\right)^{1-\beta}$, then, we obtain the following expression:

$$_0^A D_t^\beta(f(c)) = \lim_{\varepsilon \to 0} \frac{f\left(c + \varepsilon\left(t_0 + \frac{1}{\Gamma(\beta)}\right)^{1-\beta}\right) - f(c)}{\varepsilon}. \tag{5.72}$$

Then

$$_0^A D_t^\beta(f(c)) = h(\beta, c, a, b)\frac{f(b) - f(a)}{b - a}, \tag{5.73}$$

where

$$h(\beta, c, a, b) = \frac{(b - a)\left[\left(c + \frac{1}{\Gamma(\beta)}\right)^{1-\beta} + \left(c + \frac{1}{\Gamma(\beta)}\right)^{-\beta}\right]}{b\left(b + \frac{1}{\Gamma(\beta)}\right)^{1-\beta} - a\left(a + \frac{1}{\Gamma(\beta)}\right)^{1-\beta}}. \tag{5.74}$$

This completes the proof. □

Definition 5.4.4. *Let* $f : [a, b] \to \mathbb{R}$ *be a continuous function on the closed interval* $[a, b]$, *then the* 2α-*derivative of* f *is defined as:*

$$\begin{smallmatrix} A \\ 0 \end{smallmatrix} D_t^{2\beta} (f(t)) = \begin{smallmatrix} A \\ 0 \end{smallmatrix} D_t^{\beta} \left(\begin{smallmatrix} A \\ 0 \end{smallmatrix} D_t^{\beta} (f(t)) \right), 0 \le \beta \le 1. \tag{5.75}$$

In general, the $n\beta$-*derivative of* f *is given as:*

$$\begin{smallmatrix} A \\ 0 \end{smallmatrix} D_t^{n\beta} (f(t)) = \begin{smallmatrix} A \\ 0 \end{smallmatrix} D_t^{\beta} \left(\begin{smallmatrix} A \\ 0 \end{smallmatrix} D_t^{(n-1)\beta} (f(t)) \right), 0 \le \beta \le 1. \tag{5.76}$$

Remark 1. It is very important to notice that the $n\beta$-derivative of a given function gives information of the previous $n - 1$ derivatives of that function. For instance,

$$\begin{smallmatrix} A \\ 0 \end{smallmatrix} D_t^{2\beta} (f(t)) = \left(t + \frac{1}{\Gamma(\beta)} \right)^{1-\beta} \left[(1 - \beta) \left(t + \frac{1}{\Gamma(\beta)} \right)^{-\beta} f' + \left(t + \frac{1}{\Gamma(\beta)} \right)^{1-\beta} f'' \right]. \tag{5.77}$$

This gives this derivative a unique property of memory, which is not provided by any other derivative. It is also easy to verify that if $\beta = 1$, we recover the second derivative of f.

Corollary 5.4.5. *Let* $f : [a, b] \to \mathbb{R}$ *be a continuous function on the closed interval* $[a, b]$. *If* $\alpha \neq \beta$, *then*

$$\begin{smallmatrix} A \\ 0 \end{smallmatrix} D_t^{\beta} \left(\begin{smallmatrix} A \\ 0 \end{smallmatrix} D_t^{\alpha} (f(t)) \right) \neq \begin{smallmatrix} A \\ 0 \end{smallmatrix} D_t^{\alpha} \left(\begin{smallmatrix} A \\ 0 \end{smallmatrix} D_t^{\beta} (f(t)) \right). \tag{5.78}$$

Proof. In fact,

$$\left\{ \begin{smallmatrix} A \\ 0 \end{smallmatrix} D_t^{\beta} \left(\begin{smallmatrix} A \\ 0 \end{smallmatrix} D_t^{\alpha} (f(t)) \right) \right\} =$$

$$\left(t + \frac{1}{\Gamma(\beta)} \right)^{1-\beta} (1 - \alpha) \left(t + \frac{1}{\Gamma(\alpha)} \right)^{-\alpha} f'$$

$$+ \left(t + \frac{1}{\Gamma(\beta)} \right)^{1-\beta} \left(t + \frac{1}{\Gamma(\alpha)} \right)^{1-\alpha} f''.$$

On the other hand, we have

$$\begin{smallmatrix} A \\ 0 \end{smallmatrix} D_t^{\alpha} \left(\begin{smallmatrix} A \\ 0 \end{smallmatrix} D_t^{\beta} (f(t)) \right) =$$

$$\left(t + \frac{1}{\Gamma(\alpha)} \right)^{1-\alpha} (1 - \beta) \left(t + \frac{1}{\Gamma(\beta)} \right)^{-\beta} f'$$

$$+ \left(t + \frac{1}{\Gamma(\alpha)} \right)^{1-\alpha} \left(t + \frac{1}{\Gamma(\beta)} \right)^{1-\beta} f''. \qquad \square$$

Definition 5.4.6. *Let $f : [a, b] \to \mathbb{R}$ be a continuous function on the open interval (a, b), then the β-integral of f is given as:*

$$_0^A I_t^\beta (f(t)) = \int_0^t \left(x + \frac{1}{\Gamma(\beta)} \right)^{\beta-1} f(x) dx. \tag{5.79}$$

This integral was recently referred to as Atangana-beta integral.

Theorem 5.4.7. *Fundamental theorem of local β-calculus states that: First part*

$$_a^A D_t^\beta \left(_a^A I_t^\beta (f(t)) \right) = f(t), \tag{5.80}$$

with f a given continuous function. Second part:

$$_a^A I_t^\beta \left(_a^A D_t^\beta (f(t)) \right) = f(t) - f(a), \tag{5.81}$$

for all $x \geq a$ with f a given differentiable function.

Proof. We shall start with part one. Let f be a continuous function on (a, b), and let $F(t) = _0^A I_t^\beta (f(t))$, then by definition,

$$_a^A D_t^\beta \left(_a^A I_t^\beta (f(t)) \right)$$

$$= \lim_{\varepsilon \to 0} \frac{F\left(t + \varepsilon \left(t + \frac{1}{\Gamma(\beta)} \right)^{1-\beta} \right) - F(t)}{\varepsilon} \tag{5.82}$$

$$= \lim_{\varepsilon \to 0} \frac{\int_a^{\left(t + \varepsilon \left(t + \frac{1}{\Gamma(\beta)} \right)^{1-\beta} \right)} \left(x + \frac{1}{\Gamma(\beta)} \right)^{\beta-1} f(x) dx - \int_a^t f(x) \left(x + \frac{1}{\Gamma(\beta)} \right)^{\beta-1} dx}{\varepsilon} \tag{5.83}$$

$$= \lim_{h \to 0} \frac{\int_a^{(t+h)} f(x) dx - \int_a^t f(x) dx}{h} \tag{5.84}$$

$$= f(t). \tag{5.85}$$

Second part:

We shall start with part one. Let f be a differentiable function on (a, b), and let $h(t) = _0^A D_t^\beta (f(t))$, then by definition,

$$_a^A I_t^\beta \left(_a^A D_t^\beta (f(t)) \right)$$

$$= \int_a^t \left(x + \frac{1}{\Gamma(\beta)} \right)^{\beta-1} h(x) dx \tag{5.86}$$

$$= \int_a^t \left(x + \frac{1}{\Gamma(\beta)} \right)^{\beta-1} \left(\lim_{\varepsilon \to 0} \frac{f\left(x + \varepsilon \left(x + \frac{1}{\Gamma(\beta)} \right)^{1-\beta} \right) - f(x)}{\varepsilon} \right) dx \tag{5.87}$$

$$= \int_a^t \left(\lim_{h \to 0} \frac{f(x+h) - f(x)}{h} \right) dx \tag{5.88}$$

$$= f(t) - f(a). \tag{5.89}$$

This completes the proof. $\qquad\qquad\qquad\qquad\qquad\qquad\qquad\qquad\qquad\qquad\qquad\qquad\square$

Theorem 5.4.8. *Let f be a continuous real function and β-integrable on an open interval (a, b), then we can find a real number c such that:*

$$_a^A I_t^\beta (f(t)) = f(c)(b-a)h(a,b). \tag{5.90}$$

Proof. By the Extreme value theorem, we can find two real numbers N and $M \in (a, b)$ such that:

$$f(N) = \min_{t \in (a,b)} f(t) \le f(t) \le f(M) = \max_{t \in (a,b)} f(t). \tag{5.91}$$

Applying the β-integral on the above inequality (5.90), we obtain the following:

$$\int_a^b \left(t + \frac{1}{\Gamma(\beta)} \right)^{\beta-1} f(N) dt \le \int_a^b \left(t + \frac{1}{\Gamma(\beta)} \right)^{\beta-1} f(t) dt \tag{5.92}$$

$$\le \int_a^b \left(t + \frac{1}{\Gamma(\beta)} \right)^{\beta-1} f(M) dt. \tag{5.93}$$

After integration of the left- and the right-hand sides of inequality (5.92), we obtain,

$$f(N) \frac{\left(b + \frac{1}{\Gamma(\beta)} \right)^{\beta} - \left(a + \frac{1}{\Gamma(\beta)} \right)^{\beta}}{\beta} \le \int_a^b \left(t + \frac{1}{\Gamma(\beta)} \right)^{\beta-1} f(t) dt$$

$$\le \frac{\left(b + \frac{1}{\Gamma(\beta)} \right)^{\beta} - \left(a + \frac{1}{\Gamma(\beta)} \right)^{\beta}}{\beta} F(M). \tag{5.94}$$

For ease, let

$$I(a, b) = \frac{\left(b + \frac{1}{\Gamma(\beta)}\right)^{\beta} - \left(a + \frac{1}{\Gamma(\beta)}\right)^{\beta}}{\beta}. \tag{5.95}$$

Dividing (5.94) by $I(a, b)$, we obtain

$$f(N) \leq \frac{1}{I(a, b)} \int_a^b \left(t + \frac{1}{\Gamma(\beta)}\right)^{\beta - 1} f(t)dt \leq F(M). \tag{5.96}$$

However, making use of the Intermediate value theorem, we can find a real number $c \in (a, b)$ such that

$$\frac{1}{I(a, b)} \int_a^b \left(t + \frac{1}{\Gamma(\beta)}\right)^{\beta - 1} f(t)dt = f(c). \tag{5.97}$$

After manipulations, we obtain

$$^A_a I_t^{\beta}(f(t)) = f(c)(b - a)h(a, b), \qquad h(a, b) = \frac{1}{I(a, b)(b - a)}. \tag{5.98}$$

This completes the proof. $\qquad \qquad \qquad \qquad \qquad \qquad \qquad \qquad \qquad \qquad \square$

5.5 Fractal Type

In applied mathematics and mathematical analysis, the fractal derivative is a nonstandard type of derivative in which the variable such as t has been scaled according to t^{α}. The derivative is defined in fractal geometry. Porous media, aquifer, turbulence and other media usually exhibit fractal properties. The classical physical laws such as Fick's laws of diffusion, Darcy's law and Fourier's law are no longer applicable for such media, because they are based on Euclidean geometry, which doesn't apply to media of non-integer fractal dimensions. The basic physical concepts such as distance and velocity in fractal media are required to be redefined; the scales for space and time should be transformed according to (x^{β}, t^{α}). The elementary physical concepts such as velocity in a fractal space–time (x^{β}, t^{α}) can be redefined by:

$$v' = \frac{dx'}{dt'} = \frac{dx^{\beta}}{dt^{\alpha}}, \qquad \alpha, \beta > 0, \tag{5.99}$$

where $S^{\alpha, \beta}$ represents the fractal space–time with scaling indexes α and β. The traditional definition of velocity makes no sense in the non-differentiable fractal space–time.

Definition 5.5.1. *Based on the above discussion, the concept of the fractal derivative of a function $u(t)$ with respect to a fractal measure t has been introduced as follows:*

$$\frac{\partial f(t)}{\partial t^\alpha} = \lim_{t_1 \to t} \frac{f(t_1) - f(t)}{t_1^\alpha - t^\alpha}, \qquad \alpha > 0. \tag{5.100}$$

A more general definition is given by

$$\frac{\partial^\beta f(t)}{\partial t^\alpha} = \lim_{t_1 \to t} \frac{f^\beta(t_1) - f^\beta(t)}{t_1^\alpha - t^\alpha}, \qquad \alpha > 0, \beta > 0. \tag{5.101}$$

The above definition has been widely used in the field of porous media and here are some examples of application. As an alternative modeling approach to the classical Fick's second law, the fractal derivative is used to derive a linear anomalous transport-diffusion equation underlying anomalous diffusion process,

$$\frac{du(x,t)}{dt^\alpha} = D \frac{\partial}{\partial x^\beta} \left(\frac{\partial u(x,t)}{\partial x^\beta} \right), \quad -\infty < x < +\infty, \quad (1)u(x,0) = \delta(x), \tag{5.102}$$

where $0 < \alpha < 2, 0 < \beta < 1$, and $\sigma(x)$ is the Dirac Delta function.

In order to obtain the fundamental solution, the transformation of variables is applied as follows:

$$t' = t^\alpha, \quad x' = x^\beta. \tag{5.103}$$

Then Eq. (5.102) becomes the normal diffusion form equation. The solution of (5.102) has the stretched Gaussian form:

$$u(x,t) = \frac{1}{2\sqrt{\pi t^\alpha}} e^{-\frac{x^{2\beta}}{4t^\alpha}}. \tag{5.104}$$

The mean squared displacement of the above fractal derivative diffusion equation has the asymptote:

$$\left\langle x^2(t) \right\rangle \propto t^{(3\alpha - \alpha\beta)/2\beta}. \tag{5.105}$$

5.6 Caputo–Fabrizio Type

Recently Caputo and Fabrizio presented a new definition of fractional derivative with a smooth kernel which takes on two different representations for the temporal and spatial variables. The first works on the time variables; thus it is suitable to use the Laplace transform. The second definition is related to the spatial variables, by a non-local fractional derivative, for which it is more convenient to work with the Fourier transform. The interest for this new approach with a regular kernel was born from the prospect that there is a class of non-local systems, which have the ability to describe the material heterogeneities and the fluctuations of different scales, which cannot be well described by classical local theories or by fractional models with singular kernel. The new definition of fractional derivative with singular kernel is presented here:

Definition 5.6.1. *Let $f \in H^1(a, b)$, $b > a$, $a \in [0, 1]$, then, the Caputo–Fabrizio derivative with order α is defined as:*

$$^{CF}D_t^a(f(t)) = \frac{M(a)}{1-a}\int_a^t f'(x)\exp\left[-a\frac{t-x}{1-a}\right]dx \tag{5.106}$$

where $M(a)$ is a normalization function such that $M(0) = M(1) = 1$ [1]. However, if the function does not belong to $H^1(a, b)$, then the derivative can be reformulated as

$$^{CF}D_t^a(f(t)) = \frac{aM(a)}{1-a}\int_a^t (f(t) - f(x))\exp\left[-a\frac{t-x}{1-a}\right]dx. \tag{5.107}$$

Remark 2. Caputo and Fabrizio remarked that, if $s = \frac{1-a}{a} \in [0, \infty)$, $a = \frac{1}{1+s} \in [0, 1]$, then Eq. (5.107) assumes the form

$$^{CF}D_t^s(f(t)) = \frac{N(s)}{s}\int_a^t f'(x)\exp\left[-\frac{t-x}{s}\right]dx, \quad N(0) = N(\infty) = 1. \tag{5.108}$$

In addition,

$$\lim_{s\to 0}\frac{1}{s}\exp\left[-\frac{t-x}{s}\right] = d(x-t). \tag{5.109}$$

Now after the introduction of a new derivative, the associated anti-derivative becomes important. The associated integral of the new Caputo derivative with fractional order was proposed by Nieto and Losada [85–90].

Definition 5.6.2. *[85,86] Let $0 < a < 1$. The fractional integral of order a of a function f is defined by*

$$I_a^t(f(t)) = \frac{2(1-a)}{(2-a)M(a)}f(t) + \frac{2a}{(2-a)M(a)}\int_0^t f(s)\,ds, \quad t \geq 0. \tag{5.110}$$

Remark 3. [85,86]. Note that, according to the above definition, the Caputo–Frabrizio integral of a function f with order $0 < a < 1$ is an average between the same function f and its classical integral. This therefore imposes

$$\frac{2(1-a)}{(2-a)M(a)} + \frac{2a}{(2-a)M(a)} = 1. \tag{5.111}$$

The above expression yields an explicit formula for

$$M(a) = \frac{2}{2-a}, \quad 0 \le a \le 1.$$

Because of the above, Nieto and Losada proposed that the Caputo–Liouville derivative of order $0 < a < 1$ can be reformulated as

$$^{CF}D_t^a(f(t)) = \frac{1}{1-a} \int_a^t f'(x) \exp\left[-a\frac{t-x}{1-a}\right] dx. \tag{5.112}$$

Theorem 5.6.3. *For the new Caputo derivative with fractional order, if the function $f(t)$ is such that*

$$f^{(s)}(a) = 0, \quad s = 1, 2, \ldots n,$$

then we have

$$^{CF}D_t^\alpha\left(D_t^n(f(t))\right) = D_t^n\left(^{CF}D_t^a(f(t))\right).$$

For proof see [85].

The Laplace transform of the new derivative with fractional order is given as

$$L\left(^{CF}D_t^\alpha(f(t))\right) = \frac{pF(p) - f(0)}{p + \alpha(1-p)}. \tag{5.113}$$

Theorem 5.6.4. *Let $f(t)$ be a function for which the Caputo–Fabrizio derivative with fractional order exists; then the Sumudu transform is given as*

$$S\left(^{CF}D_t^\alpha(f(t))\right) = \frac{F(p) - f(0)}{1 - \alpha + \alpha p}. \tag{5.114}$$

Corollary 5.6.5. *For any natural number $n \in \mathcal{N}$, given a n-time differentiable function the following relation holds*

$$S\left(^{CF}D_t^{\alpha+n}(f(t))\right) = \left[\frac{S(f(t))}{p^n} - \sum_{k=0}^n \frac{f^k(0)}{p^{n-k}}\right] \frac{1}{1 - \alpha + \alpha p}. \tag{5.115}$$

5.7 Caputo–Fabrizio in Riemann–Liouville Sense

The application of the two most popular fractional derivatives, namely the classic Riemann–Liouville and Caputo fractional derivatives lead to a new differential operator known as Caputo–Fabrizio fractional derivative in Riemann–Liouville sense. Both concepts are subjected to some issues. The range of these issues is wide, including initialization with Caputo derivative and its observed difficulties compared to Riemann–Liouville initialization conditions, the principle and equivalence of the two definitions to solve fractional differential equations (FDE) with integer order initial states and decomposition of fractional differential equation (FDE). Being aware of these issues and reacting to the newly introduced Caputo–Fabrizio fractional derivative without singular kernel (CFFD), Atangana and Doungmo presented a new definition of fractional derivative generated by modification of the classical Riemann–Liouville definition and called here the new Riemann–Liouville fractional derivative without singular kernel.

Definition 5.7.1. *Let f be a function not necessarily differential, let α be a real number such that $0 \le \alpha \le 1$, then the Caputo–Fabrizio derivative in Caputo sense with order α is given as:*

$$^{A.G.}D_t^a\left(f\left(t\right)\right) = \frac{1}{1-a}\frac{d}{dt}\int_a^t f(x)\exp\left[-a\frac{t-x}{1-a}\right]dx. \tag{5.116}$$

If α is zero, we have the following:

$$^{A.G.}D_t^0\left(f\left(t\right)\right) = \frac{d}{dt}\int_a^t f(x)dx = f(x). \tag{5.117}$$

Using the argument by Caputo and Fabrizio, we also have that when α goes to 1, we recover the first derivative.

Theorem 5.7.2. *The Caputo–Fabrizio derivative with fractional order is connected to the A.G. derivative as follows:*

$$\theta(\alpha)^{CF}D_t^a\left(f\left(t\right)\right) =^\theta (\alpha)^{CF}A.G.D_t^\alpha\left(f\left(t\right)\right) + f(0)\exp(-f(\alpha)x) \tag{5.118}$$

where

$$\theta(\alpha) = \frac{1-\alpha}{M(\alpha)},\ f(\alpha) = \frac{\alpha}{1-\alpha}. \tag{5.119}$$

Proof. By definition, we have the following:

$$\theta(\alpha)^{CF} D_t^a(f(t)) = \frac{d}{dx} \int_0^t h(y) \exp(-f(\alpha)(t-y)) dy \qquad (5.120)$$

$$= h(t) - f(\alpha) \int_0^t h(y) \exp(-f(\alpha)(t-y)) dy \qquad (5.121)$$

$$= h(t) - f(\alpha) \left(\frac{h(t)}{f(\alpha)} - \frac{h(0)}{f(\alpha)} \exp(-f(\alpha)t) \right.$$

$$\left. - \frac{1}{f(\alpha)} \int_0^t \exp(-f(\alpha)(t-y)) \right) dy \qquad (5.122)$$

$$= \frac{h(0)}{f(\alpha)} \exp(-f(\alpha)t) + \int_0^t \frac{dh(y)}{dt} \exp(-f(\alpha)(t-y)) dy \quad (5.123)$$

$$= \theta(\alpha)^{A.G.} D_t^{\alpha}(f(t)) + f(0) \exp(-f(\alpha)x). \qquad (5.124)$$

This completes the proof. □

5.8 Atangana–Baleanu Derivatives With Fractional Order

We recall that the existing fractional derivatives have been used in many real-world problems with great success (see for example Refs. [74–82] and the references therein) but still there are many things to be done in this direction. The aim of the work by Caputo and Fabrizio [85, 86] was to introduce a new derivative with exponential kernel. Its anti-derivative was reported in [86] and it was found to be the average of a given function and the Riemann–Liouville fractional integral of the same function. The derivative introduced in [85] cannot produce the original function when α is one. However this issue was, so far, independently solved by Atangana with Goufo [90,91] and Caputo with Fabrizio [105], respectively. We believe that the main message presented in the work done by Caputo and Fabrizio [90,105] was to find a way to describe even better the dynamics of systems with memory effect. However, some issues were pointed out against the Caputo–Fabrizio derivative with fractional order:

- The kernel used in the Caputo–Fabrizio derivative is local.
- The anti-derivative associated to their derivative is not a fractional integral but the average of the function and its integral.
- When the fractional order is zero we do not recover the initial function.

It appears that such a used kernel cannot be used for many physical problems. Therefore, for a given data we ask the following question: What is the most accurate kernel which better describes it? Then Atangana and Baleanu suggested a possible answer, presented in the following sections.

5.8.1 Motivations and Definitions With New Kernel

The exponential function is the solution of the following ordinary differential equation:

$$\frac{dy}{dx} = ay. \tag{5.125}$$

However, the Mittag–Leffler function is the solution of the following fractional ordinary differential equation:

$$D_y^\alpha(y) = ay, 0 < \alpha < 1. \tag{5.126}$$

The Mittag–Leffler function and its generalized versions are therefore considered as non-local functions. Let us consider the following generalized Mittag–Leffler function:

$$E_\alpha(-t^\alpha) = \sum_{k=0}^\infty \frac{(-t)^{\alpha k}}{\Gamma(\alpha k + 1)}. \tag{5.127}$$

The above function has the following properties:

$$E_1(-t) = \sum_{k=0}^\infty \frac{(-t)^k}{\Gamma(\alpha k + 1)} = \exp(-t). \tag{5.128}$$

Additionally,

$$\lim_{\epsilon \to 0} \frac{E_\alpha - \frac{-t^\alpha}{\epsilon}}{\epsilon} = \delta(t). \tag{5.129}$$

The Taylor series of $\exp(-(t-y))$ at the point t is given by:

$$\exp(-(t-y)) = \sum_{k=0}^\infty \frac{(-a(t-y))^k}{k!}. \tag{5.130}$$

Chosing $a = \frac{\alpha}{1-\alpha}$ and replacing the above expression into Caputo–Fabrizio derivative, the following formula is obtained:

$$^{C.F.}D_t^\alpha(f(t)) = \frac{M(\alpha)}{1-\alpha} \int_b^t f'(y) \sum_{k=0}^\infty \frac{(-a(t-y))^k}{k!} dy. \tag{5.131}$$

Rearranging, they obtained

$$^{C.F.}D_t^\alpha(f(t)) = \frac{M(\alpha)}{1-\alpha} \sum_{k=0}^\infty \frac{(-a)^k}{k!} \int_b^t f'(y)(t-y)^k dy. \tag{5.132}$$

To solve the problem of non-locality, they derived the following expression. In Eq. (5.132) they replaced $k!$ by $\Gamma(\alpha k + 1)$ and also $(t-y)^k$ by $(t-y)^{\alpha k}$ to obtain

$$^{A.B.}D_t^\alpha(f(t)) = \frac{M(\alpha)}{1-\alpha} \sum_{k=0}^{\infty} \frac{(-a)^k}{\Gamma 1 + k\alpha} \int_b^t f'(y)(t-y)^{k\alpha} dy. \tag{5.133}$$

Then the following derivative was proposed.

Definition 5.8.1. *Let $f \in H^1(a, b)$, $b > a$, $\alpha \in [0, 1]$, then the definition of the Atangana–Baleanu derivative in Caputo sense is given as:*

$$^{ABC}D_t^\alpha(f(t)) = \frac{B(\alpha)}{1-\alpha} \int_b^t \frac{df(y)}{dy} E_\alpha \left[-\alpha \frac{(t-y)^\alpha}{1-\alpha} \right] dy. \tag{5.134}$$

Of course $B(\alpha)$ has the same properties as in Caputo and Fabrizio case [103]. The above definition will be helpful to real-world problems and also will have a great advantage when using Laplace transform to solve some physical problems with initial conditions. However, when α is 0 we do not recover the original function except when at the origin the function vanishes. To avoid this issue, we propose the following definition such that researchers in the field of fractional calculus have a choice when dealing with a given problem.

Definition 5.8.2. *Let $f \in H^1(a, b)$, $b > a$, $\alpha \in [0, 1]$, then the definition of the Atangana–Baleanu derivative in Riemann–Liouville sense is given as:*

$$^{ABC}D_t^\alpha(f(t)) = \frac{B(\alpha)}{1-\alpha} \frac{df}{dt} \int_b^t f(y) E_\alpha \left[-\alpha \frac{(t-y)^\alpha}{1-\alpha} \right] dy. \tag{5.135}$$

Eqs. (5.134) and (5.135) have a non-local kernel. Eq. (5.134) can be used when employing Laplace transform to solve a real-world problem with usual initial condition. Also in Eq. (5.134) when the function is constant the fractional derivative produces zero.

5.8.2 Properties of Atangana–Baleanu Derivatives With Fractional Order

In this section, we start by presenting the relation between both derivatives with Laplace transform

$$L\left(^{ABR}D_t^\alpha(f(t))\right) = \frac{B(\alpha)}{1-\alpha} \frac{p^\alpha L(f(t))(p)}{p^\alpha + \frac{\alpha}{1-\alpha}}. \tag{5.136}$$

Also we have the following relation with the one in Caputo sense

$$L\left(^{ABC}D_t^\alpha(f(t))\right) = \frac{B(\alpha)}{1-\alpha} \frac{p^\alpha L(f(t))(p) - p^{\alpha-1}f(0)}{p^\alpha + \frac{\alpha}{1-\alpha}}. \tag{5.137}$$

The following theorem can therefore be established:

Theorem 5.8.3. *Let $f \in H^1(a, b)$, $b > a$, $\alpha \in [0, 1]$, then the following relation is established*

$$^{ABC}D_t^\alpha(f(t)) = {}^{ABR}D_t^\alpha(f(t)) + K(t). \tag{5.138}$$

Definition 5.8.4. *The fractional integral associated to the new fractional derivative with non-local kernel is defined as:*

$$_a^{AB}I_t^\alpha(f(t)) = \frac{1-\alpha}{B(\alpha)}f(t) + \frac{\alpha}{B(\alpha)\Gamma(\alpha-1)}\int_a^t f(t)(t-y)^{\alpha-1}dy. \tag{5.139}$$

Remark 4. The Atangana–Baleanu fractional integral is the average between a given function and its Riemann–Liouville fractional integral, therefore this new version is more general than the Riemann–Liouville fractional derivative. In addition, if α is zero, we recover the initial function and if α is 1, we obtain the ordinary integral.

We present fractional integral of some especial functions in this section:

$$_0^{AB}I_t^\alpha\left\{E_\alpha(t^\alpha)\right\} = \frac{1}{B(\alpha)}E_\alpha(t^\alpha) - \frac{\alpha}{B(\alpha)}. \tag{5.140}$$

Proof. From the definition,

$$_0^{AB}I_t^\alpha\left\{E_\alpha(t^\alpha)\right\} = \frac{1-\alpha}{B(\alpha)}E_\alpha(t^\alpha) + \frac{\alpha}{B(\alpha)\Gamma(\alpha)}\int_0^t E_\alpha(y^\alpha)(t-y)^{\alpha-1}dy \tag{5.141}$$

where

$$\frac{1}{\Gamma(\alpha)}\int_0^t E_\alpha(y^\alpha)(t-y)^{\alpha-1}dy = E_\alpha(t^\alpha) - 1. \tag{5.142}$$

Replacing in Eq. (5.142) we obtain:

$$_0^{AB}I_t^\alpha\left\{E_\alpha(t^\alpha)\right\} = \frac{1}{B(\alpha)}E_\alpha(t^\alpha) - \frac{\alpha}{B(\alpha)}. \tag{5.143}$$

$\square$

Theorem 5.8.5. *If $\alpha, \beta, \gamma > 0$ then the following function $t^{\beta-1}E_{\alpha,\beta}(\mu t^\alpha)$ has the following integral*

$$_0^{AB}I_t^\gamma\left\{t^{\beta-1}E_{\alpha,\beta}(\mu t^\alpha)\right\} \tag{5.144}$$

$$= \frac{1-\gamma}{B(\gamma)}t^{\beta-1}E_{\alpha,\beta}(\mu t^\alpha) + \frac{\gamma}{B(\gamma)\Gamma(\gamma)}\left\{t^{\beta+\gamma-1}E_{\alpha,\beta+\gamma}(\mu t^\alpha)\right\}.$$

Proof. By definition, we have that

$$
{}^{AB}_0 I^\gamma_t \left\{ t^{\beta-1} E_{\alpha,\beta}(\mu t^\alpha) \right\} = \frac{1-\gamma}{B(\gamma)} t^{\beta-1} E_{\alpha,\beta}(\mu t^\alpha) \tag{5.145}
$$

$$
+ \frac{\gamma}{B(\gamma)\Gamma(\gamma)} \int_0^t z^{\beta-1} E_{\alpha,\beta}(\mu z^\alpha)(t-z)^{\gamma-1} dz.
$$

However,

$$
\frac{\gamma}{B(\gamma)\Gamma(\gamma)} \int_0^t z^{\beta-1} E_{\alpha,\beta}(\mu z^\alpha)(t-z)^{\gamma-1} dz = \frac{\gamma}{B(\gamma)} \left\{ t^{\beta+\gamma-1} E_{\alpha,\beta+\gamma}(\mu t^\alpha) \right\}. \tag{5.146}
$$

$\square$

Then the requested result is obtained.

Theorem 5.8.6. *If $y, z \in \mathbb{C}$ with the condition that $y \neq z$, α, β, $\gamma > 0$, the following special function $t^{\gamma-1} E_{\alpha,\gamma}(yt^\alpha) E_{\alpha,\beta}(z(h-u))^\alpha$ has the following fractional integral*

$$
{}^{AB}_0 I^\gamma_t \left\{ t^{\gamma-1} E_{\alpha,\gamma}(yt^\alpha) E_{\alpha,\beta}(z(h-u))^\alpha \right\} \tag{5.147}
$$

$$
= \frac{1-\gamma}{B(\gamma)} t^{\gamma-1} E_{\alpha,\beta}(z(h-u))^\alpha
$$

$$
+ \frac{\gamma}{B(\gamma)\Gamma(\gamma)} \left\{ \frac{y E_{\alpha,\beta+\gamma}(yt^\alpha) - z E_{\alpha,\beta+\gamma}(zt^\alpha)}{y-z} t^{\beta+\gamma-1} \right\}.
$$

Proof. By definition, we have the following:

$$
{}^{AB}_0 I^\gamma_t \left\{ t^{\gamma-1} E_{\alpha,\gamma}(yt^\alpha) E_{\alpha,\beta}(z(h-u))^\alpha \right\} \tag{5.148}
$$

$$
= \frac{1-\gamma}{B(\gamma)} t^{\gamma-1} E_{\alpha,\beta}(z(h-u))^\alpha
$$

$$
+ \frac{\gamma}{B(\gamma)\Gamma(\gamma)} \int_0^t u^{\gamma-1} E_{\alpha,\gamma}(yu^\alpha) E_{\alpha,\beta}(z(t-u))^\alpha (t-u)^{\gamma-1} du.
$$

Here the integral part is given as:

$$
\int_0^t u^{\gamma-1} E_{\alpha,\gamma}(yu^\alpha) E_{\alpha,\beta}(z(t-u))^\alpha (t-u)^{\gamma-1} du \tag{5.149}
$$

$$
= \frac{y E_{\alpha,\beta+\gamma}(yt^\alpha) - z E_{\alpha,\beta+\gamma}(zt^\alpha)}{y-z} t^{\beta+\gamma-1}.
$$

If $y, z \in \mathbb{C}$ with the condition that $y \neq z$, then $\alpha, \beta, \gamma > 0$. Now replacing Eq. (5.148) into Eq. (5.149), we obtain the requested result. $\qquad\square$

With the new fractional integral, we have the following result for exponential, sine, and cosine functions:

$$_0^{AB}I_t^\gamma \{\exp(\tau t)\} = \exp(\tau t)\frac{1-\gamma}{B(\gamma)} + \frac{\gamma}{B(\gamma)}t^\gamma E_{1,\alpha+1}(\tau t), \tag{5.150}$$

$$_0^{AB}I_t^\gamma \left\{\frac{\sinh(\sqrt{\tau}t)}{\sqrt{\tau}}\right\} = \frac{1-\gamma}{B(\gamma)}\frac{\sinh(\sqrt{\tau}t)}{\sqrt{\tau}} + \frac{\gamma}{B(\gamma)}t^{\gamma+1}E_{2,2+\gamma}(\tau t^2),$$

$$_0^{AB}I_t^\gamma \{\cosh(\sqrt{\tau}t)\} = \frac{1-\gamma}{B(\gamma)}\cosh(\sqrt{\tau}t) + \frac{\gamma}{B(\gamma)}t^\gamma E_{2,1+\gamma}(\tau t^2).$$

5.8.3 Relation With Integral Transforms

In this section, the relation between both derivatives with Laplace, Sumudu, Fourier, and Mellin transforms is presented.

Theorem 5.8.7. *The Laplace transform of Atangana–Baleanu fractional derivative in Caputo sense and in Riemann–Liouville sense produces*

$$L\left\{_0^{ABC}D_t^\alpha (f(t))\right\}(p) = L\left\{\frac{B(\alpha)}{1-\alpha}\int_0^t f'(x)E_\alpha\left[-\alpha\frac{(t-x)^\alpha}{1-\alpha}\right]dx\right\} \tag{5.151}$$

$$= \frac{B(\alpha)}{1-\alpha}\frac{p^\alpha L\{f(t)\}(p) - p^{\alpha-1}f(0)}{p^\alpha + \frac{\alpha}{1-\alpha}}$$

and

$$L\left\{_0^{ABR}D_t^\alpha (f(t))\right\}(p) = L\left\{\frac{B(\alpha)}{1-\alpha}\frac{d}{dt}\int_0^t f(x)E_\alpha\left[-\alpha\frac{(t-x)^\alpha}{1-\alpha}\right]dx\right\} \tag{5.152}$$

$$= \frac{B(\alpha)}{1-\alpha}\frac{p^\alpha L\{f(t)\}(p)}{p^\alpha + \frac{\alpha}{1-\alpha}}.$$

Theorem 5.8.8. *Let $f \in H^1(a, b)$, $b > a$, $\alpha \in [0, 1]$, then the Sumudu transform of Atangana–Baleanu fractional derivative in Caputo sense is given as:*

$$ST\left\{_0^{ABC}D_t^\alpha (f(t))\right\} = \frac{B(\alpha)}{1-\alpha}\left(\frac{\alpha\Gamma(\alpha+1)}{(1-\alpha)^k}E_\alpha(-u^\alpha)\right)(ST(f(t)) - f(0)). \tag{5.153}$$

Proof. By definition, we have the following:

$$ST \left\{ {}_{0}^{ABC} D_t^\alpha \left(f(t) \right) \right\} = ST \left\{ \frac{B(\alpha)}{1-\alpha} \int_0^t f'(x) E_\alpha \left[-\alpha \frac{(t-x)^\alpha}{1-\alpha} \right] dx \right\}. \tag{5.154}$$

By definition of Sumudu transform, we have

$$ST \left\{ {}_{0}^{ABC} D_t^\alpha \left(f(t) \right) \right\} = \int_0^\infty e^{-t} \frac{B(\alpha)}{1-\alpha} \int_0^t f'(u\tau) E_\alpha \left[-\alpha \frac{(t-u\tau)^\alpha}{1-\alpha} \right] dx d\tau. \tag{5.155}$$

Using convolution theorem for Sumudu transform, we obtain

$$ST \left\{ {}_{0}^{ABC} D_t^\alpha \left(f(t) \right) \right\} = \frac{B(\alpha)}{1-\alpha} u \, ST \left(\frac{df(t)}{dt} \right) ST \left(E_\alpha \left[\frac{-\alpha}{1-\alpha} t^\alpha \right] \right). \tag{5.156}$$

Then we have the following result

$$ST \left\{ {}_{0}^{ABC} D_t^\alpha \left(f(t) \right) \right\} = \frac{B(\alpha)}{1-\alpha} \left(\frac{\alpha \Gamma(\alpha+1)}{(1-\alpha)^k} E_\alpha(-u^\alpha) \right) \left(ST \left(f(t) \right) - f(0) \right). \tag{5.157}$$

This provides the requested result. □

Theorem 5.8.9. *Let* $f \in H^1(a, b)$, $b > a$, $\alpha \in (0, 1]$, *then the Sumudu transform of the definition of the Atangana–Baleanu fractional derivative in Riemann–Liouville sense is given as:*

$$ST \left\{ {}_{0}^{ABR} D_t^\alpha \left(f(t) \right) \right\} = \frac{B(\alpha)\alpha\Gamma(\alpha+1)}{(1-\alpha)^{k+1}} E_\alpha(-u^\alpha) ST \left(f(t) \right) - \frac{B(\alpha)}{1-\alpha} \frac{f(0) E_\alpha(0)}{u}. \tag{5.158}$$

Theorem 5.8.10. *Let* $f \in H^1(a, b)$, $b > a$, $\alpha \in [0, 1]$, *then the Fourier transform of Atangana–Baleanu fractional derivative in Caputo sense is given as:*

$$F \left\{ {}_{0}^{ABC} D_t^\alpha \left(f(t) \right) \right\} (w) \tag{5.159}$$

$$ = \frac{B(\alpha)}{1-\alpha} (jw) F(w) \left(-\frac{i(-1+e^{2i\alpha})(-1+\alpha)\alpha |w|^{\alpha-1} \left(\begin{array}{c} 1 \\ +sgn[w] \end{array} \right)}{2\sqrt{2} \left(e^{\frac{3i\alpha}{2}} \alpha - |w|^\alpha + \alpha |w|^\alpha \right) \left(\begin{array}{c} \alpha - e^{\frac{i\alpha}{2}} |w|^\alpha \\ +e^{\frac{i\alpha}{2}} \alpha |w|^\alpha \end{array} \right)} \right). $$

Proof. By definition, we have the following:

$$F\left\{_0^{ABC}D_t^\alpha\left(f\left(t\right)\right)\right\}(w) = F\left\{\frac{B(\alpha)}{1-\alpha}\int\limits_0^t f'(x)E_\alpha\left[-\alpha\frac{(t-x)^\alpha}{1-\alpha}\right]dx\right\}(w)$$

$$= \int\limits_{-\infty}^\infty \frac{B(\alpha)}{1-\alpha}\exp(-jwt)\int\limits_0^t f'(x)E_\alpha\left[-\alpha\frac{(t-x)^\alpha}{1-\alpha}\right]dxdt$$

$$= \frac{B(\alpha)}{1-\alpha}F\left(\frac{df(t)}{dt}\right)F\left(E_\alpha\left[\frac{-\alpha}{1-\alpha}t^\alpha\right]\right)$$

$$= \frac{B(\alpha)}{1-\alpha}(jw)F(w).$$

$$\left(-\frac{\begin{matrix}i(-1+e^{2i\alpha})(-1+\alpha)\\ \alpha|w|^{\alpha-1}(1+sgn[w])\end{matrix}}{2\sqrt{2}\left(\begin{matrix}e^{\frac{3i\alpha}{2}}\alpha-|w|^\alpha\\+\alpha|w|^\alpha\end{matrix}\right)\left(\begin{matrix}\alpha-e^{\frac{i\alpha}{2}}|w|^\alpha\\+e^{\frac{i\alpha}{2}}\alpha|w|^\alpha\end{matrix}\right)}\right). \qquad (5.160)$$

$\square$

Theorem 5.8.11. *Let $f \in H^1(a,b)$, $b > a$, $\alpha \in [0,1]$, then the Fourier transform of Atangana–Baleanu fractional derivative in Riemann–Liouville sense is given as:*

$$F\left\{_0^{ABR}D_t^\alpha\left(f\left(t\right)\right)\right\}(w) \qquad (5.161)$$

$$= \frac{B(\alpha)}{1-\alpha}F(w)\left(-\frac{\begin{matrix}i(-1+e^{2i\alpha})\\ (-1+\alpha)\alpha|w|^{\alpha-1}(1+sgn[w])\end{matrix}}{2\sqrt{2}\left(\begin{matrix}e^{\frac{3i\alpha}{2}}\alpha\\-|w|^\alpha+\alpha|w|^\alpha\end{matrix}\right)\left(\begin{matrix}\alpha-e^{\frac{i\alpha}{2}}|w|^\alpha\\+e^{\frac{i\alpha}{2}}\alpha|w|^\alpha\end{matrix}\right)}\right).$$

Theorem 5.8.12. *Let $f \in H^1(a,b)$, $b > a$, $\alpha \in [0,1]$, then the Mellin transform of Atangana–Baleanu fractional derivative in Caputo sense is given as:*

$$M\left\{_0^{ABC}D_t^\alpha\left(f\left(t\right)\right)\right\}(s) = \frac{B(\alpha)}{1-\alpha}(s-1).$$

$$M(f)((s-1))M\left(E_\alpha\left[-\alpha\frac{(t-x)^\alpha}{1-\alpha}\right]\right) \qquad (5.162)$$

$$M\left(E_\alpha\left[-\alpha\frac{(t-x)^\alpha}{1-\alpha}\right]\right) = \begin{cases} 0, & t > 0 \\ \sum\limits_{k=0}^\infty \frac{(-1)^{k\alpha}\left(\frac{\alpha}{1-\alpha}\right)^k}{\Gamma(\alpha k+1)(k\alpha+s)}, & 0 < t < 1 \end{cases}.$$

Proof. By definition, we have

$$
M\left\{{}_{0}^{ABC}D_{t}^{\alpha}\left(f\left(t\right)\right)\right\}(s) = M\left(\frac{B(\alpha)}{1-\alpha}\int_{0}^{t}f'(x)E_{\alpha}\left[-\alpha\frac{(t-x)^{\alpha}}{1-\alpha}\right]dx\right) \quad (5.163)
$$

$$
= \left(\int_{0}^{\infty}\frac{B(\alpha)}{1-\alpha}t^{s-1}\int_{0}^{t}f'(x)E_{\alpha}\left[-\alpha\frac{(t-x)^{\alpha}}{1-\alpha}\right]dxdt\right).
$$

Using convolution theorem for Mellin transform, we obtain

$$
M\left\{{}_{0}^{ABC}D_{t}^{\alpha}\left(f\left(t\right)\right)\right\}(s) = \frac{B(\alpha)}{1-\alpha}M\left(\frac{df(t)}{dt}\right)M\left(E_{\alpha}\left[\frac{-\alpha}{1-\alpha}t^{\alpha}\right]\right). \quad (5.164)
$$

Then, we have the result as below:

$$
M\left\{{}_{0}^{ABC}D_{t}^{\alpha}\left(f\left(t\right)\right)\right\}(s) = \frac{B(\alpha)}{1-\alpha}(s-1).
$$

$$
M(f)(s-1)M\left(E_{\alpha}\left(-\alpha\frac{(t-x)^{\alpha}}{1-\alpha}\right)\right) \quad (5.165)
$$

$$
M\left(E_{\alpha}\left[-\alpha\frac{(t-x)^{\alpha}}{1-\alpha}\right]\right) = \begin{cases} 0, & t>0 \\ \sum_{k=0}^{\infty}\frac{(-1)^{k\alpha}\left(\frac{\alpha}{1-\alpha}\right)^{k}}{\Gamma(\alpha k+1)(k\alpha+s)}, & 0<t<1 \end{cases}. \qquad \square
$$

Theorem 5.8.13. *Let $f \in H^{1}(a,b)$, $b > a$, $\alpha \in [0,1]$, then the Mellin transform of Atangana–Baleanu fractional derivative in Riemann–Liouville sense is given as:*

$$
M\left\{{}_{0}^{ABR}D_{t}^{\alpha}\left(f\left(t\right)\right)\right\}(s) = \frac{B(\alpha)}{1-\alpha}F(s)M\left(E_{\alpha}\left[-\alpha\frac{(t-x)^{\alpha}}{1-\alpha}\right]\right), \quad (5.166)
$$

$$
M\left(E_{\alpha}\left[-\alpha\frac{(t-x)^{\alpha}}{1-\alpha}\right]\right) = \begin{cases} 0, & t>0 \\ \sum_{k=0}^{\infty}\frac{(-1)^{k\alpha}\left(\frac{\alpha}{1-\alpha}\right)^{k}}{\Gamma(\alpha k+1)(k\alpha+s)}, & 0<t<1 \end{cases}.
$$

5.9 Physical Interpretation of Fractional Derivatives

In the last past years the big problem faced by researchers within the field of fractional calculus is the physical meaning of derivative with fractional order. Some fewer researchers have tried to answer this question, however, many other researchers around science were not satisfied with their demonstration. For instance Tavassoli et al. suggested that, we quote: *"We conclude that the product of fractional order derivative with the correspondent area is constant, so the fractional derivative produces the change in the area of the triangle enclosed by the tangent line at particular point and vertical line passing through this point and above X-axis with respect to fractional gradient line."* The question we ask here is: How can we then use this to portray the flow of groundwater within a geological formation? Regarding epidemiology, what can we say when we are describing the spread of the disease? Podlubny suggested that the geometry interpretation of the fractional derivative is the shadows on the walls. According to the dictionary, the shadow is, we quote: *"a region where light from a light source is obstructed by opaque object.* It occupies all of the three-dimensional volume behind an object with light in front." The shadow on the walls! It is a very big philosophical term, but the problems with the shadow are the following:

- The shadow does not always represent the real shape of an object,
- The shadow cannot tell the physical properties of an object.

Nonetheless, the mathematical design of all the definitions of fractional derivatives presented earlier is based upon the definition of a convolution. The convolution of f and g is written $f * g$, using an asterisk or star. It is defined as the integral of the product of the two functions after one is reversed and shifted. As such, it is a particular kind of integral transform:

$$(f * g)(t) \stackrel{\text{def}}{=} \int_{-\infty}^{\infty} f(\tau) g(t - \tau) d\tau = \int_{-\infty}^{\infty} f(t - \tau) g(\tau) d\tau. \qquad (5.167)$$

While the symbol t is used above, it need not represent the time domain. But in that context, the convolution formula can be described as a weighted average of the function $f(\tau)$ at the moment t where the weighting is given by $g(-\tau)$ simply shifted by amount t. As t changes, the weighting function emphasizes different parts of the input function. For functions f, g supported on only $[0, \infty)$ (that is to say, zero for negative arguments), the integration limits can be truncated, resulting in

$$(f * g)(t) = \int_0^t f(\tau) g(t - \tau) d\tau \quad \text{for } f, g : [0, \infty) \to \mathbb{R}. \qquad (5.168)$$

In this case, the Laplace transform is more appropriate than the Fourier transform and boundary terms become relevant.

In mathematics and, in particular, functional analysis, convolution is a mathematical operation on two functions f and g, producing a third function that is typically viewed as a modified version of one of the original functions, giving the integral of the pointwise multiplication of the two functions as a function of the amount that one of the original functions is translated. Convolution is similar to cross-correlation. It has applications that include probability, statistics, computer vision, natural language processing, image and signal processing, engineering, and differential equations. We present some roles of convolution here:

- Express each function in terms of a dummy variable τ.
- Reflect one of the functions: $g(\tau) \rightarrow g(-\tau)$.
- Add a time-offset, t, which allows $g(t - \tau)$ to slide along the τ-*axis*.
- Start t at $-\infty$ and slide it all the way to $-\infty$. Wherever the two functions intersect, find the integral of their product. In other words, compute a sliding, weighted-sum of function $f(\tau)$, where the weighting function is $g(-\tau)$. The resulting waveform (not shown here) is the convolution of functions f and g. If $f(t)$ is a unit impulse, the result of this process is simply $g(t)$, which is therefore called the impulse response. Formally:

$$\int_{-\infty}^{\infty} \delta(\tau) \, g(t - \tau) \, d\tau = g(t). \tag{5.169}$$

Let us look at the mathematical formulation of derivative with fractional order. With the Caputo derivative with fractional order, we have a convolution of the local derivative with the power function. With the Riemann–Liouville derivative with fractional order, we have the derivative of a convolution of a given function and the power function. With the Caputo–Fabrizio derivative with fractional order, we have a convolution of local derivative of a given function and the function exponent. Now what is the convolution? What are the applications of convolution? Convolution and associated functions are found in many applications in science, engineering, and mathematics (see [103]).

- **In image processing.** In digital image processing convolutional filtering plays an important role in many important algorithms in edge detection and related processes. In optics, an out-of-focus photograph is a convolution of the sharp image with a lens function. The photographic term for this is bokeh. In image processing applications, such as adding blurring.
- **In digital data processing.** In analytical chemistry, Savitzky–Golay smoothing filters are used for the analysis of spectroscopic data. They can improve signal-to-noise ratio with minimal distortion of the spectra. In statistics, a weighted moving average is a convolution.
- **In acoustics.** Reverberation is the convolution of the original sound with echoes from objects surrounding the sound source. In digital signal processing, convolution is used to map the impulse response of a real room on a digital audio signal. In electronic music, convolution is the imposition of a spectral or rhythmic structure on a sound. Often this envelope or structure is taken from another sound. The convolution of two signals is the filtering of one through the other [103].
- **In electrical engineering.** The convolution of one function (the input signal) with a second function (the impulse response) gives the output of a linear time-invariant system (LTI) [103]. At any given moment, the output is an accumulated effect of all the prior values of the input function, with the most recent values typically having the most influence

(expressed as a multiplicative factor). The impulse response function provides that factor as a function of the elapsed time since each input value occurred [103].

- **In physics.** Wherever there is a linear system with a "superposition principle," a convolution operation makes an appearance [102]. For instance, in spectroscopy, line broadening due to the Doppler effect on its own gives a Gaussian spectral line shape and collision broadening alone gives a Lorentzian line shape. When both effects are operative, the line shape is a convolution of Gaussian and Lorentzian, a Voigt function [103]. In time-resolved fluorescence spectroscopy, the excitation signal can be treated as a chain of delta pulses, and the measured fluorescence is a sum of exponential decays from each delta pulse [103]. In computational fluid dynamics, the large eddy simulation (LES) turbulence model uses the convolution operation to lower the range of length scales necessary in computation thereby reducing computational cost.
- **In probability theory.** The probability distribution of the sum of two independent random variables is the convolution of their individual distributions [101,102]. In kernel density estimation, a distribution is estimated from sample points by convolution with a kernel, such as an isotropic Gaussian [103].
- **In radiotherapy.** In treatment of planning systems, most parts of all modern codes of calculation apply the convolution-superposition algorithm [103].

The above applications of a convolution show that the fractional derivative as convolution has multiple purposes, it can portray the memory as in the case of theory of elasticity, it can be considered as a filter, in particular the Caputo and Caputo–Fabrizio type can be viewed as a filter of local derivative with power and exponent functions. The aim of a filter is to get rid of impurities and produce only the real product, the fractional derivative is therefore in some cases the real velocity. Atangana and Koca show that the Caputo and Fabrizio derivative is a low-pass filter for α greater than $1/2$; it is a band-pass filter when α is $1/2$, and finally it is a high-pass filter when α is greater than $1/2$. It is important to recall that the low-pass filter is a filter that passes signals with a frequency lower than a certain cutoff frequency and attenuates signal with frequencies higher than cutoff frequency [104]. The high-pass filter is the opposite of low-pass filter. The band-pass filter is the combination of a law-pass filter and a half-pass filter. Nevertheless, with Caputo derivative, authors in [104] show that the Caputo derivative is a high-pass filter for α greater than $1/2$; it is a band-pass filter when α is $1/2$, and finally it is a low-pass filter when α is greater than $1/2$. Therefore we can see that both derivatives play the same role; however, the new derivative with fractional order does not have singularity, which is a greater advantage of this derivative over existing ones [104].

5.10 Advantages and Limitations

Fractional calculus has been used to model physical and engineering processes, which are found to be best described by fractional differential equations. It is worth noting that the standard mathematical models of integer-order derivatives, including nonlinear models, do not work adequately in many cases. In the recent years, fractional calculus has played a very important role in various fields such as mechanics, electricity, chemistry, biology, economics, notably control theory, and signal and image processing. Major topics include anomalous diffusion, vibration and control, continuous time random walk, Levy statistics, fractional Brownian motion, fractional neutron point kinetic model, power law, Riesz potential, fractional derivative and fractals, computational fractional derivative equations, nonlocal phenomena, history-dependent process, porous media, fractional filters, biomedical engineering, fractional phase-locked loops, fractional variational principles, fractional transforms, fractional wavelet, fractional predator–prey system, soft matter mechanics, fractional signal and image processing, singularities analysis and integral representations for fractional differential systems, special functions related to fractional calculus, non-Fourier heat conduction, acoustic dissipation, geophysics, relaxation, creep, viscoelasticity, rheology, fluid dynamics, chaos, and groundwater problems. An excellent literature of this can be found [106]. It is very important to point out that all these fractional derivative order definitions have their advantages and disadvantages; here we will include Caputo, variational order, Riemann–Liouville, Jumarie, and Weyl. We will examine first the variational order differential operator [106].

5.10.1 Advantages of Fractional Derivatives

1. *With the Riemann–Liouville fractional derivative, an arbitrary function needs not be continuous at the origin and it needs not be differentiable.*
2. *One of the great advantages of the Caputo fractional derivative is that it allows traditional initial and boundary conditions to be included in the formulation of the problem [106]. In addition, its derivative for a constant is zero. It is customary in groundwater investigations to choose a point on the centerline of the pumped borehole as a reference for the observations and therefore neither the drawdown nor its derivatives will vanish at the origin, as required [106]. In such situations where the distribution of the piezometric head in the aquifer is a decreasing function of the distance from the borehole, the problem may be circumvented by rather using the complementary, or Weyl, fractional order derivative [106]. The Caputo fractional derivative also allows the use of the initial and boundary conditions when dealing with real-world problems. The Caputo derivative is the most appropriate fractional operator to be used in modeling real world problem.*
3. *The q-analogs find applications in a number of areas, including the study of fractals and multi-fractal measures, and expressions for the entropy of chaotic dynamical systems. The relationship with fractals and dynamical systems results from the fact that many fractal patterns have the symmetries of Fuchsian groups in general (see, for example, Indra's pearls and the Apollonian gasket) and the modular group in particular. The connection passes through hyperbolic geometry and ergodic theory, where the elliptic integrals and*

modular forms play a prominent role; the q-series themselves are closely related to elliptic integrals [106].

4. *As an alternative modeling approach to the classical Fick's second law, the fractal derivative is used to derive a linear anomalous transport-diffusion equation underlying anomalous diffusion process [106].*

5. *Besides the mathematical satisfactions of the Atangana–Baleanu derivative with fractional order, the attentiveness for the new derivative is because of the necessity of employing a model portraying the behavior of orthodox viscoelastic materials, thermal medium and other. The new approach is able to portray material heterogeneities and some structure or media with different scales. The non-locality of the new kernel allows the full describing of the memory within structure and media with different scales, which cannot be described by classical fractional derivative and also that of the Caputo–Fabrizio type. In addition to this, we rely that as for the Caputo–Fabrizio derivative of fractional order, this new derivative can play a significant role in the study of macroscopic behavior of some materials, related to nonlocal exchanges, which are predominant in defining the properties of the material [103]. Atangana–Baleanu derivatives will therefore be very useful in describing many problems of science, engineering, and technology.*

5.10.2 Disadvantages of Fractional Derivatives

Although these fractional derivatives display great advantages, they are not applicable in all the situations. We shall begin with the Liouville–Riemann type.

1. *The Riemann–Liouville derivative has certain disadvantages when trying to model real-world phenomena with fractional differential equations. The Riemann–Liouville derivative of a constant is not zero. In addition, if an arbitrary function is a constant at the origin, its fractional derivation has a singularity at the origin, for instance, exponential and Mittag–Leffler functions. These disadvantages reduce the field of application of the Riemann–Liouville fractional derivative.*

2. *Caputo's derivative demands higher conditions of regularity for differentiability: to compute the fractional derivative of a function in the Caputo sense, we must first calculate its derivative. Caputo derivatives are defined only for differentiable functions while functions that have no first-order derivative might have fractional derivatives of all orders less than one in the Riemann–Liouville sense.*

3. *With the Caputo–Fabrizio derivative, the kernel is local and its derivative when $\alpha = 0$ does not give the initial function. In addition to this the anti-derivative associate is not fractional.*

CHAPTER 6

Regularity of a General Parabolic Equation With Fractional Differentiation

We shall recall that partial differential equation can be divided into classes based on their form. The groundwater flow equations are therefore classified under parabolic type. The main aim of this chapter is to provide a comprehensive extension of some existing results surrounding the documented results of the regularity of parabolic equations with fractional order. The cases considered here are the fractional derivatives with nonlocal and singular kernel, and nonlocal and nonsingular kernel. The purpose of this chapter is to assure the regularity of the groundwater flow models that are classified under parabolic equations. The regularity of the solution with fractional order will be given in terms involving the well-known Hölder conditions, the results will be for the viscosity. The chapter will start with the regularity for parabolic equations with fractional derivative based on power law kernel and finally we will present the regularity of parabolic equations with Mittag–Leffler kernel. This chapter is made up with results obtained in [197,205,204].

6.1 Regularity With Caputo Fractional Differentiation

In this section the main focus will be devoted to the analysis of nonlocal parabolic equations of non-divergence type which involves a generalized fractional time derivative. To be specific, this part will be presentation of the work done in [205,204]:

$$
\begin{aligned}
q(x,t) \;=\; & P_t^\nu Z(x,t) - IZ(x,t) \\
: \;=\; & \int_{-\infty}^{t} [Z(x,t) - Z(x,s)]\, G(t,s)\, ds \\
& - \sup_i \inf_j \left([Z(x+y,t) + Z(x-y,t) - 2Z(x,t)]\, G_{ij}(x,y,t)\right).
\end{aligned}
\tag{6.1}
$$

The most important contribution in the work [205,204] is that viscosity solutions obtained are Hölder continuous, which is a very important fact also in the study of groundwater flow model as a parabolic equation. Prior to giving the exact assumptions associated with the kernels K and G_{ij}, we firstly provide a description of the recent history by showing Hölder continuity for parabolic equations of non-divergence type as well as the motivation for studying a

Fractional Operators with Constant and Variable Order
with Application to Geo-Hydrology
DOI: 10.1016/B978-0-12-809670-3.00006-0

parabolic equation that involves a generalized fractional time derivative which is in our work, including the class of parabolic equation describing the flow of subsurface water within geological formation called aquifers. For local linear parabolic equations of non-divergence type more information can be found also in the work done by Krylov and Safonov as presented in [214] as the theoretical proof of the Hölder continuity of solutions without the requirement of regularity assumptions equation's coefficients, which again is a more suitable approach that favors the class of parabolic equation under investigation. Later on, this method was adapted to investigate the regularity properties for fully nonlinear elliptic equations as presented in the work by [209], also in the work done by L. Wang [218] who adapted the methods for nonlinear elliptic equations as a means of proving the continuity of solutions to fully nonlinear parabolic equations; this case will be very important for the case of the groundwater model within a leaky aquifer. Recently, nonlocal equations have become of interest as a result of the applications given in physical sciences, geohydrology, mechanics, technology, and engineering. In this section, a particular focus will be on the nonlocal elliptic operators of integro-differential equations in the form of

$$LZ(x) = \int_{R^n} \left[Z(x) - Z(y) \right] G(x, y). \tag{6.2}$$

When $G(x, y) = c_{n,\omega} |x - y|^{-n-2\omega}$, then L is simply the fractional Laplacian $(-\Delta)^{\omega}$. An assumption such as $G(x, y) = G(x, -y)$ generates L, a non-divergence type operator. Alternatively, an assumption such as $G(x, y) = G(y, x)$ generates an operator of divergence type. If L is of a non-divergence type, then L becomes the nonlocal analogue of the linear elliptic operator

$$LZ = a_{ij} Z_i Z_j.$$

An often found, ellipticity assumption on $G(x, y)$ is that $\tau |x - y|^{-n-2\omega} \leq G(x, y) \leq \Upsilon |x - y|^{-n-2\omega}$, and is comparable to the assumption for local equations that $\tau |\varepsilon|^2 \leq a_{ij}\varepsilon_i\varepsilon_j \leq \Upsilon |\varepsilon|^2$. Even though regularity estimates were known for operators of type (6.2), these estimates have no uniformity as the order ω of the operator went to 1. The work done by Caffarelli and Silvestere [208] gave adaptation to the methods for complete nonlinear local equations [209] to the nonlocal setting, and found estimates of uniformity as the order of the operator ω went to 1. Also on the other hand, the work by Chang Lara and Davila [215] then gave furthermore a proof of regularity estimates for parabolic equations in the form of

$$\partial_t Z(x, t) - I Z(x, t) = 0,$$

where I is a nonlocal nonlinear elliptic operator. The estimates in [215] have uniformity as the order of the operator ω tends to 1, so the findings in [215] are not solely a nonlocal analogue, but additionally take into account many of the regularity findings for local parabolic equations.

The motivation for the equation of interest in this section is given by the following equation

$$\partial_t^\nu Z - P_{|x|}^\beta Z = q. \tag{6.3}$$

It is worth noting that the above was introduced in [210,211] with the main aim of modeling plasma transport, a study that is close related to the movement of subsurface water within a geological formation. In their work, they considered the function Z to stand for the probability density function for tracer particles within the plasma representing the probability of discovering particles at time t and position x. On the right-hand side, q was considered to be a source term. The nonlocal diffusion operator $P_{|x|}^\beta$ is one-dimensional and gives account for transport. The fractional derivative ∂_t^ν is the Caputo derivative, whereby model's context, the Caputo derivative, helps to include into mathematical formulation the trapping effect of trace particles within turbulent eddies. Even though (6.3) is an equation associated with one spatial dimension, we may give consideration to the following nonlocal parabolic equation in higher dimensions

$$\partial_t^\nu Z - LZ = f \tag{6.4}$$

where L is nonlocal elliptic operator of type (6.2) and ∂_t^ν is the Caputo derivative. As previously mentioned, certain assumptions associated with the kernel $G(x, y)$ result in an operator of either divergence or non-divergence type. The author in [204] focused on weak solutions of (6.4) of non-divergence type. The assumptions related to the kernel $G(x, y, t)$ were that

$$\frac{\tau}{|x - y|^{n+\omega}} \le G(x, y, t) \le \frac{\Upsilon}{|x - y|^{n+\omega}}, \tag{6.5}$$

for fixed $0 < \tau < \Upsilon$, and that $G(x, y, t) = G(x, -y, t)$. For divergence form equations, the author in [205] focused on weak solutions of (6.4) of divergence type and gave proof to a De Giorgi–Nash–Moser type theorem that gives Hölder continuity of solutions. The findings in [205] were associated with the assumptions of (6.5) and $G(x, y, t) = G(y, x, t)$. The latter assumption resulted in (6.4) being an equation of divergence type. The author demonstrated in [206] that the techniques used within [205] may give application to an equation of type (6.4) with a more significant general fractional time derivative. The primary aim of this section is to extend the findings for non-divergence form equations found in [204] to equations entailing the more general fractional time derivative that is now given description.

6.1.1 Some Useful Information About the Marchaud and Caputo Derivatives

The Caputo derivative is of use to modeling phenomena which takes account of interactions within the past and also problems with nonlocal properties. In this sense, one can think of the equation as having "memory." This contrasts with parabolic equations such as the heat opera-tor $\partial_t - \Delta$ that gives no account for the past, the groundwater flow equations within confined, unconfined, and leaky aquifers and also other diffusion problems. The Caputo derivative as presented earlier is defined as

$$_a\partial_t^\nu q(t) := \frac{1}{\Gamma(1-\nu)} \int_a^t \frac{f'(s)}{(t-s)^\nu} ds.$$

For C^1 functions, there may be used integration by parts for showing the equivalent formula

$$_a\partial_t^\nu q(t) := \frac{1}{\Gamma(1-\nu)} \frac{q(t)-q(a)}{(t-a)^\nu} + \frac{\nu}{\Gamma(1-\nu)} \int_a^t \frac{Z(t)-Z(s)}{(t-s)^\nu} ds.$$

Upon defining $q(t) = q(a)$ for $t < a$ then as in [205] this gives birth to the equivalent formulation

$$\partial_t^\nu q(t) := \frac{\nu}{\Gamma(1-\nu)} \int_{-\infty}^t \frac{q(t)-q(s)}{(t-s)^\nu} ds. \tag{6.6}$$

This one-sided nonlocal derivative is referred to as the Marchaud derivative [216], which was recently looked at in [205]. The formulation in (6.6) is of significant use. It is no longer of essence to know the initial point a, and subsequently label, the initial point a. Another fea-ture of use of this formulation in (6.6) is, instead of assigning initial data simply as $Z(a) = c$, assigning a more general "initial" data as $Z(t) = \varphi(t)$ for $t \le a$ with $\varphi(x)$ not necessarily being differentiable nor even continuous. The formulation found in (6.6) will be of particu-lar use for the notion of viscosity solutions in the context of non-divergence solutions, which are described later in section 6.1.3. The formulation in (6.6) appears to be similar to the one-dimensional fractional Laplacian; however, exception is given to the fact that integration occurs solely from one side. We may then give application to many of the methods devel-oped for nonlocal elliptic operators such as the fractional Laplacian to equations entailing the Caputo derivative as seen in [205,206]. Finally, the formulation given in (6.6) brings about a different type of generalization of the Caputo derivative. Rather than generalizing as

$$\frac{d}{dt}(k * (q - q(a))) \quad \text{or} \quad \int_a^t f'(s)G(t-s)ds,$$

one may generalize as

$$P_t^\nu q(t) := \int_{-\infty}^t \left[q(t)-q(s)\right] G(t,s)ds. \tag{6.7}$$

The proof of Hölder continuity in [205] for the linear divergence equation works for the more general fractional time derivative

$$\int\limits_{-\infty}^{t} \left[q(t) - q(s) \right] G(t, s, x),$$

provided that the kernel satisfies condition

$$G(t, t - s, x) = G(t + s, t, x) \tag{6.8}$$

and

$$\frac{\nu}{\Gamma(1 - \nu)} \frac{\tau}{|t - s|^{1+\nu}} \le G(t, s, x) \le \frac{\nu}{\Gamma(1 - \nu)} \frac{\Upsilon}{|t - s|^{1+\nu}}. \tag{6.9}$$

For instance, refer to [206] where a kernel $G(t, s, x)$ does not satisfy the condition in (6.8) and (6.9) is utilized. The assumption made in (6.9) is comparable to that of (6.9) for the kernel $G(x, y, t)$ of L. Similarly, the condition (6.8) is comparable to $G(x, y, t) = G(y, x, t)$ and was a requirement in [206] as a result of the equation being of divergence form.

6.1.2 The Main Focus of Investigation and Main Results

We shall give an assumption to the kernels G_{ij} satisfying (6.5) and $G_{ij}(x, y, t) = G_{ij}(x, -y, t)$. Additionally, we assume the kernel $G(t, s)$ satisfies (6.9). As a means of proving Hölder continuity of solutions to non-divergence parabolic equations which involve the Caputo derivative, the author in [204] followed the idea in [217] as a means of solving an ordinary differential equation, so that information is captured backwards in time for a solution. The requirement is the solvability of Eq. (6.10) below

$$\begin{cases} O(t) = 0 & , \text{ if } t \le -2 \\ P_t^\nu O(t) = c_0 \left| \{ x \in B_1 : Z(x, t) \le 0 \} \right| - C_1 O(t) & , \text{ if } t \in (-2, 0). \end{cases} \tag{6.10}$$

Several complications are brought up when consideration is given to solutions of (6.10). First of all, the proof is needed for the existence of solutions. Secondly, the right-hand side is not necessarily always continuous. Thirdly, a solution of (6.10) may not be regular to an extent that it may be used as a part of a test function for the concept of viscosity solution. Fourthly, it is necessary to roughly show that if $|\{ x \in B_1 \times (-2, -1) : Z(x, t) \le 0 \}| \ge \iota_1$, then $O(t) \ge \iota_2$ if $t \in (-1, 0)$. When $P_t^\nu = \partial_t^\nu$, the author in [204] made use of explicit representation formulas [212] for solutions to (6.10) to obtain the required properties. Since there is no availability of such formulas for solutions to (6.10) in general, the four mentioned complications are overcome in a different way. We first show a priori Hölder continuity estimates for such ordinary differential equations. Our class of weak solutions shall be given consideration in the viscosity sense as described in section 6.1.3.

Theorem 6.1.1. *Let q be a continuous function on* $[-1, 0]$, *and let Z be bounded and a viscosity solution to*

$$\sup_k \inf_l \left([Z(x, t) - Z(x, s)] \, K^{kl}(t, s) ds \right) = q(t)$$

on $(-1, 0]$ *with* $f \in L^\infty$. *Assume also that the kernels* $K^{kl}(t, s)$ *satisfy* (6.9). *Then if* $0 < v \leq 1$, *there is existence of two constants* $0 < \beta \leq 1$ *and* $C > 0$ *depending on* v, τ, Υ *and* $\|Z\|_{L^\infty(-\infty, -1)}$ *but uniform as* $v \to 1$ *such that*

$$\|Z\|_{C^{0,\beta}\left(\left[\frac{-1}{2}, 0\right]\right)} \leq C \|f\|_{L^\infty}. \tag{6.11}$$

We make use of Theorem 6.1.1 for proving the existence of solutions to (6.10). For accommodating the second and third complications we demonstrate in section 6.1.5 that we may approximate (6.10) in a uniform manner from below by Lipschitz sub-solutions to (6.10) which shall be sufficient for the intention of this section. The fourth complication, however, still remains. The Hölder estimate in (6.11) depending on an L^∞ norm of q is however not enough to overcome the fourth complication. Subsequently, we given consideration to a subclass of kernels $G(t, s)$ which we assume also to satisfy (6.8). To add, this last assumption lets us give proof to our second primary result.

Theorem 6.1.2. *Let Z be a bounded, continuous viscosity solution to* (6.10) *in* $B_2 \times [-2, 0]$. *Assume the kernels* $G_{ij}(x, y, t)$ *satisfy* $G_{ij}(x, y, t) = G_{ij}(x, -y, t)$ *and* (6.5). *Assume the kernel* $G(t, s)$ *satisfies* (6.9) *and* (6.8). *Then there is an existence of three constants* $C, k, \zeta_0 > 0$ *depending only on* Υ, τ, n, v, ω *but uniform as* $v \to 1$ *such that Z is Hölder continuous in* $B_1 \times [-1, 0]$, *and for* (x, t), $(y, s) \in [-1, 0]$ *the following estimates hold:*

$$|Z(x, t) - Z(y, s)| \leq C \left(\|Z\|_{L^\infty} + \zeta_0^{-1} \|f\|_{L^\infty} \right) |x - y|^k + |t - s|^{\frac{kv}{2\omega}}.$$

We give note to the estimates in Theorem 6.1.2 which remain uniform as the order v of the fractional time derivative is approaching 1. Despite this, no uniformity in the estimates remains as the order of the elliptic operator ω is approaching 1. Upon obtaining the required theory for solutions to ordinary differential equations which involve P_t^v, in the future, the author proposes to make use of the methods of Chang Lara and Davila in [215] for proving Theorem 6.1.2 with estimates uniform as ω is approaching 1. Another interest one would find is proving Hölder regularity to solutions of an equation having the form

$$\operatorname*{supinf}_{k \quad l} \left(\int_{-\infty}^{t} [Z(x,t) - Z(x,s)] \, K^{kl}(t,s,x) \right) \tag{6.12}$$

$$- \operatorname*{supinf}_{i \quad j} \left(\int_{R^n} [Z(x+y,t) - Z(x,t)] \, K^{ij}(t,x,y) \right)$$

$$= \quad q(t,x).$$

To do this, it seems relevant to prove Theorem 6.1.1 with the Hölder estimate depending on the L^p norm of the right-hand side q.

Some importation denotations: Here we give definition to the denotations which will be consistent throughout the work.

- ∂_t^ν – the Caputo derivative as defined in Eq. (6.6).
- P_t^ν – the generalized Marchaud derivative as defined in Eq. (6.7).
- ν – will always denote the order of the fractional time derivative.
- ω – will always denote the order of the nonlocal elliptic spatial operator.
- t, s – will always be reserved as time variables.
- $G(x, y, t)$ – the kernel for the elliptic operator L as defined in (6.2).
- $G(t, s)$ – the kernel for P_t^ν as defined in Eq. (6.7).
- $AB_\omega^\pm$ – Pucci's extremal operators (defined in Section 6.2) for the elliptic spatial operators.
- τ, Υ – Ellipticity constants as appearing in (6.5).
- $Q_r(x_0, t_0)$ – the space-time cylinder $B_r(x_0) \times \left(t_0 - r^{\frac{2\omega}{\nu}}, t_0\right)$.
- Q_r – the cylinder centered at the origin $Q_r(0, 0)$.

Outline. This part will be presented as in the paper [197]: To fix the mind of the readers, we suggest the following arrangement for this section: We will start with section 6.1.3 in which we will provide a detailed explanation to the notion of the viscosity solution that shall be utilized within the work. The next part we make use of a standard method for showing the comparison principle and uniqueness for ordinary differential equations which involve P_t^ν [215]. In section 6.1.4, we will stress on the analysis to provide proof to the first primary result that gives solutions to certain ordinary differential equations which involve that P_t^ν are Hölder continuous. In section 6.1.5 we provide establishment to the necessary properties for solutions of (6.10). In Section 6.1.6 we give proof to our second significant result that solutions to parabolic equations of type (6.10) are Hölder continuous [215].

6.1.3 Investigation of the Weak Solutions of Equations in Non-divergence Form

For the purpose of studying weak solutions of equations of type $P_t^\nu Z - LZ = f$ in non-divergence form we shall make use of the concept of the viscosity solution. For the elliptic operator L make reference to the concept of Pucci's extremal operators which is introduced in [215]. For fixed time t we denote the second order difference $\delta(Z, x, y, t) := Z(x + y, t) + Z(x - y, t) - 2Z(x, t)$.

We fix two constants $0 < \tau \leq \Upsilon$ and define

$$AB_\omega^+ Z(x,t) \quad : \quad = \int_{R^n} \frac{\Upsilon \delta(Z, x, y, t)_+ - \tau \delta(Z, x, y, t)_-}{|y|^{n+2\omega}},$$

$$AB_\omega^- Z(x,t) \quad : \quad = \int_{R^n} \frac{\Upsilon \delta(Z, x, y, t)_+ - \Upsilon \delta(Z, x, y, t)_-}{|y|^{n+2\omega}}.$$

We then provide the well-known definition for a Pucci-type extremal operator for fractional derivatives of type Eq. (6.7). For fixed $x \in R_n$,

$$AB_\omega^+ Z(x,t) \quad : \quad = \frac{\nu}{\Gamma(1-\nu)} \int_{-\infty}^{t} \frac{\Upsilon (Z(x,t) - Z(x,s))_+ - \tau (Z(x,t) - Z(x,s))_-}{(t-s)^{1+\nu}},$$

$$AB_\omega^- Z(x,t) \quad : \quad = \frac{\nu}{\Gamma(1-\nu)} \int_{-\infty}^{t} \frac{\tau (Z(x,t) - Z(x,s))_+ - \Upsilon (Z(x,t) - Z(x,s))_-}{(t-s)^{1+\nu}}.$$

Since ν is reserved for P_t^ν and ω is reserved for the kernel of L, no confusion should arise between $AB_\omega^\pm$ and $AB_\nu^\pm$. These operators bring fourth to the equations

$$AB_\nu^- Z(x,t) - M^+ Z(x,t) \leq q(x,t),$$
$$AB_\nu^+ Z(x,t) - M^- Z(x,t) \geq q(x,t).$$

In this work we show regularity for the parabolic equation when the kernels $G(t,s)$ satisfy (6.8); subsequently, we will only give consideration for solutions to

$$P_t^\nu Z(x,t) - M^+ Z(x,t) \leq q(x,t), \tag{6.13}$$
$$P_t^\nu Z(x,t) - M^- Z(x,t) \geq q(x,t). \tag{6.14}$$

In providing proof for regularity for ordinary differential equations we shall not assume that $G(t,s)$ satisfies (6.8); as a result, we consider the solutions to

$$AB_\nu^- Z(t) \leq q(t), \tag{6.15}$$

$$AB_v^+ Z(t) \geq q(t). \tag{6.16}$$

As was presented in the work [209], we give the following properties for Pucci's extremal operators.

Proposition 6.1.3. *For fixed Z, v evaluated at fixed (x, t) we see the following properties where $M^\pm$ denote either $AB_v^\pm$ or $AB_\omega^\pm$:*

$$
\begin{aligned}
i)\ & M^- Z \ \leq\ M^+ Z. \\
ii)\ & M^- Z \ \leq\ -M^+(-Z). \\
iii)\ & M^\pm cZ \ =\ cM^\pm Z \ \text{if } c \geq 0. \\
iv)\ & M^+ Z + M^- v \ \leq\ M^+(Z + v) \leq M^+ Z + M^+ v. \\
v)\ & M^- Z + M^- v \ \leq\ M^-(Z + v) \leq M^- Z + M^+ v.
\end{aligned}
$$

We now define a viscosity solution. We say that an upper semi-continuous function Z is a viscosity sub-solution of (6.10) (or a solution of (6.13)) if, whenever a C^2 function satisfies $\Phi \geq Z$ on $[t_0 - r, t_0] \times B_\rho(x_0)$ and $\Phi(x_0, t_0) = Z(x_0, t_0)$ and if v is defined as

$$
v(x, t) := \begin{cases} \Phi(x, t) \text{ if } (x, t) \in [t_0 - r, t_0] \times B_\rho(x_0) \\ \qquad Z(x, t) \text{ otherwise,} \end{cases}
$$

then $P_t^v v(x_0, t_0) - I v(x_0, t_0) \leq f(x_0, t_0)$ (or v is a solution to (6.13) at (x_0, t_0)).

A viscosity super-solution of (6.13) (or a solution of (6.13)) for lower semi-continuous functions is given a similar definition. We note that a viscosity sub-solution or super-solution of (6.13) is a viscosity solution of (6.13) (6.14). A solution is both a sub-solution and a super-solution, and therefore a continuous function. The notion of viscosity solutions and super-solutions for $P_t^v Z = q$ (or solutions to (6.15) and (6.16) are defined in a similar way). We point out that we may extend our class of test functions that touch from above or from below to functions Φ that are C^2 in the x-variable for fixed t and Lipschitz in time for fixed x. It is clearly seen that if function $P_t^v Z$ can be evaluated classically and solves $P_t^v Z = q$, then Z becomes a solution in the viscosity sense. This is made clear in the following two propositions.

Proposition 6.1.4. *Let Z be a continuous bounded function. Let $\Phi \in C^{0,\gamma}$ with $v < \gamma \leq 1$. If $\Phi \geq (\leq) Z$ on $[t_0 - \zeta, t_0]$ and $\Phi(t_0) = Z(t_0)$, then the integral*

$$
\int_{-\infty}^{t_0} [Z(t_0) - Z(s)] G(t, s) ds
$$

is well-defined and possibly $\infty(-\infty)$ so that $P_t^v Z(t_0)$ is well-defined.

Proof. Without losing generality we make an assumption that $\Phi \geq Z$ on $[t_0 - \zeta, t_0]$. Then

$$\int_{-\infty}^{t_0} [Z(t_0) - Z(s)] \, G(t, s) ds$$

$$= \int_{-\infty}^{t_0-\zeta} [Z(t_0) - Z(s)] \, G(t, s) ds + \int_{t_0-\zeta}^{t_0} [Z(t_0) - Z(s)] \, G(t, s) ds$$

$$\geq \int_{-\infty}^{t_0-\zeta} [Z(t_0) - Z(s)] \, G(t, s) ds + \int_{t_0-\zeta}^{t_0} [Z(t_0) - \Phi(s)] \, G(t, s) ds$$

$$= \int_{-\infty}^{t_0-\zeta} [Z(t_0) - Z(s)] \, G(t, s) ds + \int_{t_0-\zeta}^{t_0} [\Phi(t_0) - \Phi(s)] \, G(t, s) ds$$

$$\geq \int_{-\infty}^{t_0-\zeta} [Z(t_0) - Z(s)] \, G(t, s) ds - \Upsilon \|\Phi\| \, C^{0,\gamma} \int_{t_0-\zeta}^{t_0} (t_0 - s)^{\gamma-1-\nu} ds$$

$$\geq \int_{-\infty}^{t_0-\zeta} [Z(t_0) - Z(s)] \, G(t, s) ds - C_1$$

$$\geq -2\Upsilon \|Z\| \, L^\infty \int_{-\infty}^{t_0-\zeta} (t_0 - s)^{-1-\nu} ds - C_1$$

$$\geq C_2.$$

Therefore, the integral becomes well-defined and possibly ∞. □

Proposition 6.1.5. *Let Z be a continuous bounded function on $(-\infty, T]$ and assume that for some $t \in (-\infty, T]$ there exists a Lipschitz function touching Z from below (above) at t. From Proposition 6.1.4, the term $P_t^\nu Z$ is well-defined and we have that*

$$\int_{-\infty}^{t} [Z(t) - Z(s)] \, G(t, s) ds \geq (\leq) \, q(t) \tag{6.17}$$

provided $P_t^\nu Z \leq (\geq) \, q(t)$ in the viscosity sense.

Proof. Assume the inequality in (6.17). If Φ touches Z from below in $[t - \zeta, t]$,

$$q(t) \leq \int_{-\infty}^{t} [Z(t) - Z(s)] G(t, s) ds$$

$$= \int_{-\infty}^{t-\zeta} [Z(t) - Z(s)] G(t, s) ds + \int_{t-\zeta}^{t} [Z(t) - Z(s)] G(t, s) ds$$

$$\leq \int_{-\infty}^{t-\zeta} [Z(t) - Z(s)] G(t, s) ds + \int_{t-\zeta}^{t} [Z(t) - \Phi(s)] G(t, s) ds$$

$$= \int_{-\infty}^{t-\zeta} [Z(t) - Z(s)] G(t, s) ds + \int_{t-\zeta}^{t} [\Phi(t) - \Phi(s)] G(t, s) ds$$

and so $P_t^\nu Z \geq q(t)$ in the viscosity sense. Assume now that $P_t^\nu Z \geq q(t)$ in the viscosity sense. From the assumption, we may touch Z from below by a Lipschitz function Φ in some neighborhood $[t - \zeta, t]$. We may then find Lipschitz Φ_k converging uniformly to Z in $[t - \zeta, t]$ with $\Phi_k \geq Z$ in $[t - \zeta, t]$. Due to the integral in (6.17) being well-defined, we have from Lebesgue's dominated convergence theorem:

$$q(t) \leq \lim_{k \to \infty} \int_{-\infty}^{t-\zeta} [Z(t) - Z(s)] G(t, s) ds + \int_{t-\zeta}^{t} [\Phi_k(t) - \Phi_k(s)] G(t, s) ds$$

$$= \lim_{k \to \infty} \int_{-\infty}^{t-\zeta} [Z(t) - Z(s)] G(t, s) ds + \int_{t-\zeta}^{t} [Z(t) - \Phi_k(s)] G(t, s) ds$$

$$= \int_{-\infty}^{t-\zeta} [Z(t) - Z(s)] G(t, s) ds + \int_{t-\zeta}^{t} [Z(t) - Z(s)] G(t, s) ds$$

$$= \int_{-\infty}^{t} [Z(t) - Z(s)] G(t, s) ds.$$

The concept of continuity is fundamental to viscosity solutions. If we let

$$\begin{cases} Z(t) = t^{\nu-1} & \text{if } t > 0 \\ Z(t) = 0 & \text{if } t \leq 0 \end{cases},$$

then one may explicitly compute that $\partial_t^\nu Z$ for any $t \in R$. However, Z is not upper semi-continuous, and as a result it is not a viscosity sub-solution. Viscosity solutions are closed under appropriate limits.

$\square$

Lemma 6.1.6. *Let Z_k be a sequence of continuous bounded viscosity solutions to (6.13) or (6.14) in $B_{R_1} \times [-R_2, 0]$ converging in $R^n \times (-\infty, T]$ to Z_0 bounded and continuous. Then Z_0 is a viscosity solution to (6.13) (or (6.14)) in $B_{R_1} \times [-R_2, 0]$.*

Proof. The proof is standard as well as straightforward from the definition of viscosity solutions (see [209]).

$\square$

6.1.4 Derivation of an Approximating Solutions

For the purpose of showing uniqueness for an ordinary differential equation such as $P_t^\nu Z = q$, we make use of the concept of sup- and inf-convolution. For a bounded and upper-semi-continuous function Z on $(-\infty, t_0]$, for $t \le t_0$ we define

$$Z^\zeta(t) := \sup_{s \le t} \left\{ Z(s) + \frac{1}{\zeta}(s - t) + \zeta \right\}. \tag{6.18}$$

If Z is bounded and lower-semi-continuous on $(-\infty, t_0]$, for $t \le t_0$, we define

$$Z_\zeta(t) := \inf_{s \le t} \left\{ Z(s) - \frac{1}{\zeta}(s - t) - \zeta \right\}. \tag{6.19}$$

We have the following properties

Proposition 6.1.7. *For Z^ζ as defined in (6.18) and $t \le t_0$, the following hold:*

a) there exists $t^* \le t$ such that $Z^\zeta(t) = Z(t^*) + \frac{1}{\zeta}(t^* - t) - \zeta$.

b) $Z^\zeta(t) \ge Z(t) + \zeta$.

c) $Z^\zeta(t_2) - Z^\zeta(t_1) \ge \zeta^{-1}(t_2 - t_1)$ for $t_1 < t_2$.

e) $0 < \zeta_1 < \zeta_2 \Rightarrow Z^{\zeta_1}(t) \le Z^{\zeta_2}(t)$.

f) $t - t^* \le 2\zeta \sup |Z|$.

h) $0 < Z^\zeta(t) - Z(t) \le Z(t^*) - Z(t) + \zeta$.

Proof. All properties except for (c) are as in Lemma 6.5.2 in [209]. For property (c) we point out that for $t \le t_1 \le t_2$ we have

$$Z^\zeta(t_2) \geq Z(t) + \zeta^{-1}(t - t_2) + \zeta$$
$$= Z(t) + \zeta^{-1}(t - t_1) + \zeta + \zeta^{-1}(t_1 - t_2).$$

Taking the supremum over $t \leq t_1$ we obtain

$$Z^\zeta(t_2) - Z^\zeta(t_1) \geq -\zeta^{-1}(t_2 - t_1).$$

Making use of the properties found in Proposition 6.1.7, it is standard to demonstrate the following proposition which is similar to Theorem 6.5.1 in [209] and Propositions 6.5.4 and 6.5.5 in [208]. $\qquad\square$

Proposition 6.1.8. *If Z is bounded and lower-semi-continuous (upper-semi-continuous) in $(-\infty, T]$ then Z_ζ $(-Z^\zeta)$ Γ-converges to Z $(-Z)$. If Z is continuous then Z^ζ, Z^ζ converge uniformly to Z. Furthermore, if $P_t^\nu Z \geq (\leq) f$, then there exists $P_\zeta \to 0$ as $\zeta \to 0$ such that $P_t^\nu Z_\zeta \geq f - P_\zeta$ $\left(P_t^\nu Z^\zeta \leq f + P_\zeta\right)$.*

Lemma 6.1.9. *Let Z be bounded and upper-semi-continuous and v be bounded and lower-semi-continuous on $(-\infty, T]$. Let f, g be continuous functions. If $P_t^\nu Z \leq f$ and $P_t^\nu v \geq g$ in the viscosity sense, then $P_t^\nu(v - Z) \geq g - f$ in the viscosity sense.*

Proof. We take the approximating solutions Z^ζ, v_ζ with $P_t^\nu Z^\zeta \leq f + P_\zeta$ and $P_t^\nu v^\zeta \geq f - P_\zeta$. From property (3) in Proposition 6.1.7 we have that at every point Z^ζ can be touched from above by a Lipschitz function and v_ζ can be touched from below. Then also $v_\zeta - Z^\zeta$ can be touched from below by a Lipschitz function at every point. From Propositions 6.1.4 and 6.1.5, the terms $P_t^\nu Z^\zeta$, $P_t^\nu v_\zeta$, $P_t^\nu \left(v_\zeta - Z^\zeta\right)$ are well-defined and at any point t,

$$P_t^\nu \left(v_\zeta - Z^\zeta(t)\right) = P_t^\nu v_\zeta(t) - P_t^\nu Z^\zeta(t) \geq g - q - 2P_\zeta.$$

Then from Proposition 6.1.5 we conclude that $P_t^\nu \left(v_\zeta - Z^\zeta\right) \geq g - q - 2P_\zeta$. Letting $\zeta \to 0$ we obtain from Lemma 6.1.6 that $P_t^\nu(v - Z) \geq g - f$ in the viscosity sense. $\qquad\square$

With that being said, it is now possible to prove a comparison principle.

Theorem 6.1.10. *Let Z be bounded and upper-semi-continuous and v be bounded and lower-semi-continuous on $(-\infty, T_2]$. Let q be a continuous function and assume that $P_t^\nu Z \leq f \leq P_t^\nu v$ on $(T_1, T_2]$ with $Z \leq v$ on $(-\infty, T_1]$. Then $Z \leq v$ on $(-\infty, T_2]$, and if $Z(t_0) = v(t_0)$ for some $t_0 \in (T_1, T_2]$, then $Z(t) = v(t)$ for all $t \leq t_0$.*

Proof. Suppose there exists $t_1 \in (T_1, T_2]$ such that $v(t_1) - Z(t_1) \le 0$. Let $v - Z$ achieve its minimum in $[T_1, T_2]$ at $t_0 \in (T_1, T_2]$. From Lemma 6.1.9, we have $P_t^v (v - Z) \ge 0$ in the viscosity sense. Since $v - Z \ge 0$ for $t \le T_1$, then $(v - Z)$ is touched from below by the constant function $v(t_0) - Z(t_0)$ at t_0. From Propositions 6.1.4 and 6.1.5 the term $P_t^v (v - Z) (t_0)$ is well-defined and

$$\int_{-\infty}^{t_0} [(v - Z)(t_0) - (v - Z)(s)] G(t, s) ds \ge 0.$$

Since $(v - Z)$ achieves a minimum over $(-\infty, t_0]$ at t_0, we also have

$$0 \ge \int_{-\infty}^{t_0} [(v - Z)(t_0) - (v - Z)(s)] G(t, s) ds.$$

Then if $(v - Z)(t_0) \le 0$, we have that $(v - Z)(t) = 0$ for all $t \le t_0$. With the availability of a comparison principle, we may make use of Perron's method for proving the existence of solutions. $\qquad\square$

Theorem 6.1.11. *Let $\Phi(t)$ be continuous on $(-\infty, T_1]$ and q be continuous on $[T_1, T_2]$. There exists a unique viscosity solution Z to*

$$\left\{ \begin{array}{ll} P_t^v v = f & \text{for } t \in [T_1, T_2] \\ v = 0 & \text{for } t \le T_1 \end{array} \right\}$$

on $(T_1, T_2]$.

The proof of Theorem 6.1.11 is standard. We forgo this proof since the concepts and techniques are later used in the proof of Lemma 6.1.24. One may also show existence for solutions of (6.13); however, because the primary focus of this work is based on the regularity of solutions, here is no inclusion of the result. Later in the work, we shall require that the continuous divergence solutions constructed in [205,206] are also of the viscosity solutions.

Theorem 6.1.12. *Let q be continuous on $[-2, 0]$ with $q(-2) = 0$. Assume the kernel $G(t, s)$ of P_t^v satisfies (6.8) and (6.9). Then the weak divergence solution v constructed in [205] of*

$$\left\{ \begin{array}{ll} P_t^v v = f & \text{for } t \in (-\infty, 0], \\ v = 0 & \text{for } t \le 2, \end{array} \right\} \tag{6.20}$$

is also a viscosity solution and hence the unique viscosity solution.

Proof. In [205,206] the solution v is constructed with an ζ approximation. By mean of recursion we obtain the solution v_ζ to

$$\zeta \sum_{i<j} \left[v_\zeta(\zeta j) - v_\zeta(\zeta i) \right] G(\zeta j, \zeta i) = f(\zeta j), \tag{6.21}$$

where $i, j \in Z$. From the results obtained in the work by [205,206], $v_\zeta \to V$ where v gives solution to the divergence form equation

$$\int_{-\infty}^{T} \int_{-\infty}^{t} [v(t) - v(s)][\Phi(t) - \Phi(s)] G(t,s) ds dt \tag{6.22}$$

$$+ \int_{-\infty}^{T} \int_{-\infty}^{2t-T} v(t) \Phi(t) G(t,s) ds dt - \int_{-\infty}^{T} v(t) P_t^\nu \Phi(t) dt$$

$$= \int_{-\infty}^{T} q(t) \Phi(t) dt,$$

for all $T \leq 0$ and Φ bounded and Lipschitz on $(-\infty, 0]$. We extend v_ζ to all of $(-\infty, 0]$ by $v_\zeta(t) = v(\zeta i)$ where $\zeta(i-1) < t \leq \zeta j$. We now show that v is a viscosity solution. The primary notion in the following computations is that because (6.21) is a discrete equation, then v_ζ is also a viscosity-type discretized solution. Subsequently, the limit will also become a viscosity solution. Now, let ψ be a Lipschitz function touching v strictly from above at $t_0 \leq 0$, that is $v(t) < \psi(t)$ for $t < t_0$ and $v(t_0) = \psi(t_0)$. We let δ_ζ be such that $\psi + \delta_\zeta$ touches v_ζ from above at $t_\zeta \leq t_0$. Since ψ touches v solely from above at t_0, then $t_\zeta \to t_0$ as $\zeta \to 0$. Now, if $\zeta(j_\zeta - 1) < t_\zeta \leq \zeta j_\zeta$, then since $\psi + \delta_\zeta$ touches v_ζ from above and (6.21) is a discrete equation, we have that

$$\zeta \sum_{i<j_\zeta} \left[\psi(t_\zeta) - \psi(\zeta i) \right] G(\zeta j_\zeta, \zeta i) \geq f(\zeta j_\zeta).$$

Now

$$\zeta \sum_{i<j_\zeta} \left[\psi(t_\zeta) - \psi(\zeta i) \right] G(\zeta j_\zeta, \zeta i) = \zeta \sum_{i<j_\zeta} \left[\psi(t_\zeta) - \psi(\zeta i) \right] G(\zeta j_\zeta, \zeta i)$$

$$+ \zeta \sum_{i<j_\zeta} \left[\psi(\zeta j_\zeta) - \psi(\zeta i) \right] G(\zeta j_\zeta, \zeta i)$$

$$= (I) + (II).$$

Since ψ is Lipschitz, we have

$$
\begin{aligned}
|(I)| &= \left| \zeta \sum_{i<j_\zeta} \left[\psi\left(t_\zeta\right) - \psi\left(\zeta i\right) \right] G(\zeta j_\zeta, \zeta i) \right| \\
&\leq \zeta \sum_{i<j_\zeta} |C\zeta| \, G(\zeta j_\zeta, \zeta i) \\
&\leq \zeta^2 \sum_{i<j_\zeta} \Upsilon \zeta^{-1-\nu} \left(j_\zeta - i\right)^{-1-\nu} \\
&= \zeta^{1-\nu} \Upsilon \sum_{i=1}^{\infty} (i)^{-1-\nu}.
\end{aligned}
$$

Then $|(I)| \to 0$ as $\zeta \to 0$. Since ψ is Lipschitz continuous, it follows that $(II) \to P_t^\nu (t_0)$ as $\zeta \to 0$. Since q is continuous and $t_\zeta \to t_0$ as $\zeta \to 0$, we have that $f\left(\zeta j_\zeta\right) \to f(t_0)$. Then $P_t^\nu \psi(t_0) \geq f(t_0)$. The proof if ψ touches in a similar way from below. Then v becomes a viscosity solution. The next corollary will be of use if we want to demonstrate that a limit is not equivalently zero. $\qquad\square$

Corollary 6.1.13. *Let v be the viscosity solution to (6.20) as in Theorem 6.1.12. Assume also that $v \geq 0$. Then for $-2 < t \leq 0$ we have*

$$
\frac{\tau \nu}{\Gamma(1-\nu)} \int_{-2}^{t} \frac{v(s)}{(t-s)^\nu} ds \leq \int_{-2}^{t} f(s)\, ds \tag{6.23}
$$

$$
\leq \frac{\Upsilon \nu}{\Gamma(1-\nu)} \int_{-2}^{t} \frac{v(s)}{(t-s)^\nu} ds.
$$

Proof. We take $\Phi \equiv 1$ in (6.22). Then

$$
\begin{aligned}
\int_{-2}^{T} f(t)\, dt &= \int_{-\infty}^{T} \int_{-\infty}^{2t-T} v(t)\, G(t,s)\, ds \\
[-1.5pt] &\leq \frac{\Upsilon \nu}{\Gamma(1-\nu)} \int_{-\infty}^{T} \int_{-\infty}^{2t-T} \frac{v(t)}{(t-s)^{\nu+1}} ds\, dt \\
&= \frac{\Upsilon \nu}{\Gamma(1-\nu)} \int_{-2}^{T} \frac{v(t)}{(T-t)^\nu} ds.
\end{aligned}
$$

In the same manner, the bound from the other side is shown. $\qquad\square$

6.1.5 Hölder Continuity for the Time-Fractional Differentiation

We shall follow the concepts in [208] for proving the Hölder continuity. The outline and state-ments of the lemmas are with intention similar to those in Sections 6.8–6.12 in [208] for the readers who intend to make comparisons and contrasts in the operator's properties P_t^ν with properties of L. Instead of working with the concave envelope, we make use of a different envelope.

$$AB_Z(t) := \sup_{s \le t} Z(s).$$

Lemma 6.1.14. *Assume* $AB_\nu^- Z(t) \le f(t)$ *and* $AB_Z(t_0) = Z(t_0)$. *Let* $r_k := 2^{\frac{-1}{(1-\nu)}} 2^{-k}$ *and* $R_k(t) := [t - r_k, t - r_{k+1}]$. *There is a constant* C_0 *depending on* τ, *but not on* ν *such that for any* $M > 0$, *there exists* k *such that*

$$\left| R_k(t_0) \cap \{Z(s) < Z(t_0) - AB_{r_k}\} \right| \le C_0 \frac{f(t_0)}{M} |R_k(t_0)|. \tag{6.24}$$

Proof. We first point out that $|R_k(t_0)| = r_{k+1} = \frac{r_k}{2}$. Since $AB_Z(t_0) = Z(t_0)$ then Z is touched from above by the constant $AB_Z(t_0)$. Then from Proposition 6.1.5 we have

$$\int_{-\infty}^{t_0} [Z(t_0) - Z(s)] G(t_0, s) ds \le f(t_0).$$

If in the case that the conclusion is not true, it then follows that

$$
\begin{aligned}
f(t_0) \ &\ge \ AB_\nu^- Z(t_0) \ge \frac{\tau \nu}{\Gamma(1-\nu)} \int_{-\infty}^{t_0} \frac{Z(t_0) - Z(s)}{(t_0 - s)^{\nu+1}} ds \\[2mm]
&\ge \ \frac{\tau \nu}{\Gamma(1-\nu)} \int_{t_0 - 2^{\frac{-1}{(1-\nu)}}}^{t_0} \frac{Z(t_0) - Z(s)}{(t_0 - s)^{\nu+1}} ds \\[2mm]
&\ge \ \frac{\tau \nu}{\Gamma(1-\nu)} \sum_{k=0}^{\infty} \int_{R_k(t_0)} \frac{Z(t_0) - Z(s)}{r_{k+1}^{\nu+1}} ds \\[2mm]
&\ge \ \frac{\tau \nu}{\Gamma(1-\nu)} AB_{r_k} \frac{C_0 f(t_0)}{M} \frac{|R_k(t_0)|}{r_{k+1}^{\nu+1}} \\[2mm]
&= \ 2^\nu \frac{\tau \nu}{\Gamma(1-\nu)} C_0 f(t_0) \sum_{k=0}^{\infty} r_k^{1-\nu}.
\end{aligned}
$$

Now

$$\sum_{k=0}^{\infty} r_k^{1-v} = \frac{1}{2} \sum_{k=0}^{\infty} \left(\frac{1}{2^{1-v}} \right)^k = \frac{1}{2} \left[\frac{1}{1 - 2^{-(1-v)}} \right].$$

Then

$$f(t_0) \geq \frac{\tau v}{\Gamma(1-v)} \frac{C_0 f(t_0)}{2^{1-v} - 1}.$$

From the following relation on the Gamma function: $\Gamma(1-z)\Gamma(z) = \frac{\pi}{\sin(\pi z)}$ it follows that for $v_0 < v \leq 1$, the right-hand side of the above equation is bounded from below. Then for C_0 being large enough, a contradiction is obtained. $\qquad\square$

Remark 1. Using the above proof, it becomes clear that by selecting C_0 larger (say, for instance, $2C_0$) that for fixed v and M bounded from above we may choose $k \leq N$ for some large N. This N depends on v, giving a bound from below on r_k in a way that (6.24) holds.

Corollary 6.1.15. *For any $\zeta_0 > 0$, there exists C such that for Z as defined in Lemma 6.1.14, there exists $r \in \left(0, 2^{\frac{-1}{(1-v)}}\right)$ such that if $AB_Z(t) = Z(t)$ and $AB_v^- f Z(t) \leq f(t)$, then there exists a constant c_v (depending on v) and $r \in \left(c_v, 2^{\frac{-1}{(1-v)}}\right)$ such that*

$$\frac{\left|\{s \in \left[t-r, \frac{t-r}{2}\right]\} : Z(s) < Z(t) - Cf(t)r\right|}{\left|\left[t-r, \frac{t-r}{2}\right]\right|} \leq \zeta_0. \tag{6.25}$$

Proof. By choosing $M = Cf(t)\zeta_0^{-1}$ we obtain (6.25) from Lemma 6.1.14 and Remark 1. $\qquad\square$

Remark 2. Let $Z(t) \leq 0$ for $t < -2$ and $AB_v^- f Z(t) \leq f$ on $[-2, T]$. There exists a constant C which depends on τ, v_0 and there exist several finite intervals I_j with disjoint interiors and with length between c_v and $2^{-(1-v)}$ such that

$$\sup Z^+ \leq C \sum_j \max_{I_j} f^+(t) \left|I_j\right|. \tag{6.26}$$

Furthermore, if $I_j = \left[t_j - r_j, t_j\right]$ is such an interval, and $\left[t_j - 2r_j, t_j\right] \subset [-2, T]$, then

$$\left|\{s \in \left[t_j - 2r_j, t_j - r_j\right] : Z(s) \geq Z(r_j) - Cr_j\}\right| \geq \iota r_j. \tag{6.27}$$

Remark 3. We point out that as $v \to 1$, the inequality (6.26) becomes

$$\sup Z \leq C \int_{\{u=ABZ\}} f^+(s) \, ds.$$

Proof. We shall inductively select the finite sequence of intervals. Let Z^+ obtain its maximum on $[-2, T]$ at the point t_1. From Corollary 6.1.15 there exists r_1 with $c_\nu \leq r_1 \leq 2^{-(1-\nu)}$ such that (6.27) accounts for $I_1 = [t_1 - r_1, t_1]$:

$$AB_{Z^+}(t_1) - AB_{Z^+}(t_1 - r_1) \leq Cf(t_1) r_1 = Cf(t_1) |I_1|.$$

Proceeding inductively, supposing I_{j-1} has been chosen, we select

$$t_j := \max\{-2 \leq s \leq t_{j-1} - r_{j-1} : Z(s) = AB_Z(s)\}.$$

If no such t_j exists, then the process is terminated. Otherwise, by Corollary 6.1.15 we select $r_j \in \left(c_\nu, 2^{\frac{-1}{(1-\nu)}}\right)$ such that $[t_j - 2r_j, t_j - r_j]$ satisfies (6.27). We label $I_j = [t_j - r_j, t_j]$. As in the case of I_1 we achieve

$$AB_{Z^+}(t_j) - AB_{Z^+}(t_j - r_j) \leq Cf(t_j) |I_j|.$$

Due to each $r > c_\nu$ there will be an eventual termination of the process. Now

$$\{u = AB_{Z^+}\} \subseteq U_j I_j,$$

and so

$$\sup Z^+ \leq \sum_j |AB_{Z^+}(t_j) - AB_{Z^+}(t_j - r_j)| \leq C \sum_j \max_{I_j} f^+(t) |I_j|.$$

Finally, from how each I_j is selected, we have (6.27) which holds for each $[t_j - 2r_j, t_j - r_j]$ and also $I_j \cap I_k = \emptyset$ for $j \neq k$. For the next Lemma 6.1.26 for an interval $I_j = [t_j - r_j, t_j]$ we define

$$2I_j := [t_j - 2r_j, t_j]. \qquad \square$$

Lemma 6.1.16. *Let $A \subset [0, 1]$ and let c_1, c_2 be two positive constants. Let $\{I_j : j \in I\}$ be a collection of closed intervals with nonempty interior satisfying the following for every $j \in I$:*

$$i)\ 2I_j \subset [0, 1]$$
$$ii)\ I_j \cap I_i = \emptyset \text{ for every } j \neq i$$
$$iii)\ \sum_{j \in I} |I_j| > c_1$$
$$iv)\ |A \cap 2I_j| \geq c_2 |2I_j|,$$

then $|A| \geq \frac{c_1 \cdot c_2}{3}$.

Proof. Firstly, we assumed the collection $\{I_j\}$ is finite. We shall select a subset $J \subseteq I$. We recall that each interval is associated with a nonempty interior and that the interior's intersection is empty. We shall choose a sub-collection of intervals in the following way: We select and label I_1 as the interval having the farthest right-end point. If there is another interval I_k for some $k \in I$ such that

$$1) \ 2I_k \cap 2I_1 \ = \ \emptyset \ \text{ and}$$
$$1) \ 2I_k \backslash (2I_k \cap 2I_1) \ = \ \emptyset \, ,$$

then we select I_2 to be the interval satisfying (a) and (b) such that $2I_k$ has the farthest left-end point. If no $2I_k$ satisfies (a) and (b), then we select I_2 to be the interval whose right-end point is nearest to the left-end point of $2I_1$. Next, we select all the remaining intervals in the same way. It is clear from the development that each interval from the collection $\{2I_j\}$ for $j \in J$ may intersect at most two other intervals from that same collection. It is also clearly seen that

$$\bigcup_{i \in I} I_i \subset \bigcup_{j \in j} 2I_j. \tag{6.28}$$

Then

$$|A| \ \geq \ \frac{1}{2} \sum_{j \in J} |A \cap 2I_j|$$
$$\geq \ \frac{c_2}{3} \sum_{j \in J} |2I_j|$$
$$\geq \ \frac{c_2}{3} \sum_{i \in I} |2I_i|$$
$$\geq \ \frac{c_1 c_2}{3}.$$

When the collection $\{I_i\}$ is infinite, we simply select a finite sub-collection of intervals whose measure is within ζ, and then let $\zeta \to 0$. $\qquad\square$

Next, we prove the main and the more important lemma.

Lemma 6.1.17. *Let $0 < v_0 \leq v \leq 1$. There exist $\zeta_0 > 0$, $0 < \iota < 1$ and $M > 1$, all depending on v_0, τ, τ, such that if*

$$i) \ Z \ \geq \ 0 \ in \ (-\infty, 1]$$
$$ii) \ Z(1) \ \leq \ \frac{1}{2}$$
$$iii) \ AB_v^+ Z \ \geq \ -\zeta_0 \ in \ [-1, 1]$$

then

$$|\{Z \le M\} \cap [0, 1]| > \iota.$$

Proof. We make use of the function

$$\Phi(t) := \begin{cases} 0 & \text{for } t \le \frac{3}{4} \\ 4\left(t - \frac{3}{4}\right) & \text{for } t \ge \frac{3}{4}. \end{cases}$$

We note that $AB_v^- Z \le 4\tau$ for all $0 < v \le 1$. We now give application to Lemma 6.1.17 to $v := \Phi - u$. Since $Z(1) \le \frac{1}{2}$ we have that $v(1) \ge \frac{1}{2}$. From estimate (6.26) we have

$$\frac{1}{2} \le \sup v \le C \sum_j \max_{I_j} AB_v^- (\Phi - Z) |I_j| \tag{6.29}$$

$$\le C(\zeta_0 + 4\tau C_1) \sum_j |I_j|$$

$$\le C\zeta_0 + 4\tau C_1 \sum_j |I_j|.$$

Then for ζ_0 small enough we achieve

$$\frac{1}{4} \le C \sum_j |I_j|.$$

Now Φ is supported in $\left[\frac{3}{4}, 1\right]$. Then if $I_j = \left[t_j - \frac{r_j}{4}, t_j\right]$ then $t_j \in \left[\frac{3}{4}, 1\right]$ and so $2I_j := \left[t_j - r_j, t_j\right] \subset [0, 1]$. From (4.4)

$$\left|\left\{s \in 2I_j : v(s) \ge v(t_j) - Cr_j\right\}\right| \ge \iota |I_j|.$$

Now we recall that $r_j \le 2^{-(1-v)}$ and $v(t_j) \ge 0$, such that

$$\left|\left\{s \in 2I_j : \Phi(s) - Z(s) \ge -C\right\}\right| \ge \iota |I_j|.$$

Furthermore, $\Phi(s) \le 1$ and so

$$\left|\left\{s \in 2I_j : 1 + C \ge Z(s)\right\}\right| \ge \iota |I_j|.$$

Now we make use of Lemma 6.1.16 to provide the following conclusion:

$$\left|\left\{s \in [0, 1] : 1 + C \ge Z(s)\right\}\right| \ge \iota_1. \qquad \square$$

Using the results of Lemma 6.1.17 with that of Lemma 6.1.15 from the work presented in [209] we may prove the following two lemmas similarly as in [209].

Lemma 6.1.18. *Let Z be as in Lemma 6.1.17, then*

$$\left|\left\{u > M^k\right\} \cap [0, 1]\right| \leq (1 - \iota)^k.$$

Lemma 6.1.19. *Let Z be as in Lemma 6.1.17, then*

$$|\{u > t\} \cap [0, 1]| \leq dt^{-\zeta}$$

where $d, \zeta > 0$ depend only on τ, Υ, v_0.

Now we have

Theorem 6.1.20. *Let $Z \geq 0$ in $(-\infty, 0]$ and $AB_v^+ Z \geq -C_0$ in $(-2r, 0]$. Assume $v \geq v_0 > 0$. Then*

$$|\{u > t\} \cap [-r, 0]| \leq dr \left(2Z(0) + C_0 r^v \zeta_0^{-1}\right)^\zeta t^{-\zeta}$$

for every t where the constants d, ζ depend solely on Υ, Υ, and v_0 as in Lemma 6.1.19.

Theorem 6.1.1 shall follow from the following:

Theorem 6.1.21. *Let $v_0 < v < 1$ for some $v_0 > 0$. Let Z be bounded in $(-\infty, 0]$ and continuous on $[-1, 0]$. If*

$$\begin{aligned} AB_v^+ u &\geq -C_0 \text{ in } (-1, 0], \\ AB_v^- u &\leq C_0 \text{ in } (-1, 0], \end{aligned}$$

then there are $v, \beta > 0$ and $C > 0$ which depend solely on τ, Υ, v_0 so that $Z \in C^{0,\beta}\left(\left[\frac{-1}{2}, 0\right]\right)$ and

$$\|Z\|_{C^{0,\beta}\left(\left[\frac{-1}{2}, 0\right]\right)} \leq C \left(\|Z\|_{L^\infty} + C_0\right).$$

Theorem 6.1.21 follows from Theorem 6.1.20 and the following Lemma 6.1.26.

Lemma 6.1.22. *Let $\frac{-1}{2} \leq Z \leq \frac{1}{2}$ in $(-\infty, 0]$. There exists δ_0 depending solely on τ, Υ and $0 < v_0 < 1$ so that if*

$$\begin{aligned} AB_v^+ u &\geq -\delta_0 \text{ in } [-1, 0], \\ AB_v^- u &\leq \delta_0 \text{ in } [-1, 0], \end{aligned}$$

then there is a β depending solely on τ, Υ, v_0 such that if $1 > v \geq v_0$,

$$|Z(t) - Z(0)| \leq C |t|^\beta$$

for some constant C.

Proof. The proof is almost similar to [208]. We provide the details. We shall develop a sequence AB_k, AB_k with $O_k \le Z \le AB_k$ and

$$O_k \le Z \le AB_k \text{ in } \left[-4^{-k}, 0\right] \qquad (6.30)$$

$$O_k \le O_{k+1} \text{ and } AB_k > AB_{k+1} \qquad (6.31)$$

$$AB_k - O_k = 4^{-\beta k}. \qquad (6.32)$$

This shall indicate (6.30) using constant $C = 2^\beta$. For $k = 0$ we select $O_0 = -1/2$ and $O_0 = 1/2$ and with the assumption we have (6.30). We move forward by induction. Assume (6.30) holds up to k. We have

$$\left| \left\{ Z \ge \frac{AB_k + O_k}{2} \right\} \cap \left[-3.4^{-(k+3)}, -4^{-(k+2)}\right] \right| \ge 4^{-(k+4)}$$

$$\left| \left\{ Z \le \frac{AB_k + O_k}{2} \right\} \cap \left[-3.4^{-(k+3)}, -4^{-(k+2)}\right] \right| \ge 4^{-(k+4)}$$

If the first is assumed, we define

$$v(t) := \frac{Z\left(4^{-k}t\right) - O_k}{(AB_k - O_k)/2}.$$

We have that $v \ge 0$ in $[-1, 0]$ and $\left| \{v \ge 1\} \cap \left[\frac{-1}{2}, \frac{-1}{4}\right] \right| \ge \frac{1}{16}$. We also have that

$$AB_v^+ v \ge \frac{-4^{-kv}\delta_0}{(AB_k - O_k)/2} = -2\delta_0 4^{k(\beta-v)} \ge -2\delta_0$$

as long as $\beta < v$. From the inductive hypothesis, for any $0 \le j < k$ we have

$$v \ge \frac{AB_{k-j} - O_k}{(AB_k - O_k)2} \ge \frac{O_{k-j} - AB_{k-j} + AB_k - O_k}{(AB_k - O_k)2} \ge 2\left(1 - 4^{\beta j}\right),$$

and so $v \ge -2\left(|4t|^\beta - 1\right)$ outside $[-1, 0]$. We define $z := \max\{v, 0\}$. For any $t \in \left[-\frac{3}{4}, 0\right]$ we have

$$AB_v^+ z(t) - AB_v^+ v(t) = \frac{v}{\Gamma(1-v)} \int_{\{v<0\}\cap\{t<-1\}} \frac{\Upsilon v(s)}{(t-s)^{v+1}} \qquad (6.33)$$

$$\ge \frac{\Upsilon v}{\Gamma(1-v)} \int_{-\infty}^{-1} \frac{-2\left(|4s|^\beta - 1\right)}{(t-s)^{v+1}}. \qquad (6.34)$$

Thus $AB_v^+ z(t) - AB_v^+ v(t) \geq -2\delta_0$ for β small enough. We have then $AB_v^+ z \geq -4\delta_0$ and so for any $t_0 \in [-1/4, 0]$ we may apply Theorem 6.1.20 to z to achieve

$$\frac{1}{16} \leq \left| \{z > 1\} \cap \left[\frac{-3}{8}, \frac{-1}{4}\right] \right| \leq \left| \{z > 1\} \cap \left[\frac{-3}{8}, t_0\right] \right|$$

$$d\left(\frac{1}{8}\right)\left(2Z(t_0) + 4\left(\frac{1}{8}\right)^v \delta_0 \zeta_0^{-1}\right)^\zeta.$$

Thus for δ_0 small enough we have $Z(t_0) \geq \theta > 0$ for any $t_0 \in \left[\frac{-1}{4}, 0\right]$. We let $AB_{k+1} = AB_k$ and $O_{k+1} = O_k + \theta \frac{AB_k - O_k}{2}$. Then $AB_{k+1} - O_k = \left(1 - \frac{\theta}{2}\right) 4^{-\beta k}$. Selecting β and θ small enough with $\left(1 - \frac{\theta}{2}\right) = 4^{-\beta}$ we have $AB_{k+1} - O_k = 4^{-\beta(k+1)}$.

Alternatively, when $\left| \left\{ Z \leq \frac{AB_k + O_k}{2} \right\} \cap \left[-3.4^{-(k+3)}, -4^{-(k+2)}\right] \right| \geq 4^{-(k+4)}$ then we define

$$v := \frac{AB_k - Z\left(4^{-k} t\right)}{\frac{O_k - O_k}{2}}$$

and use that $AB_v^- Z \leq \delta_0$. $\qquad\square$

6.1.6 Derivation of Sub-solution to an Ordinary Differential Equation

In this section the development of a sub-solution to an ordinary differential equation which allows us to prove the H is established Hölder continuity. We start off with the following.

Lemma 6.1.23. Let $Z \in C^{0,\beta}((-\infty, 0])$ with $v < \beta < 1$. Then $AB_v^\pm Z$ is continuous on $(-\infty, 0]$.

Proof. We shall show that for fixed t and any $\zeta > 0$, there exists h_0 such that if $0 \leq h < h_0$, then

$$\left| AB_v^\pm Z(t+h) - AB_v^\pm Z(t) \right| < \zeta. \tag{6.35}$$

For fixed $\varepsilon < t$ and for $0 \leq h$, we have

$$\left| \frac{v}{\Gamma(1-v)} \int_\xi^{t+h} \frac{\Upsilon[Z(t+h) - Z(s)]_+ - \tau[Z(t+h) - Z(s)]_-}{(t+h-s)^{1+v}} ds \right| \tag{6.36}$$

$$\leq \frac{v}{\Gamma(1-v)} \Upsilon \|Z\|_{C^{0,\beta}} \frac{(t+h-\varepsilon)^{\beta-v}}{\beta-v}. \tag{6.37}$$

We may select ε earlier enough to t and select h small enough, for the above inequality to be less than $\frac{\zeta}{2}$. Now because Z is continuous, if $t > \varepsilon$, then both

$$\left| \int_{-\infty}^{\xi} \frac{[Z(t+h) - Z(s)]_-}{(t+h-s)^{1+\nu}} - \frac{[Z(t) - Z(s)]_-}{(t-s)^{1+\nu}} ds \right| \to 0 \qquad (6.38)$$

and

$$\left| \int_{-\infty}^{\xi} \frac{[Z(t+h) - Z(s)]_+}{(t+h-s)^{1+\nu}} - \frac{[Z(t) - Z(s)]_+}{(t-s)^{1+\nu}} ds \right| \to 0 \qquad (6.39)$$

as $h \to 0$. Then we may select h small enough, for (6.38) to hold. $\square$

For the next lemma, we shall require a solution to an ordinary differential equation. We recall that from Theorem 6.1.11 if f is continuous on $[-2, 0]$ with $q(-2) = 0$, then there exists a viscosity solution Z to the differential equation

$$\begin{cases} Z(t) = 0 & \text{for } t \leq -2 \\ P_t^\nu Z = q(t) & \text{on } (-\infty, 0] \end{cases} . \qquad (6.40)$$

That u is a viscosity solution on $(-2, 0]$ is a direct result from Theorem 6.3.5. Since $q(t) = Z(t) = 0$ for $t \leq -2$ it is immediate that Z also becomes a solution on $(-\infty, -2]$. From Theorem 6.1.1 the solution Z is Hölder continuous.

In the next Lemma 6.1.26 we shall also make use of the following bump function. Let $\eta \geq 0$ with support in $[0, 1]$. Let $\eta' \geq 0$ for $t \leq \frac{1}{2}$ and $\eta' \leq 0$ for $t \geq \frac{1}{2}$. We point out that $P_t^\nu \eta \leq C$ for some C that is independent of ν and $P_t^\nu \eta(t) < 0$ for $t > 1$. We shall make use of

$$\eta_\zeta := \zeta \eta \left(\frac{t}{\zeta} \right).$$

Lemma 6.1.24. *Let f be smooth on $[-2, 0]$ with $q(-2) = 0$. Let Z be the viscosity solution to*

$$\begin{cases} Z(t) = 0 & \text{for } t \leq -2 \\ P_t^\nu Z = q(t) & \text{on } (-\infty, 0] \end{cases} . \qquad (6.41)$$

Then there is existence of a sequence of Lipschitz sub-solutions $\{Z_k\}$ to the above equation with $Z_k \leq Z$ and $Z_k \to Z$ uniformly on $[-2, 0]$.

Proof. We consider for fixed $M > \sup f$ the set

$$R_1 := \left\{ z : -M \le P_t^{\nu} z \le f \text{ on } [0,2] \text{ and } z \text{ is Lipschitz} \right\}.$$

We define

$$R_2 := \left\{ \begin{array}{l} v : -M \le P_t^{\nu} v \le f \text{ on } [0,2] \text{ and} \\ Z_k \Longrightarrow v \text{ with } Z_k \le v \text{ and } Z_k \in R_2 \end{array} \right\}.$$

We point out that any $v \in R_2$ is continuous as it is the uniform limit on a compact set of Lipschitz continuous functions. Next, we now demonstrate that G_1 is nonempty, and hence G_2 is also nonempty. With the theory of ordinary differential equations [212], we solve

$$\partial_t^{\nu} g(t) = \frac{h(t)}{\tau} \le \frac{f(t)}{\tau},$$

with h smooth and strictly decreasing. Then g is strictly decreasing, and so

$$P_t^{\nu} g \le AB_{\nu}^{+} g(t) = \tau \partial_t^{\nu} g(t) = h(t) \le f(t).$$

Thus g is a smooth sub-solution to (6.41) and g can be approximated from below through itself. We select $-M < \inf \frac{h(t)}{\tau}$. Thus G_2 is nonempty.

Now we assign a partial ordering to G_2 with the natural assignment that $v1 \le v2$ if $v1 \le v2$ everywhere on $(-\infty, 0]$. From the comparison principle we find that Z as in Eq. (6.41) is an upper bound for G_2.

We shall now show that $Z \in G_2$. By Zorn's Lemma there exists a maximal element z. We shall show that $z \equiv Z$. If $P_t^{\nu} z \equiv P_t^{\nu} Z$ then by comparison and uniqueness, $z \equiv Z$. Supposing, as a means of contradiction, that z is not identical to Z, then z is not a super-solution of Eq. (6.41). Subsequently, there exists a $t_0 \in [0,2]$ and a Lipschitz function Ψ with $\Psi(t_0) = z(t_0)$ and $\Psi \le z$ on $[t_0 - \delta, t]$ for some $\delta > 0$ such that $P_t^{\nu} \Psi(t_0) < q(t_0)$ where

$$\Phi(t) := \left\{ \begin{array}{ll} \Psi(t) & \text{if } t \in (t_0 - \delta] \\ z(t) & \text{if } t < t_0 - \delta. \end{array} \right.$$

Then with Proposition 6.1.5 we may give evaluation to $P_t^{\nu} z(t_0)$ classically and $P_t^{\nu} z(t_0) < q(t_0)$. Since $z \in G_2$ there exist Lipschitz sub-solutions $z_k \to z$ uniformly from below, then there exists k which is large enough so that

$$\Phi_k(t) := \left\{ \begin{array}{ll} \Psi(t) & \text{if } t \in (t_0 - \delta] \\ z_k(t) & \text{if } t < t_0 - \delta \end{array} \right.$$

satisfies $AB_{\nu}^{+} \Phi_k < f$ on $[t_0 - \delta_1, t_0]$ for some $0 < \delta_1 < \delta$. We now consider two different situations. If $z_k(t_0) = z(t_0)$, then for k large enough, $P_t^{\nu} z_k(t_0) < q(t_0)$, as $P_t^{\nu} z(t_0) <$

$q(t_0)$ classically. Since z_k is Lipschitz, it is followed from Lemma 6.1.23 that $P_t^\nu z_k < f$ in $[t_0 - \delta_2, t_0 + \delta_2]$. Then since $P_t^\nu (z_k + \eta_\zeta) = P_t^\nu z_k + P_t^\nu \eta_\zeta$, then $z_k + \eta_\zeta$ is a Lipschitz subsolution for ζ small enough, and $z_k(t_0) + \eta_\zeta(t_0) > z(t_0)$. Now the max $\{z_k + \eta_\zeta, z\} \in G_2$ and this contradicts the maximality of z. If in the second situation, $z_k(t_0) < z(t_0)$, then we extend Ψ_k to the right of t_0 by $\Phi_k(t_0) = -AB_1(t_0 - t) + z(t_0)$. From Lemma 6.1.23, for AB_1 large enough, $P_t^\nu \Phi_k \leq f$ for all $t \in [-2, 0]$ and $P_t^\nu \Phi_k \leq f$ in $[t_0 - \delta_3, t_0 + \delta_3]$. We let $\widetilde{\Phi}_k := \max \{\Phi_k, z_k\}$ and note that $\widetilde{\Phi}_k \in G_1$ and Lipschitz, and $P_t^\nu \widetilde{\Phi}_k \leq f$ in $[t_0 - \delta_4, t_0 + \delta_4]$ for some $\delta_4 > 0$ with $\widetilde{\Phi}_k(t_0) = z(t_0)$. Then, as before, we can take $\widetilde{\Phi}_k + \eta_\zeta$ with ζ small enough and obtain a contradiction. $\square$

For proof of Hölder continuity of solutions to (6.45) the approach in the work by [217] will be used and followed. One of the primary inputs is to solve an ordinary differential equation in time. We start off with the following.

Lemma 6.1.25. *Let C_1 be a fixed constant. Let $g(t)$ be a continuous function on $[-2, 0]$ as in [217]. There is existence of a continuous viscosity solution $O(t)$ in $[-2, 0]$ to*

$$P_t^\nu O(t) = C_1 O(t) + g(t),$$

with $O(t) = 0$ for $t \leq -2$.

Proof. Let $\beta_1 < \beta$ for β as in Theorem 6.1.1. From Theorem 6.3.5 [217], for $v \in C^{\beta_1}([0, 2])$ there exists a solution $h(t)$ to

$$P_t^\nu h(t) = C_1 v(t) + g(t),$$

with $h(t) = 0$ for $t \leq -2$. From Theorem 6.1.1 we have that

$$\|h\|_{C^{0,\beta}} \leq C \|C_1 v + g\|_{L^\infty}.$$

Since $C^{0,\beta}[-2, 0]$ is compactly contained in $C^{0,\beta}[-2, 0]$, we obtain a compact mapping $M: v \to h$ from $C^{0,\beta}[-2, 0]$ onto itself. From Corollary 6.11.2 in [213] it follows that there is an occurrence of a fixed point $O(t)$ that is a viscosity solution from Lemma 6.1.6.

Just as in [208] we will make use of an ordinary differential equation for capturing information backwards in time. We consider the fractional ordinary differential equation

$$\begin{cases} m(-2) = 0 & \text{for } t \leq -2 \\ P_t^\nu O(t) = c_0 |\{x \in B_1 : Z(x, t) < 0\}| - C_1 O(t) & \text{for } -2 < t < 0 \end{cases}. \tag{6.42}$$

We would like to utilize m as a test function for a viscosity solution. However, because the right-hand side is not continuous, Lemma 6.1.25 cannot be applied to obtain the existence of m. Furthermore, the solution m may not be Lipschitz and as a result, not a valid test function. To overcome these two challenges, we obtain a Lipschitz sub-solution to (6.38). We give consideration to $|\{x \in B_1 : Z(x, t) < 0\}|$ instead of $|\{x \in B_1 : Z(x, t) \le 0\}|$ since we may with ease approximate the former from below by smooth functions. This we accomplish by giving consideration to $g(x, t, \zeta) := \min\{\zeta - 1 \max\{0, -Z\}, 1\}$. Next, we then let

$$G(t, \zeta) = \int_{B_1} g(x, t, \zeta) dz.$$

Now $G(t, \zeta)$ is continuous in t, and $0 \le G(t, \zeta) \le |\{x \in B_1 : Z(x, t) < 0\}|$ and $G(t, \zeta) \to |\{x \in B_1 : Z(x, t) < 0\}|$ as $\zeta \to 0$. As a result of $G(t, \zeta)$ being continuous, from Lemma 6.1.25 we can solve

$$\begin{cases} m(-2) = 0 & \text{for } t \le -2 \\ P_t^\nu O(t) = c_0 G(t, \zeta) - C_1 G(t, \zeta) & \text{for } -2 < t \le 0 \end{cases}. \tag{6.43}$$

$\square$

Lemma 6.1.26. *Let $\nu_0 \le \nu < 1$. Assume that the kernel for P_t^ν satisfies Eq. (6.43). Let m be a solution to $P_t^\nu O(t) = c_0 f(t) - C_1 m$ with $O(t) = 0$ for $t \le -2$, $0 \le f \le |B_1|$, and*

$$\int_{-2}^{-1} f(t) \ge \iota. \tag{6.44}$$

Then there exist two moduli of continuity $z_1(c_0\iota)$, $z_2\left(C_1^{-1}\right)$ with z_i increasing and $z_i(s) > 0$ for $s > 0$ and depending solely on $\tau, \Upsilon, B_1, \nu_0$ such that

$$O(t) \ge z_1(c_0\iota) z_2\left(C_1^{-1}\right) \text{ for } -1 \le t \le 0.$$

Proof. From Lemma 6.1.22 any such solution m is Hölder continuous. We state that $m > 0$ for $t \in [-1, 0]$. Suppose, as a means of contradiction, that there exists $t_0 \in [-1, 0]$ such that $O(t_0) \le 0$. Let m achieve its minimum at t_1. Then m is touched from below at t_1 by the constant function $O(t_1)$, and so by Propositions 6.1.4 and 6.1.5 we may give evaluation to $P_t^\nu m$ at t_1 and

$$\int_{-\infty}^{t_1} [O(t_1) - O(s)] G(t, s) ds \ge f(t_1) \ge 0.$$

However, since $O(t_1) \le 0$ and m obtains a minimum at t_1, we maintain

$$0 \ge \int_{-\infty}^{t_1} [O(t_1) - O(s)] G(t, s) ds.$$

Then $O(t) = 0$ for $t \le t_1$. But then $q(t) = 0$ for $t \le t_1$. However, this gives contradiction to the assumption (6.44). Then $O(t) > 0$ for $-1 \le t \le 0$.

Let us fix $c_0, C_1, \iota > 0$. Suppose, as a means of contradiction, that there exist O_k, v_k, f_k all satisfying Lemma 6.1.26 assumptions, but

$$\inf_{[-1,0]} O_k \to 0$$

as $k \to \infty$. From Lemma 6.1.22, we have $O_k \to O_0$ in $C^{0,\beta}[-2,0]$ for any $\beta_1 < \beta$. Then there exists $t_1 \in [-1, 0]$ such that $O_0(t_1) = 0$.

Next, we claim that $O_0(t_0) = 0$ for some $t_0 \in [-2, -1]$. Suppose, as a way of contradiction, that $O_0 \equiv 0$ in $[-2, -1]$. From Corollary 6.1.13, we have that

$$C\Upsilon \sup_{[-2,-1]} O_k \ge \frac{\Upsilon}{\Gamma(1 - v_k)} \int_{-2}^{-1} \frac{O_k(t)}{(-1 - t)^v} dt$$

$$\ge \int_{-2}^{-1} c_0 f_k(t) - C_1 O_k dt$$

$$\ge c_0 \iota - \int_{-2}^{-1} C_1 O_k dt.$$

Letting $k \to \infty$ we obtain that $0 \ge \iota$. Which is a contradiction, and subsequently the claim that O_0 is not identically zero is true.

We now give consideration to two different cases. First assume that for a subsequence $v_k \to v_1 < 1$. Let t_2 be the first point after t_0 so that $O_0(t_2) = 0$. Since $O_k \to O_0$ uniformly, and since $O_0(t_0) > 0$, we may select $\Psi \ge 0$ smooth with $\Psi(t) = 0$ in a neighborhood of t_2 and $\Psi(t_0) > 0$ and also satisfying $O_k \ge \Psi$ for all k. Now $AB_v^+ \Psi(t) \le -\delta_1$ for $t \in [t_2 - \delta_2]$ and $v \in (v_1 - \delta_3, v_1 + \delta_3)$. We let ζ_k be such that $\Psi + \zeta_k \le O_k$ on $[t_2 - \delta_2, t_2]$ and $\Psi(t_k) + \zeta_k = O_k(t_k)$ for some $t_k \in [t_2 - \delta_2, t_2]$. We define

$$\Phi_k := \begin{cases} \Psi + \zeta_k & \text{if } t \ge t_2 - \delta_2 \\ O_k & \text{if } t < t_2 - \delta_2 \end{cases}.$$

Then

$$AB_v^+ \Phi_k(t_k) \ge P_t^{v_k} \Phi_k(t_k) \ge -C_1 O_k(t_k).$$

Since $O_0(t) > 0$ for $t \in [t_2 - \delta_2, t_2]$, as $k \to \infty$, we have $t_k \to t_2$ and $\zeta \to 0$. Then

$$\lim_{k \to \infty} AB_v^+ \Phi_k(t_k) \le AB_{v_1}^+ \Psi(t_2) \le -\delta_1 < 0.$$

This is contradictory to the first case. Now, we consider the case in which $v_k \to 1$. For a further subsequence there exists f_0 such that f_k converges to f_0 in weak star L^∞, such that

$$\int_{-2}^{-1} f_0(t) \ge \iota.$$

Then from Corollary 6.1.13 we have that

$$\frac{\tau v_k}{\Gamma(1 - v_k)} \int_{-2}^{t} O_k(s)\, ds \; \le \; \int_{-2}^{t} f_k(s) - C_1 O_k(s)\, ds$$

$$\le \; \frac{\Upsilon v_k}{\Gamma(1 - v_k)} \int_{-2}^{t} O_k(s)\, ds.$$

Since $O_k \to O_0$ uniformly and since [212] for any continuous function h, we have

$$\frac{v_k}{\Gamma(1 - v_k)} \int_{-2}^{t} \frac{h(s)}{(t - s)^{v_k}}\, ds \to h(t),$$

as $v_k \to 1$, then we obtain as $k \to \infty$ the inequality

$$\tau O_0(t) \le \int_{-2}^{t} f_0(s) - C_1 O_0(s)\, ds \le \Upsilon O_0(t).$$

Then there exists $\tau \le g(t) \le \Upsilon$ such that

$$g(t) O_0(t) = \int_{-2}^{t} f_0(s) - C_1 O_0(s)\, ds.$$

Then $g(t) O_0$ is Lipschitz continuous since f_0 is bounded. Furthermore, we have that

$$g(t) O_0(t) \ge \int_{-2}^{t} f_0(s) - C_1 \tau^{-1} g(s) O_0(s)\, ds.$$

Then $g(s) O_0$ is a super-solution and from the theory of ordinary differential equations, $g(s) O_0 \geq \widetilde{g}$ with $\widetilde{g}$ solving

$$\widetilde{g} = \int_{-2}^{t} f_0(s) - C_1 \tau^{-1} g(s) \, ds.$$

Since f_0 is not identically zero, then $\widetilde{g} > 0$ on $[-1, 0]$. It follows that $g(s) O_0 > 0$ and thus $O_0 > 0$ on $[-1, 0]$ as well. This is contradictory to the second case. Then for fixed c_0, C_1, ι there exists $\delta_4 > 0$ which depends on $c_0, C_1, \iota, \tau, \nu_0$ such that for any solution which satisfies the assumptions in the statement of Lemma 6.1.26, we have that $Z \geq \delta_4$ in $[-1, 0]$. We then achieve a modulus of continuity as stated in the lemma. $\qquad \square$

6.1.7 Analysis of the Hölder Continuity Within the Framework of Fractional Differentiation

In this part, we follow the technique used in [208] for proving our primary result. We shall require the following proposition to take account of the growth in the tails.

Lemma 6.1.27. *Let Z be a continuous function, $Z \leq 1$ in $(R^n \times [-2, 0] \cup B_2 \times [-\infty, 0])$, which satisfies the inequality in the viscosity sense in $B_2 \times [-2, 0]$*

$$P_t^{\nu} Z - M^+ Z \leq \zeta_0, \tag{6.45}$$

with $\nu_0 \leq \nu < 1$. Assume also that

$$|\{Z \leq 0\} \cap (B_1 \times [-2, -1])| \geq \iota.$$

Then if ζ_0 is small enough, thus we are able to find the existence of $\Theta > 0$ such that $Z \leq 1 - \Theta$ in $B_1 \times [-1, 0]$. The maximum value of ζ_0 as well as Θ is solely dependent on ν_0, Υ, τ, n and ω.

Proof. First of all, we mention that it suffices to prove Lemma 6.1.26 under the assumption

$$|\{Z \leq 0\} \cap (B_1 \times [-2, -1])| \geq \iota.$$

For if Z satisfies (6.45) and (6.36), then $Z - c$ for any positive constant c will satisfy (6.45) as well as the inequality above. Then $Z - c \leq 1 - \Theta$ in $B_1 \times [-1, 0]$ independent of c and so letting $c \to 0$ one achieves $Z \leq 1 - \Theta$.

We give consideration to the fractional ordinary differential equation $m : (-\infty, 0] \to R$

$$\begin{cases} m(t) = 0 & \text{for } t \leq -2 \\ P_t^\nu O(t) = c_0 f(t) - C_1 m(t) & \text{for } t > -2 \end{cases}$$

where $q(t)$ is a smooth approximation from below of $|\{x \in B_1 : Z(x, t) < 0\}|$. From the hypothesis and Lemma 6.1.26, we can select an approximation q so that

$$O(t) \geq \frac{z_1(c_0 \iota) z_2 \left(C_1^{-1}\right)}{2} > 0,$$

for $t \in [-1, 0]$. By Lemma 6.1.24 we can approximate m in a uniform manner from below by a Lipschitz function g so that

$$\begin{aligned} g(-2) &= 0 \\ P_t^\nu g(t) &= c_0 f(t) - C_1 m(t), \end{aligned}$$

and

$$g(t) \geq \frac{z_1(c_0 \iota) z_2 \left(C_1^{-1}\right)}{4} > 0$$

for $t \in [-1, 0]$. We make use of the function $g(t)$ that is not just a viscosity solution but also a classical solution, as it is Lipschitz. We can then calculate $P_t^\nu g$ everywhere classically. Furthermore, g is allowed as a means of a test function for touching above or below for viscosity solutions.

We want to show that $Z \leq 1 - g(t) + \zeta_0 c_\nu 2^\nu$ if c_0 is small and C_1 is large. We can then set $\Theta = z_1(c_0 \iota) z_2 \left(C_1^{-1}\right) / 4$ for ζ_0 small to achieve the result of the lemma. We select the constant c_ν such that $\partial_t^\nu c_\nu (2 + t)_+^\nu = 1$ for $t > -2$ and note [212] that c_ν is uniform as $\nu \to 1$. Let $\beta : R \to R$ be a fixed smooth non-increasing function so that $\beta(x) = 1$ if $x \leq 1$ and $\beta(x) = 1$ if $x \geq 2$. Let $b(x) = \beta(|x|)$, where for $b = 0$ we have $M^- b > 0$. Since b is smooth, $M^- b$ is continuous and it remains positive for b small enough [208]. Thus there exists β_1 such that $M^- b \geq 0$ if $b(x) = \beta_1$. Assume that there exists some point $(x, t) \in B_1 \times [-1, 0]$

$$Z(x, t) > 1 - g(t) + \zeta_0 c_\nu \tau^{-1} (2 + t)_+^\nu.$$

We shall arrive at a contradiction by referring to the maximum of the function

$$w(x, t) = Z(x, t) + g(t) b(x) - \zeta_0 c_\nu \tau^{-1} (2 + t)_+^\nu.$$

Assume there exists a point in $B_1 \times [-1, 0]$ where $z(x, t) > 1$. Let (x_0, t_0) be the point that realizes the maximum of z:

$$z(x_0, t_0) = \max_{R^n \times (-\infty, 0]} z(x, t).$$

This maximum is greater than 1, and therefore it must be obtained when $t > -2$ and $|x| < 2$.

Let $\Phi(x, t) := z(x_0, t_0) - g(t) b(x) - \zeta_0 c_\nu \tau^{-1} (2 + t)_+^\nu$. We remark that $\Phi(x, t) = \Phi(x, -2)$ for $t \le -2$ and Φ touches Z from above at the point (x_0, t_0). We define

$$v(x, t) := \begin{cases} \Phi(x, t) & \text{if } x \in B_r \\ Z(x, t) & \text{if } x \notin B_r. \end{cases}$$

Then from the definition of viscosity solution we have

$$\partial_t^\nu v - M^+ v \le \zeta_0 \tag{6.46}$$

at (x_0, t_0). We have that

$$
\begin{aligned}
P_t^\nu v(x_0, t_0) &= P_t^\nu \left(-g(t_0) b(x_0) - \zeta_0 c_\nu \tau^{-1} (2 + t)_+^\nu \right) \\
&\ge (C_1 m(t) - c_0 f(t)) b(x_0) + \zeta_0 \tau^{-1} A B_\nu^- (2 + t)_+^\nu \\
&= (C_1 m(t) - c_0 f(t)) b(x_0) + \zeta_0 \tau^{-1} \tau \partial_t^\nu (2 + t)_+^\nu \\
&= (C_1 m(t) - c_0 f(t)) b(x_0) + \zeta_0 \\
&\ge (C_1 m(t) - c_0 |\{x \in B_1 : Z(x, t) < 0\}|) b(x_0) + \zeta_0 \\
&\ge (C_1 m(t) - c_0 |\{x \in B_1 : Z(x, t) \le 0\}|) b(x_0) + \zeta_0.
\end{aligned}
$$

Then

$$
\begin{aligned}
\zeta_0 &\ge P_t^\nu v(x_0, t_0) - M^+ v(x_0, t_0) \\
&\ge (C_1 m(t) - c_0 |\{x \in B_1 : Z(x, t) < 0\}|) b(x_0) \\
&\quad + \zeta_0 - M^+ v(x_0, t_0)
\end{aligned}
$$

or

$$
\begin{aligned}
0 &\ge (C_1 m(t) - c_0 |\{x \in B_1 : Z(x, t) \le 0\}|) b(x_0) + \zeta_0 \\
&\quad - M^+ v(x_0, t_0).
\end{aligned} \tag{6.47}
$$

Now exactly as in [15] we obtain the following bound for $G := \{x \in B_1 : Z(x, t) \le 0\}$

$$M^+ v(x_0, t_0) \le -m(t_0) M^- b(x_0, t_0) - c_0 |G \backslash B_r| \tag{6.48}$$

for some universal constant c_0. This is how we select c_0 in the fractional ordinary differential equation. We now make reference to two different cases and achieve a contradiction in both. Suppose $b(x_0) \le \beta_1$. Then $M - b(x_0) \ge 0$, and so from (6.48)

$$M^+ v(x_0, t_0) \le -c_0 |G \backslash B_r|.$$

Combining the above inequality with (6.40), we achieve

$$0 \ge (C_1 m(t) - c_0 |\{x \in B_1 : Z(x, t) \le 0\}|) b(x_0) + c_0 |G \backslash B_r|.$$

For any $C_1 > 0$ this will be a contradiction by taking r small enough. Now suppose $b(x_0) \le \beta_1$. Since b is a smooth compactly supported function, there exists C such that $|M^{-b}| \le C$. We then have from (6.6) the bound

$$M^+ v(x_0, t_0) \le C m(t_0) - c_0 |G \backslash B_r|$$

and inserting this in (6.40) we obtain

$$0 \ge (C_1 m(t) - c_0 |\{x \in B_1 : Z(x, t) \le 0\}|) b(x_0) - C m(t_0) + c_0 |G \backslash B_r|.$$

Letting $r \to 0$ we obtain

$$
\begin{aligned}
0 \quad &\ge \quad c_0 (1 - b(x_0)) |G| + (C_1 b(x_0) - C) m(t_0) \\
&\ge \quad c_0 (1 - b(x_0)) |G| + (C_1 \beta_1 - C) m(t_0).
\end{aligned}
$$

Choosing C_1 large enough we achieve a contradiction.

We now define

$$Q_r := B_r \times \left[-r^{\frac{2\omega}{\nu}}, 0 \right]$$

and note the re-scaling property that if $v(x, t) = Z(rx, -r^{\frac{2\omega}{\nu}})$, then

$$\tilde{D}_t^\nu v(x_0, t_0) - A B_\omega^\pm v(x, t) = r^{2\omega} \left(P_t^\nu Z(rx, -r^{\frac{2\omega}{\nu}}) - A B_\omega^\pm Z(rx, -r^{\frac{2\omega}{\nu}}) \right)$$

where if $G(t, s)$ is the kernel for P_t^ν then $\tilde{D}_t^\nu$ has kernel

$$\frac{G(rt, rs)}{r^{1+\nu}}$$

which will also satisfy (6.8) and (6.9). For the next three results we fix $r = \min\{4^{-1}, 4^{-\nu/2\omega}\}$. We shall require the following proposition to bound the tails.   □

Proposition 6.1.28. *Let $h(t) = \max\{2\,|rt|^{v} - 1, 0\}$ with $r = \min\{4^{-1}, 4^{-v/2\omega}\}$. If $t \leq 0$ and $v < v$ then*

$$0 \geq P_t^{\nu} h(t_1) \geq -\tau c_{v,v}$$

where $c_{v,v}$ is a constant which depends solely on v and v but for fixed v remains uniform as $v \to 1$.

Proof. Now

$$0 \geq P_t^{\nu} h(t_1) \geq A B_v^{-} h(t_1) = \tau \partial_t^{\nu} h(t_1).$$

From [1] we have

$$\partial_t^{\nu} h(t_1) \geq c_{v,v},$$

which for fixed v remains uniform as $v \to 1$. By combining the above two inequalities, the proposition is proven. $\square$

Lemma 6.1.29. *Let Z be a bounded continuous function which satisfies the following two inequalities in the viscosity sense in Q_1:*

$$\begin{aligned}
P_t^{\nu} Z - M^{+} u &\leq \frac{\zeta_0}{2}, \\
P_t^{\nu} Z - M^{-} u &\geq -\frac{\zeta_0}{2}.
\end{aligned} \tag{6.49}$$

Let the kernel $G(t, s)$ of P_t^{ν} satisfy (6.8) and (6.9) with $0 < v_0 \leq v < 1$. Then there are universal constants $\Theta > 0$ and $v > 0$ depending only on $n, \tau, \omega, \Upsilon, v_0$ such that if

$$\begin{aligned}
|Z| &\leq 1 && \text{in } B_1 \times [-1, 0] \\
|Z(x,t)| &\leq 2\,|rx|^{v} - 1 && \text{in } (R^n \times B_1) \times [-1, 0] \\
|Z(x,t)| &\leq 2\,|rt|^{v} - 1 && \text{in } B_1 \times (-\infty, -1]
\end{aligned}$$

with $r = \min\{4^{-1}, 4^{-v/2\omega}\}$, then

$$\mathrm{osc}_{Q_r} Z \leq 1 - \Theta.$$

Proof. We give consideration to the re-scaled version

$$\tilde{Z}(x,t) := Z\left(r^{-1}x, r^{-v/2\omega}t\right).$$

The function $\tilde{Z}$ will remain either positive or negative in half of the points in $B_1 \times [-2, -1]$. Let us assume that $\{\tilde{Z} \leq 0\} \cap (B_1 \times [-2, -1]) \geq |B_1|/2$. Otherwise we can repeat the proof for

$-\widetilde{Z}$. We would like to give application to Lemma 6.1.27. To do this, we would require $\widetilde{Z} \leq 1$. We give consideration to $v := \min\{1, \widetilde{Z}\}$. Inside Q_{r-1} we have $v = \widetilde{Z}$. The error comes solely from the tails in the computations. The same as in [217], we obtain for k small enough

$$-M^+ v \leq -M^+ \widetilde{Z} + \frac{\zeta_0}{4}.$$

From Proposition 6.1.28 we have for small enough k that

$$P_t^v v \leq P_t^v \widetilde{Z} + \frac{\zeta_0}{4}.$$

Thus

$$P_t^v v - M^+ v \leq \zeta_0.$$

We now apply Lemma 6.1.27 to v and re-scale back for concluding the proof. We are now able to give the proof of our primary result. $\qquad\square$

Theorem 6.1.30. *We first choose $k < v$ for v as in Lemma 6.1.29. Let $(x_0, t_0) \in Q1$. We consider the re-scaled function*

$$v(x, t) = \frac{Z(x_0 + x, t + t_0)}{\|Z\|_{L^\infty} + \zeta_0 \|f\|_{L^\infty}},$$

and note that $|v| \leq 1$ and v is a solution to

$$P_t^v v - AB_\omega^+ v \leq \zeta_0,$$
$$P_t^v v - AB_\omega^- v \geq -\zeta_0,$$

in $B_2 \times [-1, 0]$. We let $r = \min\{4^{-1}, 4^{-\frac{v}{2\omega}}\}$; the estimate will follow as soon as we show

$$osc_{Q_{r_k}} v \leq 2 r^{kK}. \tag{6.50}$$

Estimate (6.8) will be proven by developing two sequences $a_k \leq v \leq b_k$ in Q_{r_k}, $b_k - a_k = 2r^{\kappa k}$ with a_k non-decreasing and b_k non-increasing. The sequence is developed inductively. Since $|v| \leq 1$ everywhere, we can begin by selecting some $a_0 \leq \inf v$ and $b_0 \geq \sup v$ such that $b_0 - a_0 = 2$. Assuming now that the sequences have been developed up to the value k, we scale

$$w(x, t) = \left(v\left(r^k x, r^{\frac{2k\omega}{v}} t\right) - (b_k - a_k)/2\right) r^{-\kappa k}.$$

We then have

$$|z| \leq 1 \qquad \text{in } Q_1$$
$$|z| \leq 2r^{-kK} - 1 \quad \text{in } Q_{r^{-k}}.$$

And so

$$|z(x,t)| \leq 2|x|^{\nu} - 1 \quad \text{for } (x,t) \in B_1^c \times [-1,0]$$
$$|z(x,t)| \leq 2|t|^{\nu} - 1 \quad \text{for } (x,t) \in B_1 \times (-\infty, -1).$$

Notice also that z has a new right-hand side bounded by

$$\zeta_0 r^{k(k-2\omega)}$$

which is strictly smaller than ζ_0 for $k < 2\omega$. For k small enough, we may give application to Lemma 6.1.29 to obtain

$$osc_{Q_r} z \leq 1 - \Theta.$$

Then if k is chosen smaller than k in Lemma 6.1.29 and also so that $1 - \Theta \leq rk$, then this infers

$$osc_{Q_{rk+1}} z \leq r^{k(k+1)}$$

so we can find a_{k+1} and b_{k+1} and this finishes the proof. Also this section has presented a clear and detailed analysis of the general parabolic equations with fractional differentiation. The concept of fractional differentiation used here is based on the power law kernel and therefore has a singular kernel. This class of fractional parabolic equations included those groundwater flow models with Caputo and Riemann–Liouville fractional differentiation and integrations. Therefore clear details on regularity of general parabolic equations involving the concept of fractional differentiation and integration with the Mittag–Leffler kernel will be provided to include groundwater models with such differentiation. This is presented in detail in the next section.

6.2 Regularity of General Fractional Parabolic Equation With Atangana–Baleanu Derivative

We devote this section to the discussion underpinning the regularity of a general parabolic equation with fractional derivative with the newly established fractional derivative known as the Atangana–Baleanu fractional derivative in Caputo sense. Henceforth in this section, we present in detail some proofs of the well-known Hölder regularity results for viscosity solutions of fractional integro-differential equations in which the kernel is considered as the generalized Mittag–Leffler function and the spatial nonlocal operator kernel corresponds to the fractional Laplacian. The main aim of this section is to reproduce the results obtained in the above section [205,204] where the fractional parabolic equation was constructed using the power law kernel fractional differentiation, in particular the definition used here is that of Caputo. We point out that, the results presented here are the collection of the works done by Djida, Atangana and Ivan in their paper with title "Parabolic problem with fractional time derivative with nonlocal and nonsingular Mittag-Leffler kernel." The main aim is to show the regularity of a general fractional parabolic equation with Atangana–Baleanu derivative in Caputo sense, as the groundwater flow models are classified under parabolic equations.

To achieve this, we collect the notation found in the paper by [205,204] and this notation will be found in the following reference in Section 6.2.2. In this section, we will be particular and also we will not try to deviate from the original idea presented in [205,204]. We are also going to recall the results found in the literature in the following in Section 6.2.1. The generalized fractional equation that is considered in this section is given below as:

$$Hv(t, x) - Bv(t, x) = w(t, x), \quad \text{in } (-\infty, b) \times R^n. \tag{6.51}$$

For simplicity's sake, we consider the following operators

$$Jv(x) = \int_{R^n} \delta_h v(x) V(x, h) dh, \tag{6.52}$$

we shall recall that $\delta_h v(x) = v(x + h) + v(x - h) - 2v(x)$ is the second difference approximation of the function v at a point x. In addition to this, we consider $v(0, 2)$ to be the kernel. Therefore,

$$U(x, h) = C(n, v)\|h\|^{-n-v}. \tag{6.53}$$

For a suitable constant, the above $C(n, v)$ produces the following formulation $B = (-\Delta)^{v/2}$.

This has been presented in detail in the work done by [203]. The class of kernel considered here will be U because it will capture a large area $U(x, h)$ where one will not necessarily compare to $(\|h\|^{-n-v})$ as from below. The class of kernel considered in this work is known as A_{sec}; this will be found in (6.60). This will be treated without any assumption of regularity for the variable x. By means of the assumption needed to be above $\|h\|^{-n-v}$ this is considered to take place in a small set as defined below:

$$\frac{\pi}{|h|^{n+v}} \leq U(x, h) \leq \frac{\zeta}{|h|^{n+v}} \quad h \in R^n \|\{0\}.$$

Additionally to this, considering the parameter $\alpha \in (0, 1)$ for all $t < a$ and for $n \in N$, in this case we consider defining the time-fractional operator of differentiation L which is found in (6.82) as

$$Lv(t) = Y(n, \alpha) \int_{-\infty}^{t} [v(t) - v(s)] P(t, s) ds. \tag{6.54}$$

In this formulation, the used kernel $P(t, s)$ is included in the class of kernels T_{sec} as presented in

$$T_{sec} = \left\{ T : (-\infty, b) \to R : T(t, t - s) = T(t + s, t), \text{ and} \right.$$ (6.55)

$$\left. -c\pi \frac{(t - s)^{\alpha - 1}}{\Gamma(\alpha + 1)} \leq P(t, s) \leq -c\zeta \frac{(t - s)^{\alpha - 1}}{\Gamma(\alpha + 1)} \right\}.$$ (6.56)

The class of kernels considered here is chosen in order to include all the existing kernels used to build a fractional differential operator, for instance the well-known Marchaud derivative, which can be found in [193], Caputo derivative, which can be found in [205,193,194,202], and the fractional derivative with nonlocal and nonsingular Mittag–Leffler kernel that was recently established by Atangana and Baleanu in their work [196]. We shall include further properties of the fractional differential operators and integral in Subsection 6.2.3.

Remark 4. It is worth keeping in mind that the consideration of the nonlocal differential operator in Caputo sense is presented as:

$$DCv(t) = \frac{\alpha}{\Gamma(1 - \alpha)} \int_{-\infty}^{t} \left[v(t) - v(s) \right] (t - s)^{-\alpha - 1} ds.$$

We recall that the kernel used here is $(t - s)^{-\alpha - 1}$ which is constructed from repetition and providing the power law belongs to the class of kernels T_{sec} that was defined earlier. We also stress the fact that this Caputo fractional differential operator has been employed in [205,197] to present the detailed proof of the well-known Hölder regularity of some parabolic problems in the non-divergence form.

An important example considered is that of Atangana–Baleanu fractional differential operator in Caputo sense introduced in [196]:

$$ABv(t) := c_\alpha \int_{-\infty}^{t} \left[v(t) - v(s) \right] (t - s)^{\alpha - 1} E_{\alpha, \alpha} \left[c(t - s)^\alpha \right] ds.$$ (6.57)

We also stress the fact that the considered kernel $(t - s)^{\alpha - 1} E_{\alpha, \alpha} \left[c(t - s)^\alpha \right]$ is a member of the class of kernels T_{sec}. Not to point out just abstract mathematical excitement of the well-known Caputo fractional differential operator and its wider use in the field of fractional differentiation and integration, we shall point out that the attentiveness of the fractional differential operator introduced by Atangana and Baleanu is based on its properties of portraying the behavior of orthodox viscoelastic materials, thermal medium, material heterogeneities and some structure or media with different scales. In addition to this, the fractional operator is also attracting more nonlocalities than the existing fractional differential operators including the Riemann–Liouville and the Caputo–Fabrizio types [196]. In the paper where this new fractional differential operator was introduced, Atangana and Baleanu argued that the nonlocality of the new kernel allows a good description of the memory within structure and media with different scale, which cannot be described by classical fractional derivative [196]. This derivative also takes into account power and exponential decay laws for which many natural occurrences follow and also many non-localities can be captured. We could also mention that the Mittag–Leffler function allows us to describe phenomena in processes that progress or decay too slowly to be represented by classical functions like the exponential function and its successors which in practice are the dynamic of super-diffusion and also sub-diffusion. The Mittag–Leffler function arises naturally in the solution of fractional integral equations, and especially in the study of the fractional generalization of the kinetic equation, random walks, Lévy flights, and so-called super-diffusive transport.

Remark 5. Again to accommodate readers that are not aware of the background of the field of fractional differentiation and integration, there are in the nowadays literature several definitions of the concept of fractional differential operators including: Riemann, Liouville, Caputo, Grunwald–Letnikov, Marchaud, Weyl, Riesz, Feller, and others. For fractional derivatives and integrals one can find an excellent literature in the following references (see e.g. [190–193, 202] and the references therein). Due to the fact that each kernel is able to handle a class of particular physical problems, one will therefore be obliged to have several definitions of the concept of fractional differentiation. Therefore in this section, apart from what was presented earlier for the power law kernel, one will investigate the parabolic problem for different kernels. In this section, we study the problem for a large class of kernels T_{sec} satisfying the properties given by (6.66).

The solutions of these nonlocal equations, involving the fractional time derivative and the nonlocal spatial operators, are of particular interest. Numerous authors, such as [202,197,199, 200], have studied the problem of Hölder continuity for solutions to master equations and Hölder continuity for parabolic equations with Riemann–Liouville derivative and the Caputo fractional time derivative and divergence form nonlocal operator. Recently [205] proved the Hölder continuity of viscosity solutions of Eq. (6.82) in the non-divergence form, but using the generalized fractional time derivative of Marchaud or Caputo type under the appropriate assumptions, obtaining estimates which remain uniform as the order of the fractional derivative $\beta \to 1$. With the clear intention to obtain a similar result as in [197,205] where the generalized Marchaud or Caputo derivative was used to show the Hölder continuity of viscosity solutions, we propose to study the same problem, but now with another kernel (e.g. $T = (t - s)^{\beta-1} E_{\beta,\beta}[c(t - s)^{\beta}]$), where the Caputo kernel can be deduced from and which belongs to the class of kernels T_{sec}.

Remark 6. We believe our presentation of the same results but with more general kernel is a useful contribution to the field, especially since it seems that most of the previously applied kernels of the fractional time derivative belong to T_{sec}. For example, if we take the case of the

Caputo fractional derivative, it has been proven by [196] that there exists a well-defined function h that equalizes the Caputo fractional time-derivative and the fractional time-derivative with nonlocal and nonsingular Mittag–Leffler kernel with fractional order β,

$$h(t) = h(0)\left\{(1 + cB(\beta))t^{\beta-1}\right\}.$$

To study regularity properties of solutions to Eq. (6.82), one could in some sense study the solution v which simultaneously solves the two inequalities

$$\inf_{T \in T_{sec}} \left\{L_T v(t, x) - J v(t, x)\right\} \leq C \text{ and } \sup_{T \in T_{sec}} \left\{L_T v(t, x) - J v(t, x)\right\} \geq -C \text{ in } (\infty, T) \times \Omega.$$

The kernel $T = (t - s)^{\beta-1} E_{\beta,\beta}\left[c(t - s)^{\beta}\right]$ is chosen in the class of kernels T_{sec} which at least contains all the kernels involved in the fractional time derivatives in the literature so that it will be convenient if one wishes to attain further properties of the extremal operators. The program of studying existence of solutions and regularity properties of parabolic problem with fractional nonlocal space-time operators such as (6.82) was presented in [205,202,197], using respectively Riemann fractional derivative and Caputo fractional derivative. We extend those results to cover the larger class, T_{sec}. Our main results are the existence of a weak solution and the Hölder regularity estimate.

Next we state the Hölder regularity estimate result.

Theorem 6.2.1 (Hölder Regularity). *Let $\beta \in (0, 1)$, $\nu \in (\nu_0, 2)$ and let $M_A^{\pm}$ be as defined in (6.95) and (6.96). Assume also that $w \in L^\infty(-\infty, b) \times R^n$. There are positive constants $\beta \in (0, 1)$ and $C \geq 1$ depending only on $n, \pi, \zeta, \beta, \nu$ such that if v is bounded continuous viscosity solution in $R_2 \times [-2, 0]$ satisfying*

$$L_T v - M_A^- v \leq \xi_0 \ \text{ and } L_T v - M_A^+ v \geq -\xi_0, \tag{6.58}$$

then v is Hölder continuous in $R_1 \times [-1, 0]$ and for $(x, t), (y, s) \in R_1 \times [-1, 0]$ the following estimate holds

$$|v(x, t) - v(y, s)| \leq C(\|v\|_{L^\infty} + \xi_0^{-1}\|g\|_{L^\infty})|x - y|^\iota + |t - s|^{\iota\beta/(2\nu)}. \tag{6.59}$$

Furthermore, C remains bounded as $\beta \to 1$ and $\nu \to 2$.

Remark 7. We draw attention of the reader to the fact that in order to get solutions to Theorem 6.2.1, the solution of fractional differential equations involving the fractional time derivative with nonlocal and nonsingular Mittag–Leffler kernel is proposed in order to use it as a test function for of viscosity solution. The case with the Caputo derivative was well presented by the author in [205] showing that if $|\{x \in R_1 \times (-2, -1) : v(x, t) \le 0\}| \ge \mu_1$, then $v(t) \ge \mu_2$ if $t \in (-1, 0)$. So we omit to prove it here in this note in the case for the fractional time derivative with nonlocal and nonsingular Mittag–Leffler kernel since the idea of the proof is similar. So with the Hölder continuity estimates for ordinary differential equations involving the Caputo derivative [205], our class of weak solutions will be considered in the viscosity sense as described in 6.2.6. This will then allow us to get the similar result.

This section will be structured as follow: In Section 6.2.1 we review some background associated with Theorem 6.2.1. In Section 6.2.2 we collect notation, definitions, and preliminary results regarding Eq. (6.82) and Theorem 6.2.1. Section 6.2.4 is dedicated to the sketch of proof of the existence of weak solutions by mean of approximating solutions, mainly Eq. (6.82). Finally in Section 6.2.6 we discuss the point-wise estimates and put together the remaining pieces of the proof for the and Hölder Regularity.

6.2.1 Literature Review of Results Related to the Hölder Regularity

There are very few results related to Theorem 6.2.1 with both space and time fractional nonlocal operators. We will focus on the type of results solely depending on the ellipticity constants, λ and Λ, for the spatial nonlocal operator and possibly the order, β, ν. We will try to see if we recover the result obtained by [204,197,205] by making use of the fractional time derivative associated with the nonlocal and nonsingular Mittag–Leffler kernel. The requirement of regularity for the parabolic problem with the fractional time derivative has recently become a significant research focus, for example [201] applied the original method of De Giorgi for proof that solutions are bounded and have local Hölder regularity. This approach was similarly was applied by [204] to prove a priori local Hölder estimates of solutions to the fractional parabolic type equation, where the De Giorgi method is also applied but in addition accounts for the fractional time derivative in the sense of Caputo [193]. These results infer that use of the fractional nature of the derivative allows the estimates made to vary with the fractional order $\beta \to 1$. The requirement of regularity for the parabolic problem with the fractional time derivative is also investigated by [202]; however, instead of using the Caputo derivative, the Riemann–Liouville fractional time derivative and right-hand side are given consideration. It is important to note that these findings based on fractional space and time were obtained where both kernels are bounded. In the case of the fractional spatial nonlocal operator there are a few interesting distinctions that are generally made: whether or not $U(x, h)$ is assumed to be even in h; whether or not the corresponding equations are linear; and whether or not a Harnack inequality holds for [203]. Regularity results (such as Theorem 6.2.1) and also the Harnack inequality for linear equations with operators similar to (6.52) were obtained by [198,203]. Furthermore, in [205] the Hölder continuity of viscosity solutions to certain nonlocal parabolic equations that involve a generalized fractional time derivative of Marchaud or Caputo type is obtained under the assumption that the kernel of the fractional time operator satisfies the symmetry condition $T(t, t - s) = T(t + s, t)$. The estimates are uniform as the order of the operator β approaches 1, so that the results recover many of the regularity results for the local parabolic problem. Finally, higher regularity in fractional time type estimates were found in [204]. An important class of kernels are those for which the symmetry $T(t, t - s) = T(t + s, t)$ is assumed to hold and for which all the fractional time operators belong to the class T_{sec}. To add, this class may be extended to the non-symmetric case. In this note, we shall discuss the situation where the kernel of the non-local time operator belongs to a more general class of kernels T_{sec}, which contain most of the known kernels.

6.2.2 Preliminaries and Some Useful Denotations

We first collect some denotations which will be used throughout this article.

- L – the time fractional derivative in Atangana–Baleanu sense.
- $\nu \in (\nu_0, 2)$ – stands for fractional order for spatial operator.
- χ – will be considered here as an arbitrary order derivative.

$$A_{sec} = \left\{ K : R^n \to R \ : J(-h) = J(h), \qquad \text{and} \quad \frac{\pi}{|h|^{n+\nu}} \leq J(h) \leq \frac{\zeta}{|h|^{n+\nu}} \right\} \qquad (6.60)$$

$$\delta_h v(x) = v(x + h) + v(x - h) - 2v(x) \qquad (6.61)$$

$$
\begin{aligned}
J_A v(x) &= \int_{R^n} \delta_h v(x) J(h) dh \\
mv(dh) &= \|h\|^{-n-2\nu} dh \\
Q_\xi(x_0) &= \left\{ x \in R^n : x - x_{0\infty} < tfrac\xi 2 \right\} \\
R_\xi(x_0) &= \left\{ x \in R^n : x - x_0 < \xi \right\}
\end{aligned}
$$

The $\|.\|$ is the well-known absolute value operator, the Euclidean norm, and the n-dimensional Lebesgue measure at the same time. Within this section, $\Omega \subset R^n$ will be considered as a bounded domain. In case of balls and cubes, one will have that $x_0 = 0$ thus this can be written as Q_l in place of $Q_l(0)$ and this will also be similar for R_l. This will therefore follow:

$$R_{1/2} \subset F_1 \subset F_3 \subset R_3 \subset R_2 .$$

In the next part, the Atangana–Baleanu fractional differentiations and their properties will be used and also the viscosity solutions from [196] and [205], and for Hölder continuity one will rely on the works done in [204,195,197,198].

6.2.3 The Fractional Time Derivative With Nonlocal and Nonsingular Mittag–Leffler Kernel

In this section, to accomodate researchers that are not aware of the newly established fractional operators of differentiation and also integration based on the generalized Mittag–Leffler kernel as presented earlier, we recall some definitions of fractional time derivative with the nonlocal and nonsingular kernel as stated in [196] and including some new properties associated to them. It is recalled that the Atangana–Baleanu fractional differential operator is presented as

$$\,_{a}^{AB}D_t^{\chi}v(t) = \frac{B(\chi)}{1-\chi}\int_a^t E_\chi\big[-c(t-s)^\chi\big]v'(s)ds. \tag{6.62}$$

The Atangana–Baleanu fractional integral associate is given as:

$$\,_{a}^{AB}I_t^{\chi}a\chi v(t) = \frac{1-\chi}{B(\chi)}v(t) + \frac{\chi}{B(\chi)\Gamma(\chi)}\int_a^t v(y)(t-y)^{\chi-1}dy. \tag{6.63}$$

In the above formulas $B(\chi)$ is a constant depending on χ such that

$$B(\chi) = 1 - \chi + \frac{\chi}{\Gamma(\chi)}, \quad c = -\frac{\chi}{1-\chi},$$

and $E_{\chi,\chi}(z)$ the more general Mittag–Leffler function defined in terms of a series as:

$$E_{\chi,\chi}(z) = \sum_{k=0}^{\infty} \frac{z^{\chi k}}{\chi(\chi k + \chi)}, \quad \text{and} \quad \chi > 0, \quad z \in \mathbb{C}.$$

Another way of representing the Atangana–Baleanu fractional differential equation in Caputo sense is given below as for point-wise case:

$$ABv(t) = v_\chi E_\chi\big[-c(t-a)^\chi\big]\big[v(t) - v(a)\big]$$
$$+ c_\chi \int_a^t (t-s)^{\chi-1}E_{\chi,\chi}\big[-c(t-s)^\chi\big]\big[v(t) - v(s)\big]ds, \tag{6.64}$$

where $v_\chi = B(\chi)(1-\chi)^{-1}$ and $c_\chi = -cv_\chi$. We set

$$P(t,s) = (t-s)^{\chi-1}E_{\chi,\chi}\big[-c(t-s)^\chi\big]. \tag{6.65}$$

The following relation is $T(t, t-s) = T(t+s, t)$, it is satisfied by the Atangana–Baleanu kernel

$$-c\pi \frac{(t-s)^{\chi-1}}{\Gamma(\chi+1)} \leq P(t,s) \leq -c\zeta\frac{(t-s)^{\chi-1}}{\Gamma(\chi+1)}. \tag{6.66}$$

In this setting, following the idea of [204,197] we define $v(t) = v(a)$ for $t < a$,

$$Lv(t) := c_\chi \int_{-\infty}^{t} [v(t) - v(s)] U(t, s) \, ds. \tag{6.67}$$

The direct consequence of the above formulation as referred in (6.67) is the fact that it helps to drop out data, at the same time it is important for viscosity solutions as presented in [197]. The concept of a viscosity solution and that of super-solution for those classes of the initial values problems with time derivative and nonsingular Mittag–Leffler kernel as presented in (6.67) is similar to the one of (6.82). For this purpose, to start, we assume that, the following (6.67) is well-defined.

Proposition 6.2.2. *Let v a continuous bounded function and $w \in Y^{0,\chi}$ with $\chi < \chi \leq 1$. If $w \geq (\leq) u$ on $[t_0 - \xi, t_0]$ and $w(t_0) = v(t_0)$, then the integral*

$$c_\chi \int_{-\infty}^{t_0} [v(t_0) - v(s)] P(t, s) \, ds$$

is well-defined, so that $Lv(t_0)$ is well-defined.

Proof. Assume that $w \geq u$ on $[t_0 - \xi, t_0]$. This comes from Eq. (6.67).

Therefore we can confirm that the integral is well-defined. □

Here we present the estimation of the bound for the time fractional derivative $L\eta(t)$ which will be of importance for the proof of Hölder continuity in 6.2.6.

Proposition 6.2.3. *Let us consider $\eta(t) = \max\{2|rt|^\nu - 1, 0\}$ under the condition that $\nu < \chi$, and $r = \min\{4^{-1}, 4^{-\chi/2\nu}\}$. If $t_1 \leq 0$ then*

$$-d_{\chi,\nu} \leq L\eta(t_1) \leq 0$$

where the constant $d_{\chi,\nu}$ depends on χ and ν.

Proof. From (6.62) and (6.67) the re-scaled fractional time derivative takes the following form

$$-\chi \int_{a}^{t} E_\chi [c(t-s)^\chi] |s|^{\chi-1} ds \geq -\chi \int_{-\infty}^{t} E_\chi [c(t-s)^\chi] |s|^{\chi-1} ds.$$

One notices that $|s|^{\chi-1}$ and the Mittag–Leffler function $E_\chi [c(t-s)^\chi]$ are increasing functions of s, if $s < 0$, so

$$\chi \int_{-\infty}^{t} E_\chi\big[c(t-s)^\chi\big]|s|^{\chi-1}ds$$

is an increasing function of t. Furthermore, if $t \le -1$, then if follows that

$$AB\eta \ge -\chi \int_{-\infty}^{t} E_\chi\big[c(t-s)^\chi\big]|s|^{\chi-1}ds \ge -\chi \int_{-\infty}^{-1} E_\chi\big[c(-1-s)^\chi\big]|s|^{\chi-1}ds \ge -d_{\chi,\chi}.$$

Now if $t > -1$, then

$$AB\eta \ge -\chi \int_{-\infty}^{-1} E_\chi\big[c(t-s)^\chi\big]|s|^{\chi-1}ds \ge -\chi \int_{-\infty}^{-1} E_\chi\big[c(-1-s)^\chi\big]|s|^{\chi-1}ds \ge -d_{\chi,\chi}. \quad \square$$

In this case, the focus will be devoted to the investigation of the solution of the fractional differential equation constructed based on the Atangana–Baleanu fractional derivative as follows

$$Lv(t) = -c_1\, v(t) + c_0\, h(t), \quad c_1,\ c_0 < \infty. \tag{6.68}$$

Proposition 6.2.4. *Let $v \in H^1(0,b)$, $b > 0$, and $h \in \mathcal{Y}^2$, for which the Atangana–Baleanu fractional derivative exists. Then, the solution of differential equation (6.68), for $c_1 = 0$, is given by*

$$v(t) = \frac{(1-\chi)c_0}{B(\chi)}h(t) + \frac{\chi c_0}{B(\chi)\Gamma(\chi)}\int_0^t h(s)(t-s)^{\chi-1}ds, \tag{6.69}$$

and for $c_1 \ne 0$ by

$$v(t) = \zeta E_\chi\big[-\chi t^\chi\big]v(0) \tag{6.70}$$

$$+ \frac{\chi c_0 \zeta}{B(\chi)}\int_0^t \left(E_{\chi,\chi}\big[-\chi(t-s)^\chi\big] + \frac{(1-\chi)}{\chi}\chi^{-2\chi}E_{\chi,\chi}\big[-\chi^{-2\chi}(t-s)^\chi\big]\right)$$

$$\times (t-s)^{\chi-1}h(s)ds, \tag{6.71}$$

with $\chi = \dfrac{\chi c_1}{\big(B(\chi)+(1-\chi)c_1\big)}$ and $\zeta = \dfrac{B(\chi)}{\big(B(\chi)+(1-\chi)c_1\big)}$.

Proof. For $c_1 = 0$, it is obvious that the result is

$$v(t) = \frac{1-\chi}{B(\chi)}h(t) + \frac{\chi c_0}{B(\chi)\Gamma(\chi)}\int_0^t h(s)(t-s)^{\chi-1}ds.$$

The proof from Proposition 6.2.4 is considered for $c_1 \ne 0$ by using (6.63) and by applying the Laplace transform on Eq. (6.68). It comes that

$$fiab0\chi\{ABv(t)\} \quad = \quad fiab0\chi\{-c_1\,v(t) + c_0\,h(t)\} \tag{6.72}$$

$$v(t) - v(0) \quad = \quad -c_1\left\{\frac{1-\chi}{B(\chi)}v(t) + \frac{\chi}{B(\chi)\Gamma(\chi)}\int_0^t v(s)(t-s)^{\chi-1}ds\right\} \tag{6.73}$$

$$+ \quad c_0\left\{\frac{1-\chi}{B(\chi)}h(t) + \frac{\chi}{B(\chi)\Gamma(\chi)}\int_0^t h(s)(t-s)^{\chi-1}ds\right\} \tag{6.74}$$

An application of the Laplace transform operator on both sides of (6.72) yields

$$\left[\frac{B(\chi)+(1-\chi)c_1}{B(\chi)} + \frac{\chi c_1}{B(\chi)}\frac{1}{p^\chi}\right]v(p) = \frac{1}{p}v(0) + \frac{c_0(1-\chi)}{B(\chi)}h(p) + \frac{\chi c_0}{B(\chi)\Gamma(\chi)}\Gamma(\chi)p^{-\chi}h(p)$$

$$\frac{1}{B(\chi)}\left[\frac{p^\chi\big(B(\chi)+(1-\chi)c_1\big)+\chi c_1}{p^\chi}\right]v(p) = \frac{1}{p}v(0) + \frac{(1-\chi)c_0}{B(\chi)}h(p) + \frac{\chi c_0}{B(\chi)}p^{-\chi}h(p)$$

$$v(p) = \frac{B(\chi)}{p}\left[\frac{p^\chi}{p^\chi\big(B(\chi)+(1-\chi)c_1\big)+\chi c_1}\right]v(0) + \left[\frac{(1-\chi)c_0 p^\chi}{p^\chi\big(B(\chi)+(1-\chi)c_1\big)+\chi c_1}\right]h(p)$$

$$+ B(\chi)\left[\frac{p^\chi}{p^\chi\big(B(\chi)+(1-\chi)c_1\big)+\chi c_1}\right]\frac{\chi c_0}{B(\chi)}p^{-\chi}h(p)$$

$$= \frac{B(\chi)}{\big(B(\chi)+(1-\chi)c_1\big)}\left[\frac{p^{\chi-1}}{p^\chi + \frac{\chi c_1}{\big(B(\chi)+(1-\chi)c_1\big)}}\right]v(0)$$

$$+ \frac{(1-\chi)c_0}{\big(B(\chi)+(1-\chi)c_1\big)}\left[\frac{p^\chi}{p^\chi + \frac{\chi c_1}{\big(B(\chi)+(1-\chi)c_1\big)}}\right]h(p)$$

$$+ \frac{\chi}{\big(B(\chi)+(1-\chi)c_1\big)}\left[\frac{c_0}{p^\chi + \frac{\chi c_1}{\big(B(\chi)+(1-\chi)c_1\big)}}\right]h(p). \tag{6.75}$$

If we set

$$\chi = \frac{\chi c_1}{\big(B(\chi)+(1-\chi)c_1\big)} \quad \text{and} \quad \zeta = \frac{B(\chi)}{\big(B(\chi)+(1-\chi)c_1\big)},$$

it comes that

$$v(p) = \zeta\,\frac{p^{\chi-1}}{p^\chi + \chi}\,v(0) + \frac{c_0\zeta(1-\chi)}{B(\chi)}\,\frac{p^\chi}{p^\chi+\chi}\,h(p) + \frac{\chi c_0\zeta}{B(\chi)}\,\frac{1}{p^\chi+\chi}\,h(p)$$

$$= \zeta\,\frac{p^{\chi-1}}{p^\chi+\chi}\,v(0) + \frac{\zeta c_0(1-\chi)}{B(\chi)}\,\frac{1}{1+\big(\chi^{-\chi}p\big)^{-\chi}}\,h(p) + \frac{\chi c_0\zeta}{B(\chi)}\,\frac{1}{p^\chi+\chi}\,h(p) \tag{6.76}$$

with $d = \chi^{-\chi}$. It is worth noticing that:

$$\frac{1}{1+(\chi^{-\chi}p)^{-\chi}} = \sum_{k=1}^{\infty} \frac{d^{\chi k-2} t^{\chi k-1}}{\Gamma(\chi k)} = \frac{d}{dt} E_\chi \left[\chi^{-2\chi} t^\chi \right] \tag{6.77}$$

$$= d^{-2\chi} t^{\chi-1} E_{\chi,\chi} \left[\chi^{-2\chi} t^{\chi-1} \right]. \tag{6.78}$$

A direct application of the inverse Laplace transform on both sides produces

$$v(t) = \zeta E_\chi \left[-\chi t^\chi \right] v(0) \tag{6.79}$$

$$+ \frac{\chi c_0 \zeta}{B(\chi)} \int_0^t \left(E_{\chi,\chi} \left[-\chi (t-s)^\chi \right] + \frac{(1-\chi)}{\chi} \chi^{-2\chi} E_{\chi,\chi} \left[-\chi^{-2\chi} (t-s)^\chi \right] \right)$$

$$\times (t-s)^{\chi-1} h(s) ds. \tag{6.80}$$

$\square$

Corollary 6.2.5. *Assume that* $w : [-2, 0] \to R$ *is a solution to* $ABw = -c_1 w + c_0 h(t)$ *with* $w(-2) = 0$, $h \geq 0$ *and* $\int_{-2}^{-1} h(t) \geq \mu$. *Then*

$$w(t) \geq \frac{\chi}{2} E_{\chi,\chi} \left[-2c_1 \right] c_0 \mu, \quad for \quad -1 \leq t \leq 0.$$

Proof. From Proposition 6.2.4 the solution of the differential equation for w can be computed explicitly:

$$w(t) = \zeta E_\chi \left[-\chi t^\chi \right] w(-2)$$

$$+ \frac{\chi c_0 \zeta}{B(\chi)} \int_{-2}^t \left(E_{\chi,\chi} \left[-\chi (t-s)^\chi \right] + \frac{(1-\chi)}{\chi} \chi^{-2\chi} E_{\chi,\chi} \left[-\chi^{-2\chi} (t-s)^\chi \right] \right)$$

$$\times (t-s)^{\chi-1} h(s) ds.$$

If c_0 is small enough and c_1 is large, with the initial condition $w(-2) = 0$, and the fact that $E_{\chi,\chi}(t) > 0$, we have

$$w(t) \geq \frac{\chi c_0 \zeta}{2B(\chi)} E_{\chi,\chi} \left[-2\chi \right] \int_{-2}^t h(s) ds \geq \frac{\chi c_0 \zeta}{2B(\chi)} E_{\chi,\chi} \left[-2c_1 \right] c_0 \mu$$

$$\geq \frac{\chi}{2} E_{\chi,\chi} \left[-2c_1 \right] c_0 \mu. \quad\quad \square$$

6.2.4 Detailed Proof of the Existence of Weak Solutions Using Approximating Solutions

In this section the details of the proof of existence of a solution to the weak equation (6.82) by means of approximating solutions is given, following the approach of [204]. The weak formulation of the problem is first considered, and the discretization of the weak formulation will enable the proof of existence of the unique solution and Hölder continuity.

6.2.5 Numerical Approximation of the Problem

In the following, we denote by $\psi = b/\iota$ the time step which represents the subdivision of the interval (a, b), where $\iota \in N$ denotes the number of time steps. Also, for $0 \le k \le \iota$, $t = k\psi$. So the discrete form of the Atangana–Baleanu fractional derivative in the sense of Caputo holds

$$Lv(a + \psi k) = \psi^\chi c_\chi \sum_{-\infty < i < k} \frac{E_{\chi,\chi}\left[c\psi^\chi(k-i)^\chi\right]\left[v(a+\psi k) - v(a+\psi i)\right]}{(k-i)^{1-\chi}}. \qquad (6.81)$$

Using (6.81) the discrete form of (6.62) takes the form

$$\psi^\chi c_\chi \sum_{-\infty < i < k} \frac{E_{\chi,\chi}\left[c\psi^\chi(k-i)^\chi\right]\left[v(a+\psi k) - v(a+\psi i)\right]}{(k-i)^{1-\chi}}$$

$$= \int_{R^n} [v(a+\psi k, \xi) - v(a+\psi k, x)]$$
$$J(a+\psi k, x, \xi)d\xi + w(a+\psi k, x). \qquad (6.82)$$

Next we state the following lemma.

Lemma 6.2.6. *Assume $v(a) = v(a + \xi j) = 0$ for $j < 0$. Then the discrete integration by parts type estimate holds*

$$\sum_{k \le j} v(a + \psi k) Lv(a + \psi k)$$

$$\ge \frac{\psi^\chi}{2} C_\chi \sum_{0 \le i < k \le j} \frac{\left[v(a+\psi k) - v(a+\psi i)\right]^2}{(k-i)^{\chi-1}} E_{\chi,\chi}\left[c\psi^\chi(k-i)^\chi\right] \qquad (6.83)$$

$$+ \frac{\psi^\chi}{2} C_\chi \sum_{k \le j} \frac{\left[u^2(a+\psi k)\right]^2}{(j-i)^{\chi-1}} E_{\chi,\chi}\left[c\psi^\chi(j-i)^\chi\right]. \qquad (6.84)$$

Proof. The proof of this lemma follows from the direct computation of the discrete integration by parts type estimate. □

Now to get the discrete form of the weak formulation Proposition 6.2.6 we use the integration by parts type estimate given by Lemma 6.2.6.

Now we are ready to prove existence of weak solutions following the idea of the authors in [197] by using the approximating method. For this purpose, we write the operators in 6.2.6 and (6.83) as $\mathcal{H}$ and $\mathcal{H}_\psi$ respectively.

Proof. Let ϑ be a bounded an Lipschitz function on $(-\infty, b) \times R^n$. There exists a sequence of solutions v_ψ to (6.82) with $\psi \to 0$, such that

$$u_\psi \to u \in L^p\big((-\infty, b) \times R^n\big),$$

with p as defined in [204] as

$$p = 2\left(\frac{\chi n + \chi}{\chi n + (1 - \chi)\chi}\right).$$

For $\psi(k-1) < t \leq \psi k$, we let $\mathcal{B}_\psi$ be the bilinear form associated with K_ψ. Our aim is to show that for ϑ to be a bounded and Lipschitz function on $(-\infty, b) \times R^n$,

$$\mathcal{H}(v, \vartheta) + \mathcal{H}_\psi(u_\psi, \vartheta) \to 0.$$

The first part of the proof where we shall consider the fractional Laplacian, we mean

$$\lim_{\psi \to 0} \int_a^b B_\psi(v_\psi, \vartheta)\,dt = \lim_{\psi \to 0} \psi \sum_{0 < k \leq j} B(v_{\psi k}(\psi k, x), \omega(\psi, x)) \tag{6.85}$$

has been proven by the authors in [204]. So next we shall focus on the component in time. To do this we start by showing that

$$\lim_{\psi \to 0} \int_{R^n} \int_{-\infty}^b \int_{-\infty}^t [v_\psi(t, x) - v_\psi(s, x)][\omega(t, x) - \omega(s, x)]\frac{E_{\chi,\chi}[c(t-s)^\chi]}{(t-s)^{1-\chi}}\,ds\,dt \tag{6.86}$$

$$= \lim_{\psi \to 0} \sum_{0 \leq i < k \leq j} \sum \int_{\psi(k-1)}^{\psi k} \int_{\psi(i-1)}^{\psi i} [u_\psi(\psi k) - u_\psi(\psi i)][\omega(\psi k) - \omega(\psi i)]$$

$$\times \frac{E_{\chi,\chi}[c\psi^\chi(k-i)^\chi]}{(\tau(k-i))^{1-\chi}}. \tag{6.87}$$

In order to achieve our goal, since ϑ is a bounded and Lipschitz function, then $\vartheta_\psi(t) \to \omega(t)$ and $v_\psi \to u \in L^p\big((-\infty, b) \times R^n\big)$ we have that

$$\lim_{\psi \to 0}\left| \int_{R^n} \int_{-\infty}^b \int_\infty^t \frac{E_{\chi,\chi}[c(t-s)^\chi]}{(t-s)^{1-\chi}}[u_\psi(t) - u_\psi(s)] \times \right.$$

$$\left. [(\omega(t) - \omega(s)) - (\vartheta_\psi(t) - \vartheta_\psi(s))]\,ds\,dt \right| \to 0.$$

We show also that

$$\lim_{\psi \to 0} \sum_{0 \le i < k \le j} (u_\psi(\psi k) - u_\psi(\psi i))(\omega(\psi k) - \omega(\psi i)) \times \quad (6.88)$$

$$\int_{\psi(k-1)}^{\psi k} \int_{\psi(i-1)}^{\psi i} \left(\frac{E_{\chi,\chi}\left[c(t-s)^\chi\right]}{(t-s)^{1-\chi}} - \frac{E_{\chi,\chi}\left[c\psi^\chi(k-i)^\chi\right]}{\left(\tau(k-i)\right)^{1-\chi}} \right) = 0.$$

To do this we break up the integral over the sets $(t - s) \le \psi^{1/2}$ and $(t - s) > \psi^{1/2}$. Then with the relation (6.66),

$$0 = \lim_{\psi \to 0} \int_{R^n} \int\int_{t-s \le \psi^{1/2}} \frac{|(u_\psi(t) - u_\psi(s)(\vartheta_\psi(t) - \vartheta_\psi(s))|}{(t-s)^{1-\chi}} E_{\chi,\chi}\left[c(t-s)^\chi\right] \quad (6.89)$$

$$\ge \lim_{\psi \to 0} \int_{R^n} \sum\sum_{\psi(k-i) \le \psi^{1/2}} \left| (u_\psi(\psi k) - u_\psi(\psi i))(\omega(\psi k) - \omega(\psi i)) \right|$$

$$\times \int_{\psi(k-1)}^{\psi k} \int_{\psi(i-1)}^{\psi i} \frac{E_{\chi,\chi}\left[c\psi^\chi(k-i)^\chi\right]}{\left(\tau(k-i)\right)^{1-\chi}} \quad (6.90)$$

For $t - s > \psi^{1/2}$, and $\psi(i-1) \le s \le \psi i$ and $\psi(k-1) \le t \le \psi k$, we can compute the estimate

$$\left| (t-s)^{\chi-1} E_{\chi,\chi}\left[c(t-s)^\chi\right] - \left(\tau(k-i)\right)^{\chi-1} \right|$$

$$\le \left(\psi^{1/2} - \tau\right)^{\chi-1} \times \quad (6.91)$$

$$E_{\chi,\chi}\left[c\left(\psi^{1/2} - \tau\right)^\chi\right] - \left(\psi^{1/2}\right)^{\chi-1} E_{\chi,\chi}\left[c\left(\psi^{1/2}\right)^\chi\right] \quad (6.92)$$

$$\le T_{\chi,\psi}. \quad (6.93)$$

Hence the result follows for $\psi \to 0$. Next we consider the following

$$\int_{R^n} \int_{-\infty}^{b} u_\psi(t) L\omega(t) \, dt \, dx - \int_{R^n} \psi \sum_{0 < k \le j} u_\psi(\psi k) L\omega(\psi k).$$

Similarly as in the previous case, since ϑ is a bounded Lipschitz function and $v_\psi \to u \in L^p\left((-\infty, b) \times R^n\right)$, and from (6.83) one shows that this term also goes to zero. The remaining pieces in time are handled in the similar manner. Thus the theorem is proven. $\square$

6.2.6 The Derivation of the Point-Wise Estimation and Hölder Regularity

This section contains the some auxiliary results, which are the key to prove Theorem 6.2.1. The proof of Proposition 6.2.12 uses the main contributions of this note. Once Proposition 6.2.12 is established, a priori Hölder regularity estimates follow by the classical method of diminishing oscillation given by Lemma 6.2.13. Before going into this, we first collect ingredient that will be useful. As one feature, we underline the viscosity solution. One of the useful properties of viscosity solution is that viscosity sub-solutions themselves can be used to evaluate their corresponding equation classically at all of the points where the sub-solution can be touched from above by a smooth test function. In this note, we need the following property for this equation to hold:

$$Lv(t, x) - M_A^+ v(t, x) \leq w(t, x) \quad \text{in } R_1 \times [-1, 0]. \tag{6.94}$$

Next we state the following proposition to clarify that v is a solution on (6.94) and (6.82) in the viscosity sense by also making reference to the Proposition 6.2.2.

Proposition 6.2.7. *Let v be a continuous bounded function on $(-\infty, b)$ and assume that for some $t \in (-\infty, b)$ there is a Lipschitz function touching v by above at t. Then*

$$\int_{-\infty}^{t} [v(t, x) - v(l, x)] \P(t, s) \, ds \geq w(t, x)$$

if and only if $Lv(t, x) \geq w(t, x)$ in the viscosity sense.

Proof. The proof is standard and is based on the proof of Proposition 2.3 in [205]. □

Proposition 6.2.8. *Assume that v solves (6.94) in the viscosity sense. $\phi \geq u$ defined on the cylinder $Y := [t_0 - \xi, t_0] \times R_\xi(x_0)$ has a global maximum and touches v from above at $(x_0, t_0) \in Y$, and we define v as*

$$v(x, t) := \begin{cases} \phi(x, t) & \text{if } (x, t) \in Y \\ v(x, t) & \text{otherwise,} \end{cases}$$

then v is solution to (6.94) at (x_0, t_0), or

$$Lv(t_0, x_0) - Bv(t_0, x_0) \leq w(t_0, x_0).$$

So the solution is both a sub-solution and a super-solution.

For the brevity we are not going to prove this proposition since the proof is straightforward. Before we state the point evaluation of the proposition, we recall the comparison principle for which the proof is similar as in [205] and use Perron's method.

Lemma 6.2.9 (Comparison Principle). *Let v be bounded and upper-semi-continuous and w be bounded and lower-semi-continuous $(-\infty, R_2)$. Let w be a continuous function such that $Lu \leq g \leq Lw$ on $(b_1, b_2]$, with $v \leq w$ on $(-\infty, b_1)$. Then $v \leq w$ on $(-\infty, b_2)$, and if $v(t_0) = w(t_0)$ for some $t_0 \in (b_1, b_2]$, then $v(t) = w(t)$ for all $t \leq t_0$.*

Lemma 6.2.10. *Let $\phi(t)$ be continuous $(-\infty, b_1)$. Let w be a continuous function on $[b_1, b_2]$. There exists a unique viscosity solution v to,*

$$\begin{cases} Lv(t) & = w(t) & \text{for} & t \in [b_1, b_2] \\ v(t) & = \phi(t) & \text{if} & t \leq b_1 \end{cases}$$

on $(b_1, b_2]$.

We start by recalling the definitions of Pucci's extremal operators as defined in [203] for the spatial operator and next for the fractional-time derivative operator.

Lemma 6.2.11 (Extremal Formula). *Assume $v \in C^{1,1}(-\infty, b) \cap L^\infty(R^n)$. Then we have the following elliptic spacial operator*

$$M_A^+ v(t, x) = \int_{R^n} \left(\zeta \left(\delta_h v(t, x) \right)_+ - \pi \left(\delta_h v(t, x) \right)_- \right) \mu(dh), \tag{6.95}$$

$$M_A^- v(t, x) = \int_{R^n} \left(\pi \left(\delta_h v(t, x) \right)_+ - \zeta \left(\delta_h v(t, x) \right)_- \right) \mu(dh), \tag{6.96}$$

and for the fractional-time derivative operator as

$$M_{T_{sec}}^+ v(t, x) := \mathcal{Y}(n, \chi) \int_{-\infty}^t \left[\zeta \left(v(t, x) - v(l, x) \right)_+ - \pi \left(v(t, x) - v(l, x) \right)_- \right] T_{sec} \tag{6.97}$$

$$M_{T_{sec}}^- v(t, x) := \mathcal{Y}(n, \chi) \int_{-\infty}^t \left[\pi \left(v(t, x) - v(l, x) \right)_+ - Zeta \left(v(t, x) - v(l, x) \right)_- \right] T_{sec} \tag{6.98}$$

Next, in order to prove the Hölder continuity, we use essentially the same ideas as the proof in [197,198].

Proposition 6.2.12 (Point Estimate). *Let $v \leq 1$ in $(R^n \times [-2, 0]) \cup (R_1 \times [-\infty, 0])$ and assume it satisfies the following inequality in the viscosity sense in $R_2 \times [-2, 0]$:*

$$Lu - M^+ u \leq \xi_0.$$

Assume also that $\left| \{ v(x, t) \leq 0 \} \cap (R_1 \times [-2, -1]) \right| \geq \mu > 0$. Then if ξ_0 is small enough there exists $\theta > 0$ such that $v \leq (1 - \theta)$ in $R_1 \times [-2, 0]$. The maximal value of θ as well as ξ_0 depends on χ, π, ζ, n and v, but remains uniform as $\chi \to 1$.

Proof. We consider the differential equation

$$\begin{cases} Lw(t) = c_0 \left| \{ x \in R_1 : v(x, t) \leq 0 \} \right| - c_1 w(t), \\ w(-2) = 0, \quad \text{for } t \leq -2. \end{cases} \tag{6.99}$$

From Proposition 6.2.4 this ordinary differential equation can be computed explicitly and from Corollary 6.2.5 we have that

$$w(t) \geq \frac{\chi c_0 \mu}{2} E_{\chi,\chi}\big[-2c_1\big], \qquad \text{for } -1 \leq t \leq 0.$$

In the following we will show that if c_0 is small and c_1 is large, then $v \leq 1 - w(t) + \xi_0 c_\chi 2^\chi$ in $R_1 \times [-1, 0]$. The constant c_χ is chosen such that $Lc_\chi (2+t)_+^\chi = 1$ for $t \geq -2$. Since for $t \in [-1, 0]$

$$
\begin{aligned}
w(t) &\geq \frac{\chi}{2} E_{\chi,\chi}\big[-2c_1\big] c_0 \big| \{x : v(x,t) \leq 0\} \cap R_1 \times [-2, -1] \big| \\
&\geq \frac{\chi}{2} E_{\chi,\chi}\big[-2c_1\big] c_0 \mu,
\end{aligned}
$$

we set $\theta = \frac{\chi c_0 \mu}{4} E_{\chi,\chi}\big[-2c_1\big]$ for ξ_0 small and finish the proof of the lemma. Next as in [197], let $\chi : R \to R$ be a fixed smooth non-increasing function such that $\chi(x) = 0$ if $x \geq 2$. Let $\eta(x, t) = \chi(|x|)$. As a function of x, $\eta(x, t)$ looks like a bump function for every fixed t. The main strategy of the proof is to show that the function $v(x, t)$ stays below $1 - w(t)\eta(x, t) + \xi_0 c_\chi (2+t)_+^\chi$. In order to arrive to a contradiction, we assume that $\eta(x, t) > 1 - w(t) + \xi_0 c_\chi (2+t)_+^\chi$ for some point $(x, t) \in R_1 \times [-1, 0]$. We then look at the maximum of the function

$$\tilde{\eta}(x, t) = v(x, t) + w(t)\eta(x, t) - \xi_0 c_\chi (2+t)_+^\chi.$$

Assume that there is one point (x_0, t_0) in $R_1 \times [-1, 0)$ where $\tilde{\eta}(x, t) > 1$, $\tilde{\eta}$ must be larger that 1 at the point that realize the maximum of $\tilde{\eta}$. We mean by that

$$\tilde{\eta}(x_0, t_0) = \max_{R^n \times (-\infty, 0]} \tilde{\eta}(x, t).$$

Since $\tilde{\eta}(x_0, t_0) > 1$, the point (x_0, t_0) must belong to the compact support η. Hence $|x| < 2$. Now the remainder of the proof is exactly as in [197]. Next we call $\varphi(x, t)$ the function such that

$$\varphi(x, t) := \tilde{\eta}(x_0, t_0) - w(t)\eta(x, t) + c_\chi (2+t)_+^\chi$$

φ touches v from above at the point (x_0, t_0). We define

$$v(x, t) := \begin{cases} \varphi(x, t) & \text{if } x \in R_r \\ v(x, t) & \text{if } x \notin R_r. \end{cases}$$

Then at the point that $\tilde{\eta}$ realizes it maximum

$$Lv(x_0, t_0) - M^+ v(x_0, t_0) \leq \xi_0. \tag{6.100}$$

So one can have

$$Lv(x_0, t_0) - Lw(t_0)\eta(x_0) + \xi_0. \tag{6.101}$$

Hence for $G := \{x \in R_1 | v(x, t_0) \le o\}$, the following bound is obtained:

$$M^+ v(x_0, t_0) \le -w(t_0) M^- \eta(x_0, t_0) - c_0 |G \setminus R_r|. \tag{6.102}$$

Now if we insert the relations (6.102), (6.100), (6.101) into (6.99), we obtain

$$- Lw(t_0)\eta(x_0, t_0) + \xi_0 + c_0 |G \setminus R_r| \le \xi_0,$$

or in its explicit form

$$\left(-c_0 |\{x \in R_1 : v(x, t) \le 0\}| + c_1 w(t) \right)\eta(x_0) + \xi_0 + c_0 |G \setminus R_r| \le \xi_0.$$

But notice that for any $c_1 > 0$ this contradicts (6.99). We now analyze the case where $\eta(x_0, t_0) > \chi_1$. As we said previously that η is a smooth compactly supported function, then there exists some constant $\tilde{C}$ such that $|M^- \eta| \le \tilde{C}$. Then we have the bound

$$M^+ v(x_0, t_0) \le -\tilde{C} w(t_0) + c_0 |G \setminus R_r|. \tag{6.103}$$

As in the previous case, we insert (6.103) into (6.100), (6.101) into (6.102), we obtain

$$\left(-c_0 |\{x \in R_1 : v(x, t) \le 0\}| + c_1 w(t) \right)\eta(x_0) + \xi_0 - \tilde{C} w(t_0) + c_0 |G \setminus R_r| \le \xi_0.$$

We recall that $\eta(x_0, t_0) > 0$ and by letting $r \to 0$, we obtain

$$c_0 (1 - \eta(x_0))|G| + (c_1 \eta(x_0) - \tilde{C})w(t_0) \le 0.$$

Which can be written in the form

$$c_0 (1 - \eta(x_0))|G| + (c_1 \chi_1 - \tilde{C})w(t_0) \le 0.$$

Choosing c_1 large enough, we arrived to a contradiction. This ends the proof. $\square$

In the following we shall give an approach of the proof of the Hölder continuity. For this purpose, we state and prove the so-called growth lemma, which says that if a solution of Eq. (6.82) in the unit cylinder $F_1 = R_1 \times [-1, 0]$ has oscillation one, then its oscillation in a smaller cylinder F_r is less than a fixed constant $(1 - \theta)$ [197,198].

Lemma 6.2.13 (Diminish of Oscillation). *Let v be a bounded continuous function which satisfies (6.82) or the following two inequalities in the viscosity sense in F_1.*

$$Lv - M^+u \leq \xi_0/2 \tag{6.104}$$
$$Lv - M^+u \geq -\xi_0/2. \tag{6.105}$$

Then there are universal constants $\theta > 0$ and $v > 0$, depending on n, v, ζ, π and χ, such that if

$$\|g\|_{L^\infty(Q)} \leq \xi_0, \qquad |v| \leq 1 \ in \ F_1$$
$$|v(x,t)| \leq 2|rx|^v - 1 \ in \ (R^n \setminus R_1) \times [-1,0]$$
$$|v(x,t)| \leq 2|rt|^v - 1 \ in \ R_1 \times [-\infty,-1]$$

with $r = \min\{4^{-1}, 4^{-\chi/2v}\}$, then

$$osc_{F_r} v \leq (1 - \theta).$$

Proof. With the Proposition 6.2.3 in hand the reader can finish the proof by following the idea of the authors in [197,198]. □

Having the diminishing of oscillation lemma in hand, where the proof is similar to the one proposed by the author Mark Allen in [205], we are now going to prove one of our main results about Hölder continuity. The result requires the function to solve the equation only in a cylinder in order to have Hölder continuity in a smaller cylinder. To do this, we re-normalized the function v. If v satisfies (6.82), then the re-scale function $\eta(x,t) = v(rx, r^{2v/\chi}t)$ satisfies also (6.82), with $r \in (0,1)$ such that $r = \min\{4^{-1}, 4^{-\chi/2v}\}$. We then define the parabolic cylinders in terms of the scaling of the equation by

$$F_r := R_r \times [-r^{2v/\chi}, 0].$$

Now we state the result on Hölder continuity.

Proof of Theorem 6.2.1. The proof is the adaptation of the proof in [198,205]. We prove C^χ estimate of (6.59) by proving a C^χ estimate for $\eta(x,t)$ at the point $(x_0,t_0) \in F_1$. For any point (x_0,t_0) we consider the normalized function

$$\eta(x,t) = \frac{v(x_0 + x, t + t_0)}{\|v\|_{L^\infty} + \xi_0^{-1}\|w\|_{L^\infty}},$$

here ξ_0 being considered to be a constant as in Lemma 6.2.13, therefore, it is possible to show that $\operatorname{osc}_{R \times [-1,0]} \rho \leq 1$ and in $R_2 \times [-1, 0]$, η is a solution to

$$L\eta - M_v^+ \eta \leq \xi_0$$
$$L\eta - M_v^- \eta \geq -\xi_0.$$

Consider $r \in (0, 1)$, with the property that $r = \min\{4^{-1}, 4^{-\chi/2v}\}$. In this case to achieve the proof of the Hölder estimate, within a specific oscillation decay of cylinders for some ι [199], the induction principle is adopted.

$$\operatorname{osc}_{F_{r_k}} \eta \leq 2r^{\iota k} \quad \text{for } k = 0, 1, 2, \cdots \tag{6.106}$$

Two sequences are built here including $S_k \leq \eta \leq T_k$ in F_{r_k}, $T_k - S_k = 2r^{\iota k}$ with S_k which will be considered as a non-decreasing and T_k which is considered as a non-increasing. Therefore, with $k = 0$, thus, using the following suggestion $\operatorname{osc}_{F_{r_k}} \eta = 1$. For k comprises in a certain range, it is supposed that the sequence is true. We scale once more by considering

$$e(x, t) = (\eta(r^k x, r^{2kv/\chi} t) - (S_k + T_k)/2)r^{-\iota k}.$$

We then have

$$|e| \leq 1 \quad \text{in} \quad F_1$$
$$|e| \leq 2r^{-\iota k} - 1 \quad \text{in} \quad F_{r^{-k}},$$

and so

$$|e(x, t)| \leq 2|x|^v - 1 \quad \text{for} \quad (x, t) \in R_1^c \times [-1, 0]$$
$$|e(x, t)| \leq 2|t|^v - 1 \quad \text{for} \quad (x, t) \in R_1 \times (-\infty, -1).$$

Having in hand that $\iota < 2v$, e is bounded, then the following holds.

$$\left\| r^{k(\iota - 2v)} w \right\|_{L^\infty} \leq \xi_0 r^{k(\iota - 2v)} < \xi_0$$

If ι is a very small parameter, then one can easily apply Lemma 6.2.13 such that one can obtain

$$\operatorname{osc}_{F_r} e \leq 1 - \theta.$$

Additionally, if one considers ι to be a small parameter less than that of Lemma 6.2.13 at the same time with $r^\iota \geq 1 - \theta$. Then, the following holds:

$$\operatorname{osc}_{F_{rk}} e \leq r^\iota.$$

This implies obviously that, $\operatorname{osc}_{F_{r^{k+1}}} z \leq r^{\iota(k+1)}$ within this framework, one can find two different possibilities including: S_{k+1} and T_{k+1}. Employing the recursive approaches in this case, one concludes the proof. $\qquad\qquad\square$

CHAPTER 7

Applications of Fractional Operators to Groundwater Models

A problem that arises naturally in groundwater investigations is to choose an appropriate geometry for the geological system in which the flow occurs [111]. For example, one can use a model based on percolation theory to simulate the flow in a fractured rock system with a very large fracture density [110] or the parallel plate model [107] to simulate flow through a single fracture. However, there are many fractured rock aquifers where the flow of groundwater does not fit conventional geometries [107]. This is in particular the case with the Karoo aquifers in South Africa, characterized by the presence of a very few bedding parallel fractures that serve as the main conduits of water in the aquifers [108]. Attempts to fit a conventional radial flow model to the observed drawdown, see the work done in [109], always yield a fit that underestimates the observed drawdown at early times and overestimates it at later times. The deviation of observations from theoretically expected values is usually an indication that the theory is not implemented correctly, or does not fit the observations.

7.1 Theis Model With Fractional Derivatives

In order to include explicitly the possible effect of flow geometry into the mathematical model, the radial component of the gradient of piezometric head is replaced by the fractional derivatives of order. The fractional derivatives involved in this section are the Caputo, Caputo–Fabrizio, Atangana–Baleanu types. The expression of the Darcy velocity as a function of a complementary derivative of fractional order may sound to be merely some kind of highly technical mathematical concept and before drawing conclusions on this new model, one would like to attract the attention of the reader to some basic physical features implied by the model. Broadly seen the generalized version of the Darcy law is affecting conceptually the physical content of Darcy's work in mainly two ways. Firstly, the new generalized constitutive equation is not, unlike the usual Darcy relation, satisfying the popular principle of local (spatial) action that postulates that the evolution of a system at a given spatial position and at a given time is only affected by the behavior of the different variables in the direct neighborhood around that point [111]. Within the frame of Darcy's law this principle implies that the Darcy velocities may only be expressed as relations containing the value, at that point, of the piezometric head and/or its derivatives. The use of a constitutive relation based on a spatial integration of the piezometric head over the half infinite line is breaking away from this postulate and is allowing for the consideration of more general situations where the groundwater flow at a given point of the aquifer is governed not only by the properties of the piezometric field at that specific position but also depends on the global spatial distribution of that field in soil matrix [111]. It is, however, expected that the information relative to the direct vicinity of the specific point under consideration will have a greater influence on the fluid flow than the information dealing with events taking place far away from that point [111].

7.1.1 Theis Model With Caputo Derivative

The easiest sweeping statement of subsurface water flow equation, which while we are on the subject is in addition in harmony in the midst of the real physics of the observed fact, is to presume that water level is not in a balanced state but momentary state. Theis was the first to develop a formula for unsteady-state flow that introduces the time factor and the storativity as we discussed earlier. He noted that when a well-penetrating extensive confined aquifer is pumped at a constant rate, the influence of the discharge extends outward with time. The rate of decline of head, multiplied by the storativity and summed over the area of influence, equals the discharge. The unsteady-state (or Theis) equation, which was derived from the analogy between the flow of groundwater and the conduction of heat, is perhaps the most widely used partial differential equation in groundwater investigations:

$$S\frac{\partial h(r,t)}{\partial t} = \frac{\partial^2 h(r,t)}{\partial r^2} + \frac{1}{r}\frac{\partial h(r,t)}{\partial r}. \tag{7.1}$$

The aforementioned equation is classified under parabolic equation. The standard version of the partial derivative with respect to time is replaced here with time-fractional Caputo order derivative to obtain:

$$S^C D_t^\alpha(h(r,t)) = \frac{\partial^2 h(r,t)}{\partial r^2} + \frac{1}{r}\frac{\partial h(r,t)}{\partial r}. \tag{7.2}$$

The initial conditions are the same as presented earlier. The exact solution of the above equation will be obtained using the Laplace–Carson transform together with method of separation of variables. The Laplace–Carson transform is given as:

$$L_C f(t)(p) = \int_0^\infty x f(x)\exp(-sx)dx. \tag{7.3}$$

We shall start by assuming that the solution of the main equation can be separated as follows:

$$h(r,t) = g(r)f(t). \tag{7.4}$$

Thus, the separated equations become

$$^C D_t^\alpha (f(t)) - \lambda^2 f(t) = 0, \tag{7.5}$$

$$D_{rr}(g(r)) + \frac{1}{r} D_r g(r) + \lambda^2 g(r) = 0, \tag{7.6}$$

where λ is the separation constant. The first equation of (7.5) can be solved directly by applying on both sides the Laplace transform to obtain:

$$L(f(t))(p) = \frac{p^{\alpha-1}}{p^\alpha + \lambda^2}. \tag{7.7}$$

Using the inverse formula of Laplace transform of two-parameter Mittag–Leffler function, we get:

$$f(t) = c E_\alpha \left[-\frac{-S\lambda^2 t^\alpha}{T} \right]. \tag{7.8}$$

To solve the second equation, we make use of the Laplace–Carson transform presented earlier, to obtain

$$Lg(r)(s) = \frac{1}{\sqrt{s^\alpha + \lambda^2}}. \tag{7.9}$$

Now applying the inverse Laplace transform operator on both sides in the previous equation, we obtain the following in terms of the Bessel function of the first kind, $J_0(.)$:

$$g(r) = \sum_{k=0}^{\infty} \frac{(-1)^k}{k!} \frac{1}{\Gamma(k+1)} \left(\frac{r\lambda}{2} \right)^{2k}. \tag{7.10}$$

Therefore, the solution of the fractional groundwater flow equation is given as:

$$h(r,t) = c \sum_{k=0}^{\infty} E_\alpha \left[-\frac{-S\lambda_n^2 t^\alpha}{T} \right] J_0(\lambda_n r). \tag{7.11}$$

Making use of the initial and boundary conditions, we obtain the constant c to be

$$c = \frac{Q}{4\pi T}. \tag{7.12}$$

Finally, the exact solution of the Theis model with fractional derivative is given as:

$$h(r,t) = \frac{Q}{4\pi T} \sum_{k=0}^{\infty} E_\alpha \left[-\frac{-S\lambda_n^2 t^\alpha}{T} \right] J_0(\lambda_n r). \tag{7.13}$$

Eq. (7.5) can be solved by means of a method frequently used to derive some kind of parabolic partial differential equations is the so-called Boltzmann transformation defined for an arbitrary $t_0 < t$ by equation:

$$u_0 = \frac{Sr^2}{4T(t-t_0)}.$$ (7.14)

Let us consider now the following function:

$$h_0(r,t) = \frac{C}{t-t_0} E_{\alpha,1}[-u_0],$$ (7.15)

with c an arbitrary constant. If we assume that r_b is the ratio of the borehole from which the water is taken out of the aquifer, then the total volume of the water withdrawn from the aquifer is given by:

$$Q_0 \Delta t_0 = 4\pi c T.$$ (7.16)

Thus

$$h_0(r,t) = \frac{Q_0 \Delta t_0}{4\pi T(t-t_0)} E_{\alpha,1}[-u_0]$$ (7.17)

is the drawdown that will be observed at a distance, r, from the pumped borehole after the period Δt_0. Now suppose that the previous procedure is repeated many times; that is, water is withdrawn for a short period of time:

$$h(r,t) = \frac{1}{4\pi T} \sum_{k=0}^{n} \frac{Q_k \Delta t_k}{t-t_k} E_{\alpha,1}[-u_k].$$ (7.18)

If we take $\Delta t_k \to 0$, the sum can be converted to an integral as:

$$h(r,t) = \frac{1}{4\pi T} \int_{t_0}^{t} \frac{Q(\tau)}{t-\tau} E_{\alpha,1}\left[-\frac{Sr^2}{4T(t-\tau)}\right].$$ (7.19)

However, using the Boltzmann variable, we arrive at the following expression:

$$y = \frac{Sr^2}{4T(t-\tau)},$$ (7.20)

such that

$$h(r,t) = \frac{1}{4\pi T} \int_{y}^{\infty} \frac{Q(\tau)}{t-\tau} E_{\alpha,1}[-\tau]d\tau.$$ (7.21)

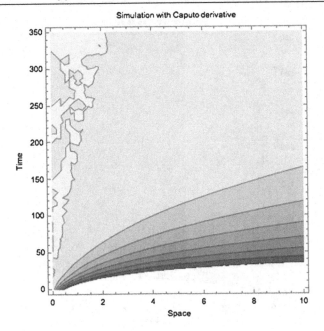

Figure 7.1: Numerical simulation for $\alpha = 0.95$.

The previous solution will be called the most generalized solution of fractional groundwater equation. However, this solution can be simplified somewhat under certain conditions. A particularly important solution arises when t_0 is taken at zero and $Q(t)$ is a constant independent of time, and then we arrive at:

$$h(r, t) = \frac{Q}{4\pi T} \int_y^\infty \frac{1}{\tau} E_{\alpha,1}[-\tau] d\tau = \frac{Q}{4\pi T} W_\alpha(u), \qquad (7.22)$$

which will be called the generalized exponential integral. It is worth pointing out that if alpha is equal to 1, we recover the exact analytical solution of the groundwater flow equation proposed by Theis. Some numerical simulations are presented here for different values of the fractional order alpha (see Figs. 7.1, 7.2, 7.3, 7.4).

7.1.2 Theis Model With Caputo–Fabrizio Derivative

The Theis groundwater flow equation was proposed under some conditions that do not correspond to the real world observations [111,108,107]. The Theis groundwater equation assumes that, the medium by mean of which the flow is taken place is homogeneous, there is no variation of the medium in space and time. On the other hand, the main mathematical tool used to derive the Theis groundwater flow equation is the well-known Darcy's law, which was also

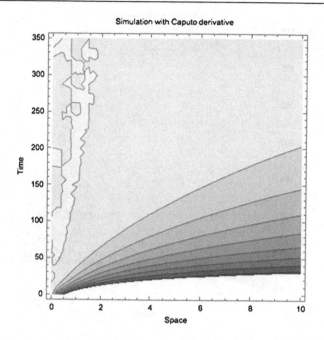

Figure 7.2: Numerical simulation for $\alpha = 0.65$.

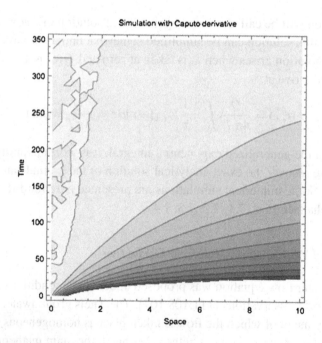

Figure 7.3: Numerical simulation for $\alpha = 0.35$.

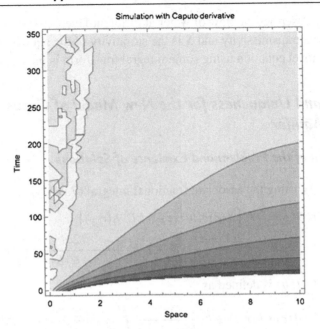

Figure 7.4: Numerical simulation for $\alpha = 0.035$.

obtained experimentally using the sand; this simply implies that there is no form of heterogeneity, or layers, this is of course in contradiction with what is observed in the real field
observation, therefore a new flow equation is needed. The newly proposed derivative with
fractional order has the properties to model the movement of water in different layers or
scales within a geological formation called aquifer. Therefore, in order to well replicate the
movement of water by means of porous media in different layers in the aquifer, we reformulate the well-known Darcy law in order to produce a new groundwater flow equation within
a confined aquifer [111,108,107]. One of the most important facts in the model of the flow
within a confined aquifer is that, the difference between the rate inflow and outflow from annular cylinder is the change of the volume of water within the annular space. Now, in order to
take into account the effect of different scale in the annular space, we proposed the mathematical model to be:

$$_{0}^{CF} D_t^\alpha (V(r,t)) = 2\pi r S dr _{0}^{CF} D_t^\alpha (h(r,t)) = Q_1 - Q_2. \tag{7.23}$$

With Q_1 and Q_2 are inflow and outflow rate respectively and V is the volume. Thus following
the discussion presented in the above section, the following equation is obtained after some
simplification:

$$_{0}^{CF} D_t^\alpha (h(r,t)) \frac{S}{T} = \frac{\partial h(r,t)}{r \partial r} + \frac{\partial^2 h(r,t)}{\partial r^2} \tag{7.24}$$

The above equation describes the flow of water at different layers in time within the confined aquifer. Here T is the transmissivity and S is the storativity of the aquifer. The derivation of the solution of the novel equation using some integral transform is presented below.

7.2 Existence and Uniqueness for the New Model of Groundwater Flow in Confined Aquifer

7.2.1 Formulation of the Problem and Existence of Solutions

Integrating Eq. (7.23), using the associate fractional integral (7.24), yields

$$
\begin{aligned}
h(r,t) - h(r,0) &= I_t^\alpha \big[\vartheta(r)\partial_r h(r,t) + kL^\gamma h(r,t)\big] \\
&= \frac{2(1-\alpha)}{(2-\alpha)B(\alpha)} K(r,t,h) + \frac{2\alpha}{(2-\alpha)B(\alpha)} \int_0^t K(r,s,h)ds, \quad (7.25)
\end{aligned}
$$

where the kernel $K(r,t,h)$ is defined as

$$
\begin{aligned}
K(r,t,h) &= \vartheta(r)\partial_r h(r,t) + k\frac{\gamma}{(1-\gamma)\sqrt{\pi}} \int_0^r \exp\big[-\lambda^2(r-\tau)^2\big]\partial_{\tau\tau}h(\tau,t)d\tau \\
&= \vartheta(r)\partial_r h(r,t) + \eta\partial_r h(r,t) + \eta \int_0^r \exp\big[-\lambda^2(r-\tau)^2\big]h(\tau,t)d\tau \\
&\quad - 2\lambda^2\eta \int_0^r (r-\tau)^2 \exp\big[-\lambda^2(r-\tau)^2\big]h(\tau,t)d\tau
\end{aligned}
$$

and where $\eta = k\gamma\big[(1-\gamma)\sqrt{\pi}\big]^{-1}$. Now let us show that the nonlinear kernel $K(r,t,h)$ satisfies the Lipschitz condition.

Theorem 7.2.1. *Let $\alpha \in (0,1)$, $\beta \in (0,1)$, $T > 0$, and $K : (0,T) \times H_0^1(\Omega)$ be a continuous function such that there exists $\Lambda > 0$ satisfying,*

$$
\|K(r,t,h) - K(r,t,\varphi) \le \Lambda\|h - \varphi\|, \qquad \text{for all } h, \varphi \in (0,T) \times H_0^1(\Omega).
$$

If $(\varepsilon_1 + \varepsilon_2 + \varepsilon_3) < 1$, the operator K is a contraction.

Proof. We consider two bounded functions h and φ in $H_0^1(\Omega)$. We have by triangular inequality

$$
\begin{aligned}
\|K(r,t,h) - K(r,t,\varphi)\| &\le |\vartheta(r)|\|\partial_r\big[h(r,t) - \varphi(r,t)\big]\| + |\eta|\|\partial_r\big[h(r,t) - \varphi(r,t)\big]\| \\
&\quad + |2\lambda^2\eta|\Big\|\int_0^r (r-\tau)^2 \exp\big[-\lambda^2(r-\tau)^2\big]\big[h(\tau,t) - \varphi(\tau,t)\big]d\tau \\
&\quad + |\eta|\Big\|\int_0^r \exp\big[-\lambda^2(r-\tau)^2\big]\big[h(\tau,t) - \varphi(\tau,t)\big]d\tau
\end{aligned}
$$

$$\|K(r,t,h) - K(r,t,\varphi) \leq |\vartheta(r) + \eta|\|h(r,t) - \varphi(r,t)\|$$

$$+ |\eta| \| \int_0^r \exp\left[-\lambda^2(r-\tau)^2\right]\left[h(\tau,t) - \varphi(\tau,t)\right]d\tau$$

$$+ |2\lambda^2\eta| \| \int_0^r (r-\tau)^2 \exp\left[-\lambda^2(r-\tau)^2\right]\|h(\tau,t) - \varphi(\tau,t)\|d\tau$$

Now, by applying Cauchy–Schwartz inequality, we have

$$\|K(r,t,h) - K(r,t,\varphi)\| \leq \varepsilon_1 \|h(r,t) - \varphi(r,t)\|$$

$$+ |2\lambda^2\eta|\left(\int_0^r \|(r-\tau)^2 \exp\left[-\lambda^2(r-\tau)^2\right]^2 d\tau\right)^{\frac{1}{2}}\left(\int_0^r \|h(\tau,t) - \varphi(\tau,t)\|d\tau\right)^{\frac{1}{2}}$$

$$+ |\eta|\left(\int_0^r \|\exp\left[-\lambda^2(r-\tau)^2\right]^2 d\tau\right)^{\frac{1}{2}}\|\left(\int_0^r \|h(\tau,t) - \varphi(\tau,t)^2 d\tau\right)^{\frac{1}{2}}$$

Thus we obtain

$$\|K(r,t,h) - K(r,t,\varphi)\| < \Lambda\|h(r,t) - \varphi(r,t)\|,$$

where $\Lambda < 1$ and it can be estimated as

$$\Lambda = \varepsilon_1 + \varepsilon_2 + \varepsilon_3 = \frac{S}{T} + \eta + k\gamma\left[(1-\gamma)\sqrt{\pi}\right]^{-1} + \frac{3\sqrt{2}}{16\lambda^5}\eta.$$

We conclude that the operator K is a contraction. The statement follows now from Banach's Fixed Point Theorem. □

In the following we show that the solution of our problem (7.23), given by Eq. (7.25), can be written as an iteration for a given subsequence $h_m(r,t) \in (0, T) \times H_0^1(\Omega)$.

Theorem 7.2.2. *Assume that a bounded sequence $h_m(r,t)$ on $(0, T) \times H_0^1(\Omega)$ converges to the exact solution of the problem (7.23), then any bounded subsequences on $(0, T) \times H_0^1(\Omega)$ converge to the exact solution and are a Cauchy sequence with respect to the norm in $H_0^1(\Omega)$.*

Proof. The kernel K being bounded on $(0, T) \times H_0^1(\Omega)$, there exists a subsequence $K(r,t,h_m)$ on $(0, T) \times H_0^1(\Omega)$ that converges to $(0, T) \times L^2(\Omega)$ by the Rellich–Kondrachov theorem; furthermore, the difference between two consecutive subsequences $K(r,t,h_m)$ and $K(r,t,\varphi_m)$ also converges to $(0, T) \times L^2(\Omega)$. Thus the subsequence $K(r,t,h_m)$ is a Cauchy sequence with respect to the norm in $H_0^1(\Omega)$. One can reformulate the previous statement as following: Let $h_m(r,t)$ and $h_{m-1}(r,t)$ be two successive subsequences on $(0, T) \times H_0^1(\Omega)$. From Eq. (7.25) it follows that

$$h_m(r,t) - h_{m-1}(r,t) = \frac{2(1-\alpha)}{(2-\alpha)B(\alpha)}\left[K(r,t,h_{m-1}) - K(r,t,h_{m-2})\right] \tag{7.26}$$

$$+ \frac{2\alpha}{(2-\alpha)B(\alpha)}\int_0^t \left[K(r,s,h_{m-1}) - K(r,s,h_{m-2})\right]ds. \tag{7.27}$$

Next we want to control the difference between the two subsequences $h_m(r,t)$ and $h_{m-1}(r,t)$. A direct application of the triangular inequality yields

$$\|h_m(r,t) - h_{m-1}(r,t)\| \le |\frac{2(1-\alpha)}{(2-\alpha)B(\alpha)}| \|K(r,t,h_{m-1}) - K(r,t,h_{m-2})\|$$

$$+ |\frac{2\alpha}{(2-\alpha)B(\alpha)}| \| \int_0^t \left[K(r,s,h_{m-1}) - K(r,s,h_{m-2})\right]ds. \|$$

$$\le \frac{2(1-\alpha)}{(2-\alpha)B(\alpha)} \|K(r,t,h_{m-1}) - K(r,t,h_{m-2})\|$$

$$+ \frac{2\alpha}{(2-\alpha)B(\alpha)}\int_0^t \|K(r,s,h_{m-1}) - K(r,s,h_{m-2})\|ds.$$

Now since the nonlinear kernel given by the operator K is a contraction, it follows that

$$\frac{2(1-\alpha)}{(2-\alpha)B(\alpha)}\|K(r,t,h_{m-1}) - K(r,t,h_{m-2})\|$$

$$+ \frac{2\alpha}{(2-\alpha)B(\alpha)}\int_0^t$$

$$|K(r,s,h_{m-1}) - K(r,s,h_{m-2})ds\|$$

$$\le \frac{2(1-\alpha)}{(2-\alpha)B(\alpha)}\|K(r,t,h_{m-1}) - K(r,t,h_{m-2})\|$$

$$+ \frac{2\alpha}{(2-\alpha)B(\alpha)}\int_0^t \|K(r,s,h_{m-1}) - K(r,s,h_{m-2})\|ds.$$

Hence there exists a solution h to the problem state by Eq. (7.23).   □

7.2.2 Uniqueness of the Exact Solution of the Problem

In this section we propose to study the uniqueness of the exact solution given by the problem (7.23). To do this, we assume that there exists another solution of the problem (7.23), namely $\varphi(r,t)$. Before stating the theorem of uniqueness of the problem (7.23), we recall the following Lemma 7.2.3.

Lemma 7.2.3. *Let $\alpha \in (0,1)$ and h be a solution of the following fractional differential equation:*

$$\partial_t^\alpha h(r,t) = 0, \qquad t > 0. \tag{7.28}$$

Then, h is a constant function.

Next we can now state the theorem of uniqueness of solution of the problem (7.23).

Theorem 7.2.4. *Let $\alpha \in (0, 1)$, $\gamma \in (1, 1)$. Then, the solution of the problem of groundwater flow given by Eq. (7.23) is unique.*

Proof. The approach of our proof comes from [207]. Suppose that the problem (7.23) has two solutions $h(r, t)$ and $\varphi(r, t)$ such that they can be written in the form of Eq. (7.25). This means that

$$\partial_t^\alpha h(r, t) - \partial_t^\alpha \varphi(r, t) = \partial_t^\alpha \left[h(r, t) - \varphi(r, t) \right] = 0, \quad \text{and} \quad \left[h(r, 0) - \varphi(r, 0) \right] = 0. \quad (7.29)$$

Thus it comes out for all $t \in (0, T)$, from Lemma 7.2.3, and Eq. (7.25) that $h(r, t) = \varphi(r, t)$.

One can then conclude that the problem of groundwater flow described by Eq. (7.23) has a unique solution given by Eq. (7.25). $\qquad \square$

Derivation of an analytic solution

In this section, a substitute analytical method is used to obtain the exact solution of the novel model of groundwater flowing within a confined aquifer with Caputo–Fabrizio derivative with fractional order. In this section, the method of separation of variable is used to provide exact solution to the new model. For this, it is assumed that the solution is

$$h(r, t) = h_1(r) h_2(t) \quad (7.30)$$

Replacing Eq. (7.30) in Eq. (7.23) we obtain the following:

$$_0^{CF} D_t^\alpha (h_1(r, t)) \frac{S}{T} = \lambda^2 h_1(t) \quad (7.31)$$

$$D_{rr}(h_2(r)) + \frac{1}{r} D_r(h_2(r)) + \lambda^2 h_2(r) = 0 \quad (7.32)$$

where λ is the separation constant. The first equation of (7.5) can be solved directly by applying to both sides the Laplace transform to obtain:

$$\frac{M(\alpha)}{1 - \alpha} \frac{p F(p) - h_2(0)}{p + \alpha(1 - p)} = \lambda^2 \quad (7.33)$$

$$\left(\frac{ap}{p + \alpha(1 - p)} - \lambda^2 \right) F(p) = a \frac{h_2(0)}{p + \alpha(1 - p)} \quad (7.34)$$

$$F(p) = \frac{a h_2(0)}{(a - \lambda^2 + \alpha\lambda^2) \left(p - \frac{\alpha\lambda^2}{a - \lambda + alpha\lambda^2} \right)}. \quad (7.35)$$

Using the inverse formula of Laplace transform in Eq. (7.33),

$$h_2(t) = \frac{ah_2(0)}{a - \lambda^2 + \alpha\lambda^2} \exp\left(\frac{\alpha\lambda^2 t}{a - \lambda^2 + \alpha\lambda^2}\right) \tag{7.36}$$

$$a = \frac{M(\alpha)}{1 - \alpha}. \tag{7.37}$$

To satisfy the initial condition, $h_2(0) = h_2(0)$, the following condition must be satisfied

$$\lambda^2 = \frac{M(\alpha)}{(1 - \alpha)^2}. \tag{7.38}$$

To solve the second equation, we make use of the Laplace–Carson presented earlier, to obtain

$$Lh_1(r)(s) = \frac{1}{\sqrt{s^\alpha + \lambda^2}}. \tag{7.39}$$

Now applying the inverse Laplace transform operator to both sides in the previous equation, we obtain the following in terms of the Bessel function of the first kind, $J_0(.)$:

$$h_1(r) = \sum_{k=0}^{\infty} \frac{(-1)^k}{k!} \frac{1}{\Gamma(k + 1)} \left(\frac{r\lambda}{2}\right)^{2k}. \tag{7.40}$$

Therefore, the solution of the fractional groundwater flow equation is given as:

$$h(r, t) = c \sum_{k=0}^{\infty} E_\alpha\left[\frac{-S\lambda_n^2 t^\alpha}{T}\right] J_0(\lambda_n r). \tag{7.41}$$

Making use of the initial and boundary conditions, we obtain the constant c to be

$$c = \frac{Q}{4\pi T}. \tag{7.42}$$

Finally, the exact solution of the Theis model with fractional derivative, since λ is known, is given as:

$$h(r, t) = \frac{Q}{4\pi T} J_0(r) \frac{ah_2(0)}{a - \lambda^2 + \alpha\lambda^2} \exp\left(\frac{\alpha\lambda^2 t}{a - \lambda^2 + \alpha\lambda^2}\right). \tag{7.43}$$

7.3 Numerical Analysis of the New Groundwater Model

In this section, the numerical approximation of the problem (7.23) is analyzed. The stability of the problem using the Fourier method is also presented. The problem will be solved using the Crank–Nicholson scheme.

7.3.1 Discretization of the Problem Using Crank–Nicholson Scheme

We consider some positive integers M and N. The grid points in time and space are defined respectively by $t_k = k\tau$, $k = 0, 1, 2, \ldots, N$, and $r_j = \xi j$, $j = 0, 1, 2, \ldots, M$. We also denote by $h_j^k = h(r_j, t_k)$ the values of the function h at the grid points. The first and second order approximations of the local derivative in the sense of Crank–Nicholson is given as

$$\partial_r h(r_j, t_k) = \frac{1}{2}\left[\frac{\left(h_{j+1}^{k+1} - h_{j-1}^{k+1}\right) + \left(h_{j+1}^k - h_{j-1}^k\right)}{2\xi} \right]$$

and

$$\partial_{rr} h(r_l, t_k) = \frac{1}{2}\left\{ \frac{\left(h_{l+1}^{k+1} - 2h_l^{k+1} + h_{l-1}^{k+1}\right) + \left(h_{l+1}^k - 2h_l^k + h_{l-1}^k\right)}{2\xi^2} \right\}$$

For discrete version of the Caputo–Fabrizio derivative and Laplacian, we recall that this has been already done by Atangana and Nieto in [221]. Thus the following theorems.

Theorem 7.3.1 (Atangana and Nieto [221]). *Let $h(t)$ be a function in $H_0^1(\Omega) \times (0, T)$, then the first order fractional derivative of the Caputo–Fabrizio derivative of order $\alpha \in (0, 1)$ at a point t_k is*

$$\partial_t^\alpha \left[h(t_k) \right] = \frac{B(\alpha)}{\alpha} \sum_{s=1}^k \left(\frac{h_j^{s+1} - h_j^s}{\tau} + O(\tau) \right) d_{s,\tau}, \tag{7.44}$$

where

$$E_{k,s,\tau} = \exp\left[-\lambda\tau(k - s + 1) \right] - \exp\left[-\lambda\tau(k - s) \right].$$

Proof. Let us put the following fractional integral to be

$$F_1 = S\frac{M(\alpha)}{1 - \alpha} \int_0^t \frac{\partial h(r, x)}{\partial x} \exp[\alpha \frac{(t - x)}{1 - \alpha}]. \tag{7.45}$$

Thus, using the numerical approximation for the local derivative within the integral, we obtain the following

$$F_1 = \frac{M(\alpha)}{1 - \alpha} \sum_{j=1}^m \int_{(j-1)\tau}^{j\tau} \left(\frac{h_i^{j+1} - h_i^j}{\tau} + O(\tau) \right) \exp\left[-\frac{m\tau - y}{1 - \alpha} \right] dy \tag{7.46}$$

$$= \frac{M(\alpha)}{1 - \alpha} \sum_{j=1}^m \left(\frac{h_i^{j+1} - h_i^j}{\tau} + O(\tau) \right) \int_{(j-1)\tau}^{j\tau} \exp\left[-\frac{m\tau - y}{1 - \alpha} \right] dy. \tag{7.47}$$

$$= \frac{M(\alpha)}{1-\alpha} \sum_{j=1}^{m} \left(\frac{h_i^{j+1} - h_i^j}{\tau} + O(\tau) \right) \left(\exp\left[-\lambda\tau(k-s+1) \right] - \exp\left[-\lambda\tau(k-s) \right] \right).$$

(7.48)

This is the completion of the proof. □

From the above proof, one will therefore provide the discretization of the following component that represents the unconfined part of the groundwater flow equation:

$$F_2 = \Omega S_y \int_0^t \frac{\partial h(r,y)}{\partial y} \exp\left[-\Omega(t-y) \right]$$

(7.49)

$$= \Omega \sum_{j=1}^{m} \left(\frac{h_i^{j+1} - h_i^j}{\tau} + O(\tau) \right) \left(\exp\left[-\Omega\tau(k-s+1) \right] - \exp\left[-\Omega\tau(k-s) \right] \right).$$

(7.50)

Thus, by replacing the above numerical approximation F_2 and also the second approximation of local derivative in the generalized groundwater flow model in an unconfined aquifer, we obtain

$$T \left(\frac{1}{2} \left(\frac{h_{j+1}^{k+1} - 2h_j^{k+1} + h_{j-1}^{k+1}}{h^2} + \frac{h_{j+1}^{k+1} - 2h_j^{k+1} + h_{j-1}^{k+1}}{h^2} \right) \right) +$$

(7.51)

$$\frac{1}{2r_i} \left(\frac{h_{j+1}^{k+1} - h_{j-1}^{k+1}}{2h} + \frac{h_{j+1}^k - h_{j-1}^k}{2h} \right)$$

(7.52)

$$= \frac{SM(\alpha)}{\alpha} \sum_{l=1}^{k} \left(\frac{h_j^{l+1} - h_j^k}{\tau} \right) d_{k,\tau} + S_y \sum_{l=1}^{k} \left(\frac{h_j^{l+1} - h_j^l}{\tau} \right) \rho_{l,\tau}$$

(7.53)

7.4 Application to Groundwater Flowing Within an Unconfined Aquifer

In general, an unconfined aquifer is strongly coupled with the vadose zone and the two cannot be separated. The topmost parts of the subsurface typically belong to the soil water zone where the pore space is filled with temporally and spatially varying fractions of water and air. This is often referred to as vadose or unsaturated zone. Its thickness varies greatly, from practically zero in swamps to hundreds of meters in arid regions. This zone is typically followed by the topmost aquifer, the so-called phreatic aquifer. It is unconfined, which means that the water table is free to move. Because of these properties of this specific medium, the model based on the local derivative cannot accurately replicate the physical observation, therefore the model can be extended to the scope of fractional order derivatives.

Groundwater Flowing Within an Unconfined Aquifer With Caputo Derivative

In this section, we examine the model of groundwater flowing within an unconfined aquifer with the Caputo fractional derivative. To achieve this, the time derivative is replaced by the time fractional Caputo derivative to obtain:

$$T\left(\frac{\partial^2 s}{\partial r^2} + \frac{\partial s}{r \partial r}\right) = S\left({}_0^C D_t^\alpha(s)\right) + DS_y \int_0^t \frac{\partial s}{\partial z} \exp\left(-D(t-z)\right) dz. \qquad (7.54)$$

The Laplace transform operator is used in time component and the Fourier transform is used in space component to obtain first:

$$T\left(\frac{\partial^2 s(r,p)}{\partial r^2} + \frac{\partial s(r,p)}{r \partial r}\right) = s(r,p)p^\alpha - p^{\alpha-1}s(r,0) + DS_y \frac{p^\alpha s(r,p) - p^{\alpha-1}s(r,0)}{p^\alpha + D}. \qquad (7.55)$$

The above equation can be reduced to

$$\frac{\partial^2 s(r,p)}{\partial r^2} + \frac{\partial s(r,p)}{r \partial r} + s(r,p)f(p) = g(r,p) \qquad (7.56)$$

$$Tg(r,p) = -\left(p^{\alpha-1} + DS_y \frac{p^{\alpha-1}}{p^\alpha + D}\right) \qquad (7.57)$$

$$Tf(p) = -\left(p^\alpha + DS_y \frac{p^{\alpha-1}}{p^\alpha + D}\right) \qquad (7.58)$$

The above equation (7.56) can be solved using the Green function technique. However, the associated Green function of the above equation is obtained by solving the homogeneous part with the Laplace transform. The Green function associated in Laplace is given as:

$$J_0(\sqrt{-f(p)}r). \qquad (7.59)$$

Thus, using the Green function technique, we construct the solution of (7.56) in Laplace space as follows:

$$s(r,p) = \int J_0(\sqrt{-f(p)}r - l)g(l,p)dl. \qquad (7.60)$$

Therefore, taking the inverse Laplace transform of (7.60) we obtain

$$s(r,t) = L(s(r,p)) = L\left[\int J_0(\sqrt{-f(p)}(r-l))g(l,p)dl\right]. \qquad (7.61)$$

The above is the exact solution of the time-fractional groundwater flow equation within an unconfined aquifer.

Groundwater Flowing Within an Unconfined Aquifer With Caputo–Fabrizio

In this section, the model of groundwater flowing within an unconfined aquifer is developed using the time fractional Caputo–Fabrizio derivative. The advance of this model with this derivative is that, the model will be able to portray the flow in different scales and also there is no singularity in this case.

$$T\left(\frac{\partial^2 s}{\partial r^2} + \frac{\partial s}{r \partial r}\right) = S\left({}^{CF}_0 D_t^\alpha(s)\right) + DS_y \int_0^t \frac{\partial s}{\partial z} \exp\left(-D(t-z)\right) dz. \tag{7.62}$$

7.5 Existence and Uniqueness of the Solution for the New Model of Groundwater Flow Within an Unconfined Aquifer

7.5.1 Existence

In this section we use the fixed-point theorem to prove the existence of the solution for the new model of groundwater flow equation.

We can rewrite Eq. (2.83) as

$$ {}^{CF}_0 D_t^\alpha(h(r,t)) = \frac{T}{S}\left[\frac{\partial^2 h(r,t)}{\partial r^2} + \frac{1}{r}\frac{\partial h(r,t)}{\partial r}\right] - \frac{\chi S_y}{S}\int_0^t \frac{\partial h(r,k)}{\partial k}\exp\left[-\chi(t-k)\right]dk. \tag{7.63}$$

We recall that the Caputo–Fabrizio integral is given by

$$ {}^{CF}_0 I_t^\alpha(f(t)) = \frac{2(1-\alpha)}{(2-\alpha)M(\alpha)}f(t) + \frac{2\alpha}{(2-\alpha)M(\alpha)}\int_0^t f(k)dk, \qquad t \geq 0. \tag{7.64}$$

We first transform Eq. (7.63) to an integral form by multiplying both sides by Caputo–Fabrizio integral and we get

$$\begin{cases} h(r,t) - h(r,0) = {}^{CF}_0 I_t^\alpha\left\{\frac{T}{S}\left[\frac{\partial^2 h(r,t)}{\partial r^2} + \frac{1}{r}\frac{\partial h(r,t)}{\partial r}\right] - \frac{\chi S_y}{S}\int_0^t \frac{\partial h(r,k)}{\partial k}\exp\left[-\chi(t-k)\right]dk\right\}, \\ h(r,0) = h_0. \qquad r \in [a,b] \end{cases}$$

$$\tag{7.65}$$

Let define the kernel operator $K : H^1([a,b] \times I) \longrightarrow L^2([a,b] \times I)$ taking the value

$$K(r,t,h) = \frac{T}{S}\left[\frac{\partial^2 h(r,t)}{\partial r^2} + \frac{1}{r}\frac{\partial h(r,t)}{\partial r}\right] - \frac{\chi S_y}{S}\int_0^t \frac{\partial h(r,k)}{\partial k}\exp\left[-\chi(t-k)\right]dk$$

Using the Caputo–Fabrizio integral, we have

$$
\begin{cases}
h(r,t) - h(r,0) = \frac{2(1-\alpha)}{(2-\alpha)M(\alpha)} K(r,t,h) + \frac{2\alpha}{(2-\alpha)M(\alpha)} \int_0^t K(r,t,h)\,ds, \\
h(r,0) = h_0 \quad r \in [a,b].
\end{cases}
$$

Theorem 7.5.1. *The kernel operator K satisfies the Lipschitz condition and also the contraction if*

$$
0 < \lambda_1 \left| \frac{T}{S} \right| + \lambda_2^2 \|\frac{1}{r}\| + \lambda_3 \sqrt{2\chi \exp(-2\chi T')} \left| \frac{\chi S_y}{S} \right| < 1, \quad r > 1; \lambda_1, \lambda_2, \lambda_3, T' > 0.
$$

Proof. Let h_1 and h_2 be two functions in H^1, then we have

$$
\begin{aligned}
\|K(r,t,h_2) - K(r,t,h_1)\| &= \left\| \frac{T}{S}\left[\frac{\partial^2 h_2(r,t)}{\partial r^2} + \frac{1}{r}\frac{\partial h_2(r,t)}{\partial r} \right] \right. \\
&\quad - \frac{\chi S_y}{S} \int_0^t \frac{\partial h_2(r,k)}{\partial k} \exp\left[-\chi(t-k) \right] dk \\
&\quad - \frac{T}{S}\left[\frac{\partial^2 h_1(r,t)}{\partial r^2} + \frac{1}{r}\frac{\partial h_1(r,t)}{\partial r} \right] \\
&\quad - \left. \frac{\chi S_y}{S} \int_0^t \frac{\partial h_1(r,k)}{\partial k} \exp\left[-\chi(t-k) \right] dk \right\| \\
&= \left\| \frac{T}{S}\left[\frac{\partial^2 \{h_2(r,t) - h_1(r,t)\}}{\partial r^2} + \frac{1}{r}\frac{\partial \{h_2(r,t) - h_1(r,t)\}}{\partial r} \right] \right. \\
&\quad - \left. \frac{\chi S_y}{S} \int_0^t \frac{\partial \{h_2(r,k) - h_1(r,k)\}}{\partial k} \exp\left[-\chi(t-k) \right] dk \right\|.
\end{aligned}
$$

Using the triangle inequality, we have

$$
\begin{aligned}
\|K(r,t,h_2) - K(r,t,h_1)\| &\leq \left| \frac{T}{S} \right| \left\| \frac{\partial^2 \{h_2(r,t) - h_1(r,t)\}}{\partial r^2} \right\| + \left\| \frac{1}{r}\frac{\partial \{h_2(r,t) - h_1(r,t)\}}{\partial r} \right\| \\
&\quad - \left| \frac{\chi S_y}{S} \right| \left\| \int_0^t \frac{\partial \{h_2(r,k) - h_1(r,k)\}}{\partial k} \exp\left[-\chi(t-k) \right] dk \right\|
\end{aligned}
$$

Now, the derivative operator satisfies the Lipschitz condition in H^1, hence there exist two positive constants λ_1 and λ_2 such that

$$\|K(r,t,h_2) - K(r,t,h_1)\| \leq A\|h_2(r,t) - h_1(r,t)\| + B\|h_2(r,t) - h_1(r,t)\|$$
$$- \left|\frac{\chi S_y}{S}\right| \left|\int_0^t \left\|\frac{\partial\{h_2(r,k) - h_1(r,k)\}}{\partial k} \exp[-\chi(t-k)]\right\| dk\right|,$$

where $A = \lambda_1\left|\frac{T}{S}\right|$ and $B = \lambda_2^2\|\frac{1}{r}\|$, $r > 1$. Now, by applying the Cauchy–Schwartz inequality to the third term, we have

$$\|K(k,t,h_2) - K(k,t,h_1)\| \leq A\|h_2(r,t) - h_1(r,t)\| + B\|h_2(r,t) - h_1(r,t)\|$$
$$- \lambda_3\left|\frac{\chi S_y}{S}\right| \|h_2(r,t) - h_1(r,t)\| \left(\int_0^{T'} |\exp[-\chi(t-k)]|^2 dk\right)^{\frac{1}{2}}.$$

Integrating the second part of the third term yields

$$\|K(r,t,h_2) - K(r,t,h_1)\| \leq A\|h_2(r,t) - h_1(r,t)\| + B\|h_2(r,t) - h_1(r,t)\|$$
$$- \lambda_3\sqrt{2\chi}\exp(-2\chi T')\left|\frac{\chi S_y}{S}\right| \|h_2(r,t) - h_1(r,t)\|$$
$$\leq A\|h_2(r,t) - h_1(r,t)\| + B\|h_2(r,t) - h_1(r,t)\|$$
$$- C\|h_2(r,t) - h_1(r,t)\|,$$

where $C = \lambda_3\sqrt{2\chi}\exp(-2\chi T')\left|\frac{\chi S_y}{S}\right|$. Now taking $L = A + B + C$, i.e.

$$L = \lambda_1\left|\frac{T}{S}\right| + \lambda_2^2\|\frac{1}{r}\| + \lambda_3\sqrt{2\chi}\exp(-2\chi T')\left|\frac{\chi S_y}{S}\right|,$$

we have

$$\|K(r,t,h_2) - K(r,t,h_1)\| \leq L\|h_2(r,t) - h_1(r,t)\|, \tag{7.66}$$

showing that the kernel operator K satisfies the Lipschitz condition. If also $0 < L < 1$ then K is a contraction. Now let us show that the solution is bounded. Let's consider the recursive sequence as follows

$$h_n(r,t) = \frac{2(1-\alpha)}{(2-\alpha)M(\alpha)}K(r,t,h_{n-1}) + \frac{2\alpha}{(2-\alpha)M(\alpha)}\int_0^t K(r,t,h_{n-1})ds.$$

The idea is to show that for a given sequence h_n that satisfies our problem, the sequence converges.

Whenever the difference between the two consecutive terms of the sequence is given by

$$h_n(r, t) - h_{n-1}(r, t) = \frac{2(1-\alpha)}{(2-\alpha)M(\alpha)} \left[K(r, t, h_{n-1}) - K(k, t, h_{n-2}) \right]$$
$$+ \frac{2\alpha}{(2-\alpha)M(\alpha)} \int_0^t [K(r, t, h_{n-1}) - K(r, t, h_{n-2})]ds,$$

if we take the norm on both sides of the above equation, we have

$$\|h_n(r, t) - h_{n-1}(r, t)\| = \left\| \frac{2(1-\alpha)}{(2-\alpha)M(\alpha)} \left[K(r, t, h_{n-1}) - K(k, t, h_{n-2}) \right] \right.$$
$$\left. + \frac{2\alpha}{(2-\alpha)M(\alpha)} \int_0^t [K(k, t, h_{n-1}) - K(k, t, h_{n-2})]ds \right\|.$$

By applying the triangle inequality we get

$$\|h_n(r, t) - h_{n-1}\| \le \frac{2(1-\alpha)}{(2-\alpha)M(\alpha)} \|K(r, t, h_{n-1}) - K(r, t, h_{n-2})\|$$
$$+ \frac{2\alpha}{(2-\alpha)M(\alpha)} \left\| \int_0^t [K(r, t, h_{n-1}) - K(r, t, h_{n-2})]ds \right\|.$$

Now, let put

$$H_n(r, t) = h_n(r, t) - h_{n-1}(r, t).$$

Then we have

$$h_n(r, t) = \sum_{j=0}^n H_j(r, t)$$

with $H_0(r, t) = h_0$. Using the fact that the kernel K satisfies the Lipschitz condition, we get

$$\|h_n(r, t) - h_{n-1}(r, t)\| \le$$
$$\frac{2(1-\alpha)L}{(2-\alpha)M(\alpha)} \|h_{n-1}(r, t)) - h_{n-2}(r, t)\| + \frac{2\alpha L}{(2-\alpha)M(\alpha)} \int_0^t \|h_{n-1} - h_{n-2}(r, s)\|ds.$$

We can then put our expression as follows

$$\|H_n(r, t)\| \le \Lambda \|H_{n-1}\| + \Theta \int_0^t \|H_{n-1}\|ds,$$

where $\Lambda = \dfrac{2(1-\alpha)L}{(2-\alpha)M(\alpha)}$ and $\Theta = \dfrac{2\alpha L}{(2-\alpha)M(\alpha)}$.

By induction with respect to $n \geq 1$ we get the following:

$$\|H_1\| \leq \Lambda\|H_0\| + \Theta \int_0^t \|H_0\|ds\| \leq \Lambda\|H_0\| + \Theta T'\|H_0\| = (\Lambda + \Theta T')\|H_0\|$$

$$\|H_2\| \leq \Lambda\|H_1\| + \Theta \int_0^t \|H_1\|ds\| \leq \Lambda(\Lambda + \Theta T')\|H_0\|) + \Theta \int_0^t (\Lambda + \Theta T')\|H_0\|ds$$

$$= (\Lambda + \Theta T') \times (\Lambda + \Theta T')\|H_0\| = (\Lambda + \Theta T')^2\|H_0\|$$

$$\|H_2\| \leq (\Lambda + \Theta T')^2\|H_0\|$$

$$\vdots \qquad \vdots$$

$$\|H_n(r,t)\| \leq (\Lambda + \Theta T')^n\|H_0\|.$$

Finally, we have

$$\|H_n(r,t)\| \leq \|h(r,0)\| \left(\frac{2(1-\alpha)L}{(2-\alpha)M(\alpha)} + \frac{2\alpha LT'}{(2-\alpha)M(\alpha)} \right)^n. \tag{7.67}$$

We use the following theorem

Theorem 7.5.2 (Rellich). *[10] Let $\Omega \subset \mathbb{R}^d$ be a bounded domain with Lipschitz boundary. Then the embedding $H^1(\Omega) \hookrightarrow L^2(\Omega)$ is compact. That is, there exists for every bounded sequence in $H^1(\Omega)$ a subsequence that converges with respect to the norm in $L^2(\Omega)$.*

Using the fact that the initial condition of our problem is $h_0 = 0$, the inequality (7.67) shows that the sequence h_n is bounded in $H_0^1([a,b] \times I)$. Applying Theorem 7.5.2, there exists a subsequence, again relabeled h_n, that converges in $L^2([a,b] \times I)$. Therefore, our boundary value problem has a solution that is continuous. This completes the proof of the existence.

□

7.5.2 Uniqueness of the Solution

We need to show, however, that the solution is unique. Let h and h_1 be two candidate solutions of our initial boundary problem. Then, using Eq. (7.66), we will have

$$h(r,t) - h'(r,t) = \frac{2(1-\alpha)}{(2-\alpha)M(\alpha)}[K(r,t,h) - K(k,t,h')]$$

$$+ \frac{2\alpha}{(2-\alpha)M(\alpha)} \int_0^t [K(k,t,h) - K(k,t,h')]ds. \tag{7.68}$$

Applying the norm to both sides, we have

$$\|h(r,t) - h'(r,t)\|$$

$$= \left\| \frac{2(1-\alpha)}{(2-\alpha)M(\alpha)}[K(r,t,h) - K(r,t,h')] + \frac{2\alpha}{(2-\alpha)M(\alpha)} \int_0^t [K(r,t,h) - K(r,t,h')]ds \right\|$$

$$\leq \frac{2(1-\alpha)}{(2-\alpha)M(\alpha)} \|K(r,t,h) - K(r,t,h')\| + \frac{2\alpha}{(2-\alpha)M(\alpha)} \int_0^t \|K(r,t,h) - K(r,t,h')\| \, ds$$

using the Lipschitz condition of the operator kernel K

$$\leq \frac{2(1-\alpha)L}{(2-\alpha)M(\alpha)} \|h - h'\| + \frac{2\alpha L}{(2-\alpha)M(\alpha)} \int_0^t \|h - h'\| \, ds$$

We then have

$$\|h(r,t) - h'(r,t)\| \leq \left(\frac{2(1-\alpha)L}{(2-\alpha)M(\alpha)} + \frac{2\alpha L}{(2-\alpha)M(\alpha)} \right) \|h(r,t) - h'(r,t)\|. \tag{7.69}$$

This implies that

$$\left(1 - \frac{2(1-\alpha)L}{(2-\alpha)M(\alpha)} + \frac{2\alpha L}{(2-\alpha)M(\alpha)} \right) \|h(r,t) - h'(r,t)\| \leq 0. \tag{7.70}$$

Assuming that

$$\left(1 - \frac{2(1-\alpha)L}{(2-\alpha)M(\alpha)} + \frac{2\alpha L}{(2-\alpha)M(\alpha)} \right) \neq 0,$$

we have

$$\|h(r,t) - h'(r,t)\| = 0.$$

This implies that

$$h(r,t) = h'(r,t).$$

We conclude that the solution is unique.

Here we present the numerical solution of Eq. (7.62). To do this, the presentation of the first approximation of local derivative and Caputo–Fabrizio derivatives is shown together with the second approximation.

$$\frac{\partial S}{\partial x} = \frac{1}{2} \left[\frac{\left(s_{i+1}^{n+1} - s_{i-1}^{n+1} \right) - \left(s_{i+1}^{n} - s_{i-1}^{n} \right)}{2(\Delta x)} \right] \tag{7.71}$$

$$\int_0^t \frac{\partial s}{\partial z} \exp \left(-\frac{\alpha}{1-\alpha}(t-z) \right) dz = \frac{1}{2} \sum_{k=0}^n \left(s_i^{k+1} - s_i^k \right) d_{i,n,k}^1 \tag{7.72}$$

$$d_{i,n,k}^1 = \exp \left(-\frac{k\alpha}{1-\alpha}(n-j+1) \right) - \exp \left(-\frac{k\alpha}{1-\alpha}(n-j) \right). \tag{7.73}$$

Also,

$$\int_0^t \frac{\partial s}{\partial z} \exp\left(-\frac{\alpha}{1-\alpha}(t-z)\right) dz = \frac{1}{2}\sum_{k=0}^n \left(s_i^{k+1} - s_i^k\right) d_{i,n,k}^2 \tag{7.74}$$

$$d_{i,n,k}^2 = \exp\left(-kD(n-j+1)\right) - \exp\left(-kD(n-j)\right). \tag{7.75}$$

Then with the second approximation for local derivative, we have the following

$$\frac{\partial^2 s}{\partial x^2} = \frac{(s_{i+1}^{n+1} - 2s_i^{n+1} + s_{i-1}^{n+1}) + (s_{i+1}^n - 2s_i^n + s_{i-1}^n)}{2(\Delta x)^2}. \tag{7.76}$$

Thus, putting together Eqs. (7.76), (7.74), (7.72) and (7.71) into Eq. (7.62) yields:

$$T\left(\frac{(s_{i+1}^{n+1} - 2s_i^{n+1} + s_{i-1}^{n+1}) + (s_{i+1}^n - 2s_i^n + s_{i-1}^n)}{2(\Delta x)^2}\right. \tag{7.77}$$
$$\left. + \frac{1}{r_i}\left(\frac{1}{2}\left[\frac{\left(s_{i+1}^{n+1} - s_{i-1}^{n+1}\right) - \left(s_{i+1}^n - s_{i-1}^n\right)}{2(\Delta x)}\right]\right)\right)$$
$$= S\left(\frac{1}{2}\sum_{k=0}^n \left(s_i^{k+1} - s_i^k\right) d_{i,n,k}^1\right) +$$
$$DS_y\left(\frac{1}{2}\sum_{k=0}^n \left(s_i^{k+1} - s_i^k\right) d_{i,n,k}^2\right)$$

Rearranging, we obtain the following:

$$\left(-\frac{T}{(\Delta x)}^2 - \frac{Sd_{i,n,n}^1}{2} - \frac{DS_y d_{i,n,n}^2}{2}\right) s_i^{n+1} \tag{7.78}$$
$$= \left(\frac{T}{(\Delta x)}^2 - \frac{Sd_{i,n,n}^1}{2} - \frac{DS_y d_{i,n,n}^2}{2}\right) s_i^n -$$
$$-\frac{1}{r_i}\left(\frac{1}{2}\left[\frac{\left(s_{i+1}^{n+1} - s_{i-1}^{n+1}\right) - \left(s_{i+1}^n - s_{i-1}^n\right)}{2(\Delta x)}\right]\right) + S\left(\frac{1}{2}\sum_{k=0}^n \left(\beta_i^k\right) d_{i,n,k}^1\right) +$$
$$DS_y\left(\frac{1}{2}\sum_{k=0}^{n-1} \left(\beta_i^k\right) d_{i,n,k}^2\right)$$
$$\beta_i^k = s_i^{k+1} - s_i^k. \tag{7.79}$$

The above recursive formula can be used to generate the numerical solution of groundwater equation for unconfined aquifer with Caputo–Fabrizio fractional derivative.

Groundwater flowing within an unconfined aquifer with Atangana–Baleanu fractional derivative

In this section, the flow of groundwater within an unconfined aquifer is addressed using the newly established fractional differentiation based on a non-local and non-singular kernel. This modification is motivated with the fact that not all natural occurrences can be modeled using the concept of power law that is the basic assumption while using Caputo and Riemann–Liouville derivative with fractional order. Therefore in this section, the kernel based on power $x^{-\alpha}$ will be replaced by the generalized Mittag–Leffler function. It is important to notice that the data obtained within the vicinity of the borehole from which the water is being taking out in order to estimate the aquifer parameters are more reliable than those taken far from the borehole. However, the model based on the Caputo and Riemann–Liouville has a singular kernel at the origin, therefore it may not be able to accurately replicate the observed facts. On the other hand, the Atangana–Baleanu derivative with fractional order has as kernel the Mittag–Leffler function that has a fair representation in the vicinity of the origin, therefore it could possibly give a clear and accurate information about the observed facts within the vicinity of the borehole. Here the revisited groundwater flow equation within an unconfined aquifer can be represented as follows:

$$T\left(\frac{\partial^2 s}{\partial r^2} + \frac{\partial s}{r \partial r}\right) = S\left({}^{ABC}_0 D_t^\alpha (s)\right) + DS_y \int_0^t \frac{\partial s}{\partial z} \exp\left(-D(t-z)\right) dz, \qquad (7.80)$$

where the time derivative is replaced by the Atangana–Baleanu derivative in Caputo sense. The analysis will consist of presenting first the existence of unique solution, by using some fixed-point theorem, then followed by the derivation of an approximate solution using some recursive method. We shall start with the existence of the unique solution. We note that Eq. (7.80) is equivalent to

$$\frac{\alpha}{AB(\alpha)\Gamma(\alpha)} \int_0^t (t-y)^{1-\alpha} \left(T\left(\frac{\partial^2 s}{\partial r^2} + \frac{\partial s}{r \partial r}\right) - DS_y \int_0^t \frac{\partial s}{\partial z} \exp\left(-D(t-z)\right) dz\right) dy \quad (7.81)$$

$$+ \frac{1-\alpha}{AB(\alpha)} \left(T\left(\frac{\partial^2 s}{\partial r^2} \frac{\partial s}{r \partial r}\right) - DS_y \int_0^t \frac{\partial s}{\partial z} \exp\left(-D(t-z)\right) dz\right) = Ss(r, t) - Ss(r, 0). \quad (7.82)$$

For simplicity we consider the following mapping

$$V(r, t, s) = T\left(\frac{\partial^2 s}{\partial r^2} + \frac{\partial s}{r \partial r}\right) - DS_y \int_0^t \frac{\partial s}{\partial z} \exp\left(-D(t-z)\right) dz. \qquad (7.83)$$

Existence and Uniqueness of the Solution

To prove the existence of groundwater flowing within an unconfined aquifer with Atangana–Baleanu fractional derivative, we first show that the function

$$V(r, t, U) \quad \text{is Lipschitz, meaning}$$

$$\|V(r,t,U_1) - V(r,t,U_2)\| \leq \Phi \|U_1 - U_2\|.$$

Let U_1 and U_2 be two different functions

$$V(r,t,U_1) - V(r,t,U_2) = T\left[\frac{\partial^2}{\partial r^2}(U_1 - U_2) + \frac{1}{r}(U_1 - U_2)\right]$$
$$- DU_y\int_0^t \frac{\partial}{\partial z}(U_1 - U_2)\exp[-D(t-z)]dz,$$

applying the norm on both sides of the above relation yields

$$\|V(r,t,U_1) - V(r,t,U_2)\| \leq T\|\frac{\partial^2}{\partial r^2}(U_1 - U_2)\| + T\|\frac{1}{r}\frac{\partial}{\partial r}(U_1 - U_2)\|$$
$$+ DU_y\int_0^t \|\frac{\partial}{\partial z}(U_1 - U_2)\|\exp[-D(t-z)]dz.$$

Using the relationship between norm and differential operator, we obtain

$$\|V(r,t,U_1) - V(r,t,U_2)\| \leq T\Theta_1^2\|U_1 - U_2\| + \frac{1}{r}\delta_1\|U_1 - U_2\|$$
$$+ DU_y\Omega_1\|U_1 - U_2\|\int_0^t \exp[-D(t-z)]dz$$
$$\leq T\Theta_1^2\|U_1 - U_2\| + \frac{1}{r}\delta_1\|U_1 - U_2\| + DU_y\Omega_1 K_1\|U_1 - U_2\|$$
$$\leq \left(T\Theta_1^2 + \frac{1}{\Gamma_{min}}\delta_1 + DU_y\Omega_1 K_1\right)\|U_1 - U_2\|$$
$$\leq K\|U_1 - U_2\|.$$

This shows that V is Lipschitz. Let consider the following recursive formula,

$$U_{n+1}(r,t) = U_n(r,t) + \frac{\alpha}{AB(\alpha)\Gamma(\alpha)}\int_0^t (t-y)^{-\alpha}V(r,y,U_n)dy + \frac{1-\alpha}{AB(\alpha)}V(r,t,Un).$$

We define a function such that $\forall n \geq 1$

$$FU_n = U_{n+1},$$

and we aim to prove that F is a contraction.

Let $n \geq 1$, then

$$\|F.U_n - FU_{n+1}\| = \|U_{n+1} - U_n\|.$$

However,

$$
\begin{aligned}
U_{n+1} - U_n \;=\; & \frac{\alpha}{AB(\alpha)\Gamma(\alpha)} \int_0^t (t-y)^{-\alpha} V(r, y, U_n) dy + \frac{1-\alpha}{AB(\alpha)} V(r, t, U_n) \\
& - \frac{\alpha}{AB(\alpha)\gamma(\alpha)} \int_0^t (t-y)^{-\alpha} V(r, y, U_{u-1}) dy - \frac{1-\alpha}{AB(\alpha)} V(r, t, U_{n-1}).
\end{aligned}
$$

Thus,

$$
\begin{aligned}
\|U_{n+1} - U_n\| \;=\; & \left\| \frac{\alpha}{AB(\alpha)\Gamma(\alpha)} \int_0^t (t-y)^{-\alpha} \Big(V(t, t, U_n) - V(r, t, U_{n-1}) \right. \\
& \left. + \frac{1-\alpha}{AB(\alpha)} V(t, t, U_n) - V(r, t, U_{n-1}) \Big) \right\|.
\end{aligned}
$$

Using the triangular inequality, we obtain

$$
\begin{aligned}
\|U_{n+1} - U_n\| \;\leq\; & \frac{\alpha}{AB(\alpha)\Gamma(\alpha)} \int_0^t \|V(t, t, U_n) - V(r, t, U_{n-1})\| (t-y)^{-\alpha} dy \\
& + \frac{1-\alpha}{AB(\alpha)} \|V(t, t, U_n) - V(r, t, U_{n-1})\|.
\end{aligned}
$$

Using the Lipschitz condition of V, we obtain

$$
\|U_{n+1} - U_n\| \leq \left(\frac{\alpha\, T^{1-\alpha}}{AB(\alpha)\Gamma(\alpha)(1-\alpha)} K + \frac{1-\alpha}{AB(\alpha)} K \right) \|U_n - U_{n-1}\|,
$$

and, by recursion, we obtain

$$
\|U_{n+1} - U_n\| \leq \left(\frac{\alpha\, T^{1-\alpha}}{AB(\alpha)\Gamma(\alpha)(1-\alpha)} K + \frac{1-\alpha}{AB(\alpha)} K \right)^{n-1} \|U_1 - U_0\|
$$

where

$$
U_1 - U_0 = \frac{\alpha}{AB(\alpha)\Gamma(\alpha)} \int_0^t (t-y)^{-\alpha} V(r, y, U_0) dy + \frac{1-\alpha}{AB(\alpha)} V(r, y, U_0)
$$

with

$$
V(y, y, U_0) = T \left(\frac{\partial^2 U_0}{\partial r^2} + \frac{1}{r}\frac{\partial}{\partial r} U_0 \right) - D U_0 \int_0^t \frac{\partial U_0}{\partial z} \exp[-D(t-z)] dz.
$$

However, $U_0 = 0$, so $|V(r, y, U_0)| = |f(r)| < \beta$. So

$$
\|U_{n+1} - U_n\| \leq \left(\frac{\alpha\, T^{1-\alpha}}{AB(\alpha)\Gamma(\alpha)(1-\alpha)} K + \frac{1-\alpha}{AB(\alpha)} K \right)^{n-1} \beta.
$$

We chose K such that

$$\frac{\alpha\, T^{1-\alpha}}{AB(\alpha)\Gamma(\alpha)(1-\alpha)}K + \frac{1-\alpha}{AB(\alpha)}K < 1,$$

thus under this condition, when $n \to \infty$, $\|U_{n+1} - U_n\| \to 0$. Thus $(U_{n+1})_{n\in N}$ is a Cauchy sequence in a Banach space, therefore it converges. Therefore there exists a unique function such that $\lim_{n\to\infty} U_n = U(r,t)$, which is the unique solution of the model of groundwater flowing within an unconfined aquifer with Atangana–Baleanu derivative with fractional order.

Numerical Analysis of the Modified Model

Let us recall that the modified equation is given as:

$$T\left(\frac{\partial^2 h}{\partial r^2} + \frac{1}{r}\frac{\partial h}{\partial r}\right) = U_0^{ABC}D_t^\alpha h + DU_y \int_0^t \frac{\partial h}{\partial z}\exp[-D(t-z)]dz,$$

here

$$\frac{\partial^2 h}{\partial r^2} = \frac{1}{2(\Delta r)^2}\left((h_{i+1}^{j+1} - 2h_i^{j+1} + h_{i-1}^{j+1}) + (h_{i+1}^j - 2h_i^j + h_{i-1}^j)\right),$$

$$\frac{\partial h}{\partial r} = \frac{1}{2(\Delta r)}\left((h_{i+1}^{j+1} - h_{i-1}^{j+1}) + (h_{i+i}^j - h_{i-1}^j)\right),$$

$$\frac{\partial h}{\partial t} = \frac{1}{\Delta t}\left(h_i^{j+1} - h_i^j\right)$$

$$_0^{ABC}D_t^\alpha h(r,t) = \frac{AB(\alpha)}{1-\alpha}\sum_{k=0}^j \frac{h_i^{k+1} - h_i^k}{\Delta t}\left\{(t_n - t_k)E_{\alpha,z}\left(-\frac{\alpha}{1-\alpha}(t_n - t_k) - (t_n - t_{k+1})\right)\right.$$
$$\left. E_{\alpha,z}\left(-\frac{\alpha}{1-\alpha}(t_n - t_{k+1})\right)\right\}$$

$$\int_0^t \frac{\partial h}{\partial z}\exp[-D(t-z)]dz = \frac{1}{2}\sum_{k=0}^j\left(h_i^{k+1} - h_i^k\right)\{\exp[-D(t_n - t_k)] - \exp[-D(t_n - t_{k+1})]\}.$$

Replacing all in the equation, we obtain

$$T\left\{\frac{1}{2(\Delta r)^2}\left[\left(h_{i+1}^{j+1} - 2h_i^{j+1} + h_{i-1}^{j+1}\right) + \left(h_{i+1}^j - 2h_i^j + h_{i-1}^j\right)\right] + \frac{1}{2(\Delta r)}\left[\left(h_{i+1}^{j+1} - h_{i-1}^{j+1}\right)\right.\right.$$
$$\left.\left. + \frac{1}{r_i}\left(h_{i+1}^j - h_{i-1}^j\right)\right]\right\} = U\frac{AB(\alpha)}{1-\alpha}\sum_{k=0}^j\frac{h_i^{k+1} - h_i^k}{\Delta t}\delta_\alpha^{j,k} + DU_y\sum_{k=0}^j\frac{h_i^{k+1} - h_i^k}{\Delta t}d_\alpha^{j,k},$$

where

$$d_\alpha^{j,k} = \exp[-D(t_n - t_k)] - \exp[-D(t_n - t_{k+1})]$$

$$\delta_\alpha^{j,k} = (t_n - t_k)E_{\alpha,z}\left(-\frac{\alpha}{1-\alpha}(t_n - t_k)\right) - (t_n - t_{k+1})E_{\alpha,z}\left(-\frac{\alpha}{1-\alpha}(t_n - t_{k+1})\right).$$

For simplicity, we put

$$\alpha_1 = \frac{T}{2(\Delta r)^2}, \quad \alpha_2 = \frac{T}{2(\Delta r)r_i}, \quad \alpha_3 = U\frac{AB(\alpha)}{(1-\alpha)\Delta t}U_\alpha^{j,k}, \quad \alpha_4 = \frac{DU_y}{\Delta t}.$$

Thus the equation can be written as

$$\alpha_1\left\{\left(h_{i+1}^{j+1} - 2h_i^{j+1} + h_{i-1}^{j+1}\right) + \left(h_{i+1}^j - 2h_i^j + h_{i-1}^j\right)\right\}$$
$$+ \alpha_2\left\{\left(h_{i+1}^{j+1} - h_{i-1}^{j+1}\right) + \left(h_{i+1}^j - h_{i-1}^j\right)\right\}$$
$$= \alpha_3\sum_{k=0}^{j}\left(h_i^{j+1} - h_i^j\right) + \alpha_4\sum_{k=0}^{j}\left(h_i^{j+1} - h_i^j\right).$$

By rearranging, we obtain the following recursive formula

$$(-2\alpha_1 - \alpha_3 - \alpha_4)h_i^{j+1} = (2\alpha_1 - \alpha_3 - \alpha_4)h_i^j + f(\alpha, h_i^j)$$

where

$$f(\alpha, h_i^j) = (\alpha_1 - \alpha_2)h_{i+1}^{j+1} + (\alpha_1 - \alpha_2)h_{i-1}^{j+1} + \alpha_3\sum_{k=0}^{j-1}\left(h_i^{j+1} - h_i^j\right) + \alpha_4\sum_{k=0}^{j-1}\left(h_i^{j+1} - h_i^j\right).$$

Here we present the numerical solution with explicit scheme. For explicit scheme,

$$\frac{\partial h}{\partial t} = \frac{h_i^j - h_i^{j-1}}{\Delta t}, \quad \frac{\partial^2 h}{\partial r^2} = \frac{h_{i-1}^{j-1} - 2h_i^{j-1} + h_{i+1}^{j-1}}{(\Delta r)^2}, \quad \frac{\partial h}{\partial r} = \frac{h_{i+i}^{j-1} - h_{i-1}^{j-1}}{2(\Delta r)},$$

$${}_0^{ABC}D_t^\alpha h(t) = \frac{AB(\alpha)}{1-\alpha}\sum_{k=1}^{j}\frac{h_i^k - h_i^{k-1}}{(\Delta t)}\delta_\alpha^{j,k},$$

$$\int_0^t \frac{\partial h}{\partial t}\exp(-D(t-z))dz = \sum_{k=1}^{j}\frac{h_i^k - h_i^{k-1}}{\Delta t}d_\alpha^{i,k},$$

replacing all in the main equation, we obtain

$$T\left\{\frac{h_{i-1}^{j} - 2h_i^{j-1} + h_{i+1}^{j-1}}{(\Delta r)^2} + \frac{h_{i+1}^{j-1} - h_{i-1}^{j-1}}{2r_i(\Delta r)}\right\} = U\frac{AB}{1-\alpha}\sum_{k=1}^{j}\frac{h_i^k - h_i^{k-1}}{\Delta t}\delta_\alpha^{j,k}$$

$$+ DU_y\sum_{k=1}^{j}\frac{h_i^k - h_i^{k-1}}{\Delta t}d_\alpha^{j,k}.$$

Therefore with implicit scheme, we have

$$T\left\{\frac{h_{i-1}^{j} - 2h_i^{j} + h_{i+1}^{j}}{(\Delta r)^2} + \frac{h_{i+1}^{j} - h_{i-1}^{j}}{2r_i(\Delta r)}\right\} = U\frac{AB}{1-\alpha}\sum_{k=0}^{j}\frac{h_i^{k+1} - h_i^{k}}{\Delta t}\delta_\alpha^{j,k}$$

$$+ DU_y\sum_{k=0}^{j}\frac{h_i^{k+1} - h_i^{k}}{\Delta t}d_\alpha^{j,k}.$$

7.6 Hantush Model With Fractional Derivatives

The ongoing issue of groundwater flow through leaky aquifers is a significant complex issue which grasped the attention of several different researchers in the field of science and technology over the last few decades. The Hantush model which deals with the aforementioned issue relates to a groundwater flow equation dealing with a confined leaky aquifer. Confined leaky aquifers are also known as leaky phreatic aquifers or aquifers having a semi-permeable boundary such as an aquitard. This boundary is generally an upper boundary which has lower permeability than the underlying aquifer material. Additionally, this upper boundary is normally significantly thin. Along this boundary, which as mentioned is semi-permeable, leakage is expected to occur from or into the phreatic aquifer which overlies a saturated body. In this section, we refer to the concepts of heterogeneity and viscoelastic material through which leakage occurs. For an accurate representation of what is physically observed, in a mathematical sense, we replace the time derivative by the derivative having a non-local and non-singular kernel. In order to do this, the time derivative is considered to be the time fractional derivative based on the Mittag–Leffler function. Furthermore, the new model will be analyzed both analytically and numerically; and for the analytical investigation, we will concentrate on the analysis of both the existence and uniqueness of the new model's solution. Thereafter, application is given to the new numerical scheme for derivation of the numerical solution of the new model.

7.6.1 Analytical Solution of the Flow Within the Leaky Aquifer With the Atangana–Baleanu Fractional Derivative

Let $\Omega = (a, b)$ be an open set and bounded subset of $\mathbb{R}^n (n \geq 1)$ with boundary $\partial\Omega$. For a given $\alpha \in (0, 1)$, and a function $u(x, t) \in H^1(\Omega) \times [0, T]$, which represents the head, we seek

u such that the flow of water within the leaky aquifer is governed by

$$\gamma^{2ABC}D_t^\alpha u = \partial_{xx}u + \frac{1}{x}\partial_x u - \frac{u}{\phi^2}, \tag{7.84}$$

where $u = \sigma D_\rho$ and $\gamma^2 = \xi\rho\phi^2$, ξ represents the coefficient of storage, σ stands for the conductivity. Eq. (7.84) of groundwater flow within the leaky aquifer will be solved analytically and numerically with the implicit scheme.

7.6.2 Existence and Uniqueness of the Solution of the Problem

In what follows, we discuss the existence and uniqueness of solutions of the direct problem.

$$\left. \begin{aligned} {}^{ABC}D_t^\alpha u(x,t) &= f(u(x,t),t), \\ u(x_\rho, 0) &= v(x) && \text{on} \quad [0,T] \times \partial\Omega, \\ u(x_\rho, t) &= u_\rho && \text{in} \quad \{0\} \times \Omega, \end{aligned} \right\} \tag{7.85}$$

for the initial datum $u_\rho \in L^\infty(\Omega)$, $u\rho > 0$, and where

$$f(r,t,u(x,t)) = \gamma^2\left(\partial_{xx}u(x,t) + \frac{1}{x}\partial_x u(x,t) - \frac{u(x,t)}{\phi}\right). \tag{7.86}$$

Given $u_\rho \in L^\infty(\Omega)$, $u_\rho > 0$, a solution of (7.85) is a positive function $u(x,t) \in H^1(\Omega) \times [0,T]$ such that if we apply the fractional integral given by the Atangana–Baleanu integral of order α with base point a, given as

$$^{AB}I_t^\alpha u(t) = \frac{1-\alpha}{B(\alpha)\Gamma(\alpha)}\int_a^t u(s)(t-s)^{\alpha-1}ds,$$

in conjunction with (7.85) yields

$$u(x,t) = u(0) + \frac{1-\alpha}{B(\alpha)}f(u(x,t),t) + \frac{\alpha}{B(\alpha)(\alpha)}\int_0^t f(u(x,y),y)(t-y)^{\alpha-1}dy \tag{7.87}$$

for all $t \in [0,T]$. We want to use the contraction mapping theorem, so for this purpose we need to build a closed set of $H^1(\Omega) \times [0,T]$ such that the nonlinear operator g be a concentration which maps ϵ onto itself. In what follows, we first show that f is a contraction mapping.

Proposition 7.6.1. *The nonlinear operator $f \in H^1(\Omega) \times [0, T]$ is locally Lipschitz.*

Proof. We consider two bounded functions u and w in $H^1(\Omega) \times [0, T]$. Then,

$$\|f(u) - f(w)\|_{H^1} = \gamma^{-2}\|\partial_{xx}u - \partial_{xx}w + x^{-1}(\partial_x u - \partial_x w) - \phi^{-2}(u - w)\|.$$

We have by the triangular inequality

$$\|f(u) - f(w)\|_{H^1} = \gamma^{-2}\left\{\|\partial_{xx}u - \partial_{xx}w\| + \|x^{-1}(\partial_x u - \partial_x w)\| - \phi^{-2}\|(u - w)\|\right\}.$$

As the derivative operator satisfies the Lipschitz conditions in H^1, hence there exist two positive constants ρ_1 and ρ_2 such that

$$\|f(u) - f(w)\|_{H^1} \leq \gamma^{-2}\rho_1\|u - w\| + \gamma^{-2}\rho_2\|x^{-1}\|\|u - w\| - \gamma^{-2}\phi^{-2}\|u - w\| \leq C\|u - w\|,$$

where $C = |\gamma^{-2}(\rho_1 + \rho_2)\|x^{-1}\| - \phi^{-2}| < 1$. Then f is a contraction mapping. $\square$

Theorem 7.6.2. *For a given initial datum $u_\rho \in L^\infty$, $u_\rho > 0$, there exists a unique positive solution u of (7.85) on $H^1(\Omega) \times [0, T]$ for all $t < T < \infty$ and*

$$\lim_{t \to T} \|u(x, t)\|_{L^\infty} \to \infty.$$

Proof. We shall follow the proof as presented in previous sections. Since the nonlinear operator f is locally Lipschitz, for $\bar{u}_\rho = \|u_\rho\|_{H^1}$ there exists $C_{\bar{u}_\rho}$ such that $0 < C_{\bar{u}_\rho} < \infty$ and

$$\|f(u) - f(w)\|_{H^1} \leq C_{\bar{u}_\rho}\|u - w\|_{H^1}.$$

Let $T_1 > 0$ be a constant such that $T_1 < \frac{1}{C_{\bar{u}_\rho}}$. Set

$$\epsilon = \left\{u \in H^1(\Omega) \times [0, T]; \|u\|_{H^1} \leq \bar{u}_\rho, \text{ for all } t \in [0, T_1]\right\},$$

endowed with the norm

$$\|u\|_\epsilon = \sup_{0 \leq t \leq T_1} \|u\|_{H^1},$$

then ϵ is a closed convex subset of $H^1(\Omega) \times [0, T]$. Now we consider the following associated problem of (7.87) defined on ϵ:

$$\varpi(u) = u_\rho + \frac{1 - \alpha}{B(\alpha)} f(u(x, t), t) + \frac{\alpha}{B(\alpha)\Gamma(\alpha)} \int_0^t f(u(x, y), y)(t - y)^{\alpha-1} dy. \qquad (7.88)$$

For all $u \in \epsilon$ and $t \geq 0$, we have

$$\|\varpi(u)\|_\epsilon = \sup_{0 \leq t \leq T_1} \left\| u_\rho + \frac{1-\alpha}{B(\alpha)} f(u(x,t),t) + \Upsilon(\alpha) \int_0^t f(u(x,y),y)(t-y)^{\alpha-1} dy \right\|_{H^1}$$

$$\leq \|u_\rho\|_{H^1} + \left\| \frac{1-\alpha}{B(\alpha)} f(u(x,t),t) \right\|_{H^1} + \frac{\alpha}{B(\alpha)\Gamma(\alpha)} \int_0^t \|f(u(x,y),y)\|_{H^1} dy$$

$$\leq \|u_\rho\|_{H^1} + \int_0^t \|f(u(x,y),y)\|_{H^1} dy.$$

But since f is locally Lipschitz for all $y \in [0, T_1]$, it follows that

$$\|\varpi(u)\|_\epsilon \leq \|u_\rho\|_{H^1} + \int_0^t (C_{\bar{u}_\rho} \|u(x,y)\|_{H^1} + \rho_3) dy.$$

Thus for all $u_1, u_2 \in \epsilon$,

$$\|\varpi(u_1) - \varpi(u_2)\|_\epsilon = \sup_{0 \leq t \leq T_1} T_1 C_{\bar{u}_\rho} \|u_1 - u_2\|_\epsilon.$$

This shows that ϖ is a contraction mapping in ϵ. Thus ϖ has a fixed point which is a solution to (7.85). Next, we show that problem (7.85) has a unique solution. Let $u_1, u_2 \in H^1$ be two solutions of (7.85) and let $u = u_1 - u_2$. Then

$$u = \frac{1-\alpha}{B(\alpha)} \big[(u_1(x,t),t) - f(u_2(x,t),t)\big] + \frac{\alpha}{B(\alpha)\Gamma(\alpha)} \int_0^t \big[f(u_1(y,t),t) - f(u_2(y,t),t)\big] dy$$

By taking the error norm on both sides,

$$\begin{aligned} \|u\| &\leq \frac{1-\alpha}{B(\alpha)} \|(u_1(x,t),t) - f(u_2(x,t),t)\| \\ &+ \frac{\alpha}{B(\alpha)\Gamma(\alpha)} \int_0^t \|[f(u_1(y,t),t) - f(u_2(y,t),t)]\| dy \\ &\leq T_1 C_{\bar{u}_\rho} \int_0^t \|u(x,y)\|_{H^1} ds. \end{aligned}$$

By the Gronwall inequality the result follows. $\qquad\square$

7.6.3 Exact Solution of the Flow Within the Leaky Aquifer Based Upon Atangana–Baleanu Fractional Derivative

In the model here discussed for water flow in the leaky aquifer, the head $u(x,t)$ which appears in (7.84) is assumed to be governed by the one-dimensional time-fractional differential equation involving the Atangana–Baleanu fractional derivative. Applying Laplace transform to (7.84), the fundamental solution $u(x, \tau)$, $\tau \in \mathbb{N}$, results in:

$$-\gamma^2 L_t \left[^{ABC} D_t^\alpha u\right](\tau) + L_t \left[\partial_{xx} u\right](\tau) + L_t \left[\frac{1}{x}\partial_x u\right](\tau) - \phi^{-2} L_t[u](\tau) = 0,$$

where $L_t := \bar{u}$ stands for the Laplace transform. By replacing each term by its value, we obtain

$$-\frac{B(\alpha)\gamma^2[\tau^\alpha u - \tau^{\alpha-1} u_\rho]}{(1-\alpha)\tau^\alpha + \alpha} + \partial_{xx}\bar{u} + \frac{1}{x}\partial_x\bar{u} - \phi^{-2}\bar{u} = 0.$$

Hence the following differential equation holds in the form

$$x^2\partial_{xx}\bar{u} + x\partial_x\bar{u} - \hbar x^2 \bar{u} = 0, \tag{7.89}$$

with $u_\rho \approx 0$, where

$$\hbar = \frac{B(\alpha)\gamma^2\tau^\alpha}{(1-\alpha)\tau^\alpha + \alpha} + \phi^{-2}.$$

Since $\hbar$ is positive, the exact solution of the differential equation (7.89) is given in terms of Bessel function of the fist kind, G_0, and modified kind, H_0, as

$$\bar{u}(x, \tau) = A G_0(x\sqrt{\hbar} + B H_0(x\sqrt{\hbar})), \tag{7.90}$$

where A and B are constants and G_0, H_0 are given respectively as

$$G_v(z) = \sum_{k=0}^{\infty} \frac{(-1)^k}{k!\Gamma(k+v+1)}\left(\frac{z}{2}\right)^{2k+v} \quad \text{and} \quad H_v(z) = \frac{\pi}{2}(-i)^v\left(\frac{-G_v(z) + G_{-v}(z)}{\sin(v\pi)}\right).$$

Using the boundary condition in (7.90), we get $B = 0$, then the solution is reduced to

$$\bar{u}(x, \tau) = A G_0(x\sqrt{\hbar}) = A \sum_{k=0}^{\infty} \frac{(-1)^k}{k!\Gamma(k+v+1)}\left(\frac{x\sqrt{\hbar}}{2}\right)^{2k+v}. \tag{7.91}$$

Due to the difficulties to obtain the inverse Laplace transform of (7.91), we therefore propose to obtain the approximate solution of (7.85) by using the proposed numerical approximation of the Atangana–Baleanu integral.

7.6.4 Numerical Analysis of the Groundwater Flowing Within the Leaky Aquifer Based Upon Atangana–Baleanu Fractional Derivative

To achieve this, we revert the fractional differential equation to the fractional integral equation using the link between the Atangana–Baleanu derivative and the Atangana–Baleanu integral to obtain

$$u(x,t) - u(x_\rho, 0) = {}_0^{AB} I_t^\alpha f(u(x,t), x, t).$$

For some positive and large integers $M = N = 350$, the grid sizes in space and time are denoted respectively by $\chi = 1/M$ and $x = 1/N$. The grid points in the space interval $(0, 1]$ are the numbers $x_i = i\chi$, $i = 1, 2, \ldots, M$, and the grid points in the time interval $(0, 1]$ are the numbers $t_k = kx$, $k = 0, 1, 2, \ldots, N$. The value of the function u at the grid points is denoted by $u_i^k = u(x_i, t_k)$. Using the implicit finite difference method, a discrete approximation of $f(u(x,t), x, t)$ given by (7.86) can be obtained as follows

$$f(u(x_i, t_k), x_i, t_k) = \frac{\gamma^{-2}}{\chi^2}\left\{(1 + \frac{1}{2i})u_{i+1}^k - (2 + \frac{\chi^2}{\phi^2})u_i^k + (1 - \frac{1}{2i})u_{i-1}^k\right\}.$$

In order to use the numerical approximation proposed in this work, discrete solution of (7.85) is then given as

$$u_i^{k+1} - u_\rho^0 = {}^{AB} I_t^\alpha f(u(x_i, t_{k+1}), x_i, t_k).$$

$$\begin{aligned}
{}_0^{AB} I_t^\alpha f &= \frac{1-\alpha}{B(\alpha)} f(u(x_i, t_{k+1}), x_i, t_{k+1}) + \Upsilon(\alpha) \sum_{j=0}^{k} \frac{b_j^\alpha}{2}\left[f(u_i(x_i, t_{k-j}), x_i, t_{k-j})\right. \\
&\quad + \left. f(u_i(x_i, t_{k-j+1}), x_i, t_{k-j+1})\right].
\end{aligned}$$

Therefore the numerical approximation can be given as follows

$$\begin{aligned}
u_i^{k+1} - u_\rho^0 &= \frac{(1-\alpha)\gamma^{-2}}{B(\alpha)\chi^2}\left\{(1 + \frac{1}{2i})u_{i+1}^{k+1} - (2 + \frac{\chi^2}{\phi^2})u_i^{k+1} + (1 - \frac{1}{2i})u_{i-1}^{k+1}\right\} \\
&\quad + \Upsilon(\alpha)\frac{\gamma^{-2}}{\chi^2}\sum_{j=0}^{k}\left\{(1 + \frac{1}{2i})u_{i+1}^{k-j} - (2 + \frac{\chi^2}{\phi^2})u_i^{k-j} + (1 - \frac{1}{2i})u_{i-1}^{k-j}\right. \quad (7.92) \\
&\quad + \left. (1 + \frac{1}{2i})u_{i+1}^{k-j+1} - (2 + \frac{\chi^2}{\phi^2})u_i^{k-j+1} + (1 - \frac{1}{2i})u_{i-1}^{k-j+1}\right\}.
\end{aligned}$$

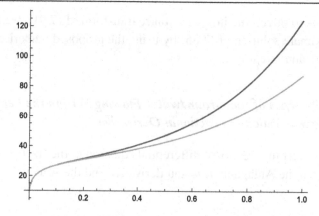

Figure 7.5: Numerical solution for $\alpha = 0.5$.

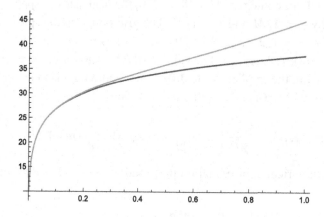

Figure 7.6: Numerical solution for $\alpha = 0.5$.

In order obtain the plots of the solution given by Eq. (7.92), we shall consider 350 equidistant nodes in [0, 1]. For the value $\alpha = 1/2$, the approximate solution found by using the numerical method of the Atangana–Baleanu integral proposed is obtained. Fig. 7.5 shows the approximate solution for the different time steps $k = 0$, $k = 20$ and $k = 50$ respectively. Moreover, to have an overview of the variation of flow or the behavior of the function in a finite time in terms of the parameter α, we consider two different time steps $k = 0$ and $k = 50$. For this purpose, Fig. 7.6 and Fig. 7.7 give us the approximate result using the method proposed. We would like to notice that for larger number of nodes in [0, 1] the better approximated solution is obtained in the whole interval. We would like to notice that for larger number of nodes in [0; 1] the better approximated solution is obtained in the whole interval. We show the approximate solution of (7.92) by using the numerical approximation of the Atangana–Baleanu

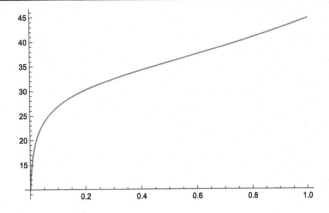

Figure 7.7: Numerical solution for $\alpha = 0.5$.

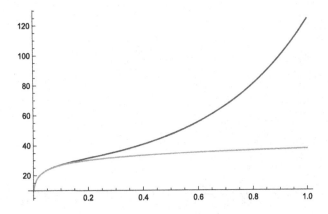

Figure 7.8: Numerical solution for $\alpha = 0.9$.

method proposed for different time steps $k = 0$ and $k = 50$ for $\alpha = 1/2$ and $\alpha = 9/10$ in $[0, 1]$ (see Figs. 7.8, 7.9, 7.10, 7.11).

7.7 Steady-State Model With Fractional Derivatives

It is important to note that groundwater flow models portray their capabilities as either steady and/or transient. In the previous section, we presented a detailed analysis of the groundwater models in transient state. We shall recall that, the groundwater model is in the steady-state if the magnitude and the direction of the flow are independent of time throughout the entire domain. This therefore implies, the hydraulic head will not change with time within the steady-state system, however does change in space. This condition renders the model of

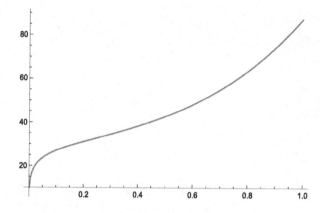

Figure 7.9: Numerical solution for $\alpha = 0.9$ and $k = 50$.

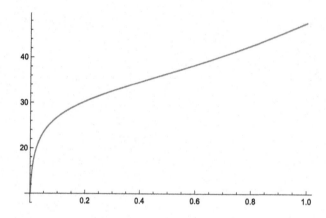

Figure 7.10: Numerical solution for $\alpha = 0.9$ and $k = 20$.

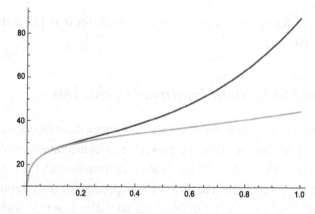

Figure 7.11: Numerical solution for $\alpha = 0.9$ and $k = 20$.

groundwater flow very simple with the confined and the unconfined aquifers. We shall start with the Theim's equilibrium formulas in the case of the unconfined aquifers.

7.7.1 Theim's Equilibrium Model Within an Unconfined Aquifer With Fractional Derivative

Within an unconfined aquifer, to establish the Theim's equilibrium formula, a non-artesian well is driven and the water is removed heavily in order to create sufficient drawdown. Then, the water level in the neighborhood will also go down symmetrically around the well. The bottom part of the cone appears to be a perfect circle with radius R which is considered as circle of influence. The formulas are therefore derived by placing two observation wells lying within the so-called circle of influence of the main well from which the water is being taken out of the unconfined aquifer. It is therefore assumed that both circles have radii, say, r_1 and r_2. It is thus assumed that the main well pumped at a sufficient rate so as to create heavy drawdown. Then the pumping is set such that the equilibrium conditions are obtained. In this case we considered that, S_1 and S_2 are the drawdowns in the two associated observed well1 and well2 respectively, at this equilibrium state. Keep in mind the formula obtained from Darcy's law stating that

$$Q = KIA. \tag{7.93}$$

In the above formula, I accounts for the hydraulic gradient using the cylindrical coordinates, r is considered the radium of a given cylinder, and h represents the height of the cone of depression at a distance r from the main borehole. Thus the mathematical formula is therefore given as

$$I = \frac{dh}{dr} \tag{7.94}$$

$$Q = KIA = K\frac{dh}{dr}2\pi rh \tag{7.95}$$

$$Q = K\frac{dh}{dr}2\pi rh. \tag{7.96}$$

Within the framework of fractional differentiation, the space local derivative is replaced by a non-local fractional operators. Thus with the power law, we have

$$Q = \frac{K}{\Gamma(1-\alpha)} \int_{r_1}^{r_2} \frac{h(y)}{dy}(r_2-y)^{-\alpha}dy2\pi rh(r). \tag{7.97}$$

With the exponential decay law or the Caputo–Fabrizio fractional derivative, we have:

$$Q = K\frac{M(\alpha)}{1-\alpha} \int_{r_1}^{r_2} \frac{h(y)}{dy}\exp\left(-\frac{\alpha}{1-\alpha}(r_2-y)\right)dy2\pi rh(r). \tag{7.98}$$

With the generalized Mittag–Leffler law or the Atangana–Baleanu fractional derivative in Caputo sense, we have:

$$Q = K \frac{AB(\alpha)}{1-\alpha} \int_{r_1}^{r_2} \frac{h(y)}{dy} E_\alpha \left(-\frac{\alpha}{1-\alpha}(r_2 - y)^\alpha \right) dy 2\pi r h(r). \qquad (7.99)$$

7.7.2 Theim's Equilibrium Model Within a Confined Aquifer With Fractional Derivative

Within a confined aquifer, the following assumptions are made to establish the Thiem equilibrium point. It is always assumed that the geological formation is homogeneous, isotropic and infinite, and areal extended. These conditions are put in place such that the coefficient of transmissivity or permeability is being constant all over the aquifer. An important assumption is that the borehole has been sunk through the full depth of aquifer and moreover has received water from the entire thickness of the geological formation. It is also important to suppose that the removal of water is continuously performed for a sufficient time at the uniform rate, thus the equilibrium point or in geohydrological terms steady-state flow criteria are reached. It is also suggested that the flow lines are radial and horizontal with a laminar flow. It is also suggested that the inclination of the table water is small so that its tangent can be used in place of sine for the hydraulic gradient in Darcy's equation. Therefore under these conditions, the mathematics employed for the unconfined aquifer is modified slightly in the case of an artesian aquifer, here the flow is of course radial and horizontal.

$$Q = KIA. \qquad (7.100)$$

In the above formula, I accounts for the hydraulic gradient using the cylindrical coordinates, r is considered the radium of a given cylinder, and h represents the height of the cone of depression at a distance r from the main borehole. Thus the mathematical formula is therefore given as

$$I = \frac{dh}{dr} \qquad (7.101)$$

$$Q = KIA = K\frac{dh}{dr}2\pi r \qquad (7.102)$$

$$Q = K\frac{dh}{dr}2\pi r. \qquad (7.103)$$

Within the framework of fractional differentiation, the space local derivative is replaced by a non-local fractional operators. Thus with the power law, we have

$$Q = \frac{K}{\Gamma(1-\alpha)} \int_{r_1}^{r_2} \frac{h(y)}{dy}(r_2 - y)^{-\alpha} dy 2\pi r. \qquad (7.104)$$

With the exponential decay law or the Caputo–Fabrizio fractional derivative we have:

$$Q = K \frac{M(\alpha)}{1-\alpha} \int_{r_1}^{r_2} \frac{h(y)}{dy} \exp\left(-\frac{\alpha}{1-\alpha}(r_2 - y)\right) dy 2\pi r. \qquad (7.105)$$

With the generalized Mittag–Leffler law or the Atangana–Baleanu fractional derivative in Caputo sense, we have:

$$Q = K \frac{AB(\alpha)}{1-\alpha} \int_{r_1}^{r_2} \frac{h(y)}{dy} E_\alpha\left(-\frac{\alpha}{1-\alpha}(r_2 - y)^\alpha\right) dy 2\pi r. \qquad (7.106)$$

The solution of the above can be obtained by mean of iteration methods.

7.8 Limitations of Fractional Derivative for Modeling Groundwater Problems

Ordinary geological formation with exceedingly contrasting permeability may structure mobile and comparatively non-movable zones, where the potential swap over connecting the mobile and immobile zones results in a wide time distribution fluid trapping. The flow process, putting together with the distinct particle status can be pigeonholed efficiently by fractional differentiation only if this property is observed all over the geological formation, thus the concept of fractional differentiation with constant order cannot be used if the properties are different from one point to another. If the high-permeable material tends to construct preferential paths, in the way that the interconnected paleochannels experimental in alluvial depositional system, then in that particular zone of the aquifer, the groundwater flow path may present a heavy leading edge, this can be portrayed by the space fractional differential operator with maximally positive skewness, also if this is observed all over the geological formation. Space fraction arises when the velocity variations are heavy tailed and describe the flow motion that takes into consideration for the variation in the flow field over the entire system. Although the time fractional groundwater flow model arises as a result of a power law particle residence time distributions but it does not describes particle motion with memory in time but the whole aquifer due to the singularity induced by the power law. Anomalous flow phenomena are thoroughly in geohydrology. To distinguish unbalanced diffusion phenomena, fractional differentiation with constant order cannot be used, in particular were initial condition have received overwhelming success. Nonetheless, it has been recognized that the fractional order process of flow is not knowledgeable of typifying some comprehensive flow processes. In addition, if the aquifer or the geological formation structure or external field changes with time in these circumstances, the concept of fractional differentiation with constant order is not suitable for modeling purposes.

CHAPTER 8

Models of Groundwater Pollution With Fractional Operators

An outstanding literature review exposed that the concept of fractional differentiation advection convection equation has been pointed out to be a very powerful mathematical tool to portray the movement of pollution within a given heterogeneous porous media. This model of movement of pollution within the geological formation is known to be a special case of a wide-ranging transport equation with sophistication flux. Indeed, the model is also a limit case of the continuous time random walk with the power law particle jumps. It is an obvious matter to draw from the fractional advection dispersion equation from the fractional conversation of mass equation employing the harmonized structure at the plume center of the mass, in exactly the same way that the usual advection–dispersion equation follows from long-established conservation of mass equation.

8.1 Time Fractional Convection–Dispersion Equation With Caputo Type

In attempt to explicitly include the effect of flow geometry into the mathematical model, we replaced the Cartesian component of the gradient of concentration $\partial_x c(x, t)$ with the Riemann–Liouville fractional derivatives of order β, $\partial_x^\beta c(x, t)$; we also replaced $\partial_x^2 c(x, t)$ with $\partial_x^\alpha c(x, t)$ in the interval $0 < \beta \leq 1 < \alpha < l$, as follows:

$$D\frac{\partial^\alpha c(x, t)}{\partial x^\alpha} - v\frac{\partial^\beta c(x, t)}{\partial x^\beta} - \lambda Rc = R\frac{\partial c(x, t)}{\partial t}. \tag{8.1}$$

This gives a generalized form of the standard equation that characterizes the transport of its solute. In fractional sense, the integro-differential equation here consists of additional parameters α and β, that are seen as new physical parameters characterizing the movement through the geological formations. This transformation provides a more general form for the boundary condition at the extreme end of the model:

$$D\frac{\partial^\alpha c(x, t)}{\partial x^\alpha} = 0, \quad x \to \infty. \tag{8.2}$$

The relations in (8.1), (8.2), and the initial condition

$$c(0, t) = c_0 \exp(-\gamma t) \quad 0 < t \leq t_0 \tag{8.3}$$

Fractional Operators with Constant and Variable Order
with Application to Geo-Hydrology
DOI: 10.1016/B978-0-12-809670-3.00008-4

make up a complete set of equations for which solution exists. The integro-differential term in the equation makes an analytical solution more difficult to obtain. An analytical solution will be discussed in what follows.

8.1.1 Analytical Solution With the Riemann–Liouville Derivative

This method involves the use of Laplace transform on both sides of (8.1) to yield

$$D\frac{\partial^\alpha c(x,s)}{\partial x^\alpha} - v\frac{\partial^\beta c(x,s)}{\partial x^\beta} - R(\lambda+s)c(x,s)c = Rc(x,0) \tag{8.4}$$

with initial data in (8.3) and further transformation; one can rewrite the above equation in the form:

$$D\frac{\partial^\alpha c(x,s)}{\partial x^\alpha} - \mu\frac{\partial^\beta c(x,s)}{\partial x^\beta} - \tau(x,s), \tag{8.5}$$

where s is the Laplace variable for the time-component, $\mu = v/D$, and $\tau = R(\lambda+s)/D$. Let $c(x,s) = y(x)$, then (8.5) results in

$$D\frac{\partial^\alpha y(x)}{\partial x^\alpha} - \mu\frac{\partial^\beta y(x)}{\partial x^\beta} - \tau y(x) = 0. \tag{8.6}$$

Taking the Laplace transform of (8.4), in the space direction, we obtain the following equation [120]:

$$L(y)(p) = \sum_{i=1}^{l} h_i \frac{p^{i-1}}{p^\alpha - \mu p^\beta - \tau}, \tag{8.7}$$

where p is the Laplace variable for the space component and $h_i = \partial_x^{\alpha-i}c(0^+)$. For $p \in \mathbb{C}$ and $|\tau p^{-\beta}/(p^{\alpha-\beta} - \mu)| < 1$, we obtain the expression $1/(p^\alpha - \mu p^\beta - \tau)$ written in series form [120]:

$$\frac{p^{i-1}}{p^\alpha - \mu p^\beta - \tau} = p^{i-1}\sum_{j=0}^{\infty}\frac{\tau^n p^{-\beta-\beta n}}{(p^{\alpha-\beta} - \mu)^{n+1}}. \tag{8.8}$$

Replacing (8.8) in (8.6) leads to the representation

$$(Ly)(p) = \sum_{i=1}^{2} h_i \sum_{j=0}^{\infty}\frac{\tau^n p^{-\beta-\beta n+i-1}}{p^{\alpha-\beta} - \mu}^{n+1}.$$

By simplifying further the above expression, for $p \in \mathbb{C}$ and $|\mu p^{\beta-\alpha}| < 1$, we first obtain

$$\frac{\tau^n p^{-\beta-\beta_{n+i-1}}}{(p^{\alpha-\beta}-\mu)^{n+1}} = \frac{\tau^n p^{(\alpha-\beta)-(\alpha+\beta_{n-i+1})}}{(p^{\alpha-\beta}-\mu)^{n+1}}.$$

Also, the second expression can be given as

$$= \frac{1}{n!} L\left\{ x^{\alpha n+\alpha-i} \left(\frac{\partial}{\partial x}\right)^n E_{\alpha-\beta,\alpha+n\beta+1-i}(\mu x^{\alpha-\beta}) \right\},$$

where

$$\left(\frac{\partial}{\partial x}\right)^n E_{\alpha,\beta}(x) = \sum_{j=0}^{\infty} \frac{\Gamma(n+j+1)x^j}{\Gamma(n\alpha+\beta+\alpha j)j!}.$$

Hence, solutions of (8.6) can be given as follows

$$y_i(x) = \sum_{n=0}^{\infty} \frac{\tau^n}{n!} x^{\alpha n+\alpha-i} \left(\frac{\partial}{\partial x}\right)^n E_{\alpha-\beta,\alpha+n\beta+1-i}(\mu x^{\alpha-\beta}). \tag{8.9}$$

Thus it follows that the solution can be written as

$$y(x) = \sum_{i=1}^{2} h_i y_i(x),$$

so that

$$c(x,s) = \sum_{i=1}^{2} h_i c_i(x,s).$$

Thus, we can give the series solution of (8.1) by applying the inverse Laplace transform operator on $c(x,s)$ to obtain

$$c_i(x,t) = L^{-1}\left(\sum_{n=0}^{\infty} \frac{\tau^n}{n!} x^{\alpha n+\alpha-i} \left(\frac{\partial}{\partial x}\right)^n E_{\alpha-\beta,\alpha+n\beta+1-i}(\mu x^{\alpha-\beta}) \right).$$

Since the inverse Laplace operator is of a linear type, it directly follows that

$$c_i(x,t) = \sum_{n=0}^{\infty} \frac{L^{-1}(\tau^n)}{n!} x^{\alpha n + \alpha - i} \left(\frac{\partial}{\partial x}\right)^n E_{\alpha-\beta,\alpha+\beta n+1-i}(\mu x^{\alpha-\beta}).$$

By substituting $r^n = (R(\lambda+s)/D)^n = (R/D)^n(\lambda+s)^n$, we have

$$L^{-1}\{\tau^n\} = \left(\frac{R}{D}\right)^n L^{-1}\{(\lambda+s)^n\} = \left(\frac{R}{D}\right)^n \frac{\exp[-\lambda t]t^{-1-n}}{\Gamma(-n)},$$

$$c_i(x,t) = \sum_{n=0}^{\infty} \frac{(R/D)^n \exp[-\lambda t]t^{-1-n}/\Gamma(-n)}{n!} x^{\alpha n + \alpha - i} \times \left(\frac{\partial}{\partial x}\right)^n E_{\alpha-\beta,\alpha+\beta n+1-i}(\mu x^{\alpha-\beta}),$$

$$c(x,t) = \sum_{i=1}^{2} h_i c_i(x,t),$$

$$c_1(x,t) = \sum_{n=0}^{\infty} \frac{(R/D)^n \exp[-\lambda t]t^{-1-n}/\Gamma(-n)}{n!} x^{\alpha n + \alpha - 1} \times \left(\frac{\partial}{\partial x}\right)^n E_{\alpha-\beta,\alpha+\beta n}(\mu x^{\alpha-\beta}),$$

$$c_2(x,t) = \sum_{n=0}^{\infty} \frac{(R/D)^n \exp[-\lambda t]t^{-1-n}/\Gamma(-n)}{n!} x^{\alpha n + \alpha - 2} \times \left(\frac{\partial}{\partial x}\right)^n E_{\alpha-\beta,\alpha+\beta n-1}(\mu x^{\alpha-\beta}).$$

To obtain the coefficient h_i, $i = 1, 2$, we apply the initial and boundary conditions on $c(x,t)$ to yield

$$h_i = \frac{c_0}{2}.$$

■ **Example 1**

Our concern is to solve (8.1) when $\alpha = 2$ and $0 < \beta \leq 1$. Following our earlier discussion, we give the two solutions of space–time fractional derivative of hydrodynamic advection–dispersion equation as

$$c_1(x,t) = \sum_{n=0}^{\infty} \left(\frac{R}{D}\right)^n \frac{\exp[-\lambda t]t^{-1-n}}{\Gamma(-n)n!} x^{2n+1} \times \left(\frac{\partial}{\partial x}\right)^n E_{2-\beta,n\beta+1}\left(\frac{R}{D}x^{2-\beta}\right),$$

$$c_2(x,t) = \sum_{n=0}^{\infty} \left(\frac{R}{D}\right)^n \frac{\exp[-\lambda t]t^{-1-n}}{\Gamma(-n)n!} x^{2n} \times \left(\frac{\partial}{\partial x}\right)^n E_{2-\beta,n\beta+1}\left(\frac{R}{D}x^{2-\beta}\right).$$

(8.10)

When $\beta < 1$, the above solutions could be referred to as the fundamental system of solution.

8.1.2 Analytical Solution With the Caputo Derivative

The Riemann–Liouville derivative has some disadvantages when applied to model real-world phenomena with fractional differential equations [121,124,125,131]. Therefore, we study the solution of space–time Caputo fractional derivative of hydrodynamic advection–dispersion equation. For the Caputo derivative, the Laplace transform is based on the expression

$$(Lc D^\alpha y)(s) = s^\alpha (Ly)(s) - \sum_{i=0}^{1} h_i s^{\alpha - i - 1}$$

with

$$h_i = y^i(0) \quad (i = 0, 1).$$

Taking the Laplace transform of both sides of (8.1) in both time and space directions yields

$$L(y)(p) = \sum_{i=0}^{2-1} h_i \frac{p^{\alpha - i - 1}}{p^\alpha - \mu p^\beta - \tau} - \mu \sum_{i=0}^{1-1} h_i \frac{p^{\beta - i - 1}}{p^\alpha - \mu p^\beta - \tau}.$$

For $p \in \mathbb{C}$ and $|\tau p^{-\beta} / (p^{\alpha - \beta} - \mu)| < 1$, in analogy with earlier discussion for the case of Riemann–Liouville, it follows that:

$$L(y)(p) = \sum_{i=0}^{2-1} h_i \sum_{n=0}^{\infty} \tau^n \frac{p^{(\alpha - \beta) - (\beta n + i + 1)}}{(p^{\alpha - \beta} - \mu)^{n+1}} - \mu \sum_{n=0}^{1-1} \tau^n \frac{p^{(\alpha - \beta) - (\beta n + i + 1 + \alpha - \beta)}}{(p^{\alpha - \beta} - \mu)^{n+1}}.$$

Hence, for $p \in \mathbb{C}$ and $|\mu p^{\beta - \alpha}| < 1$, we have that

$$\frac{p^{(\alpha - \beta) - (\beta n + j + 1)}}{(p^{\alpha - \beta} - \mu)^{n+1}} = \frac{1}{n!} \left(L\left[x^{n\alpha + i} \left(\frac{\partial}{\partial x} \right)^n E_{\alpha - \beta, \beta n + i + 1}(\mu x^{\alpha - \beta}) \right] \right),$$

$$\frac{p^{(\alpha - \beta) - (\beta n + j + 1 + \alpha - \beta)}}{(p^{\alpha - \beta} - \mu)^{n+1}} = \frac{1}{n!} L\left[x^{n\alpha + i + \alpha - \beta} \left(\frac{\partial}{\partial x} \right)^n E_{\alpha - \beta, \beta n + i + 1 + \alpha - \beta}(\mu x^{\alpha - \beta}) \right]. \tag{8.11}$$

From this expression we derive the solution to the space–time Caputo fractional derivative of hydrodynamic advection–dispersion equation (8.1) as:

$$c(x, t) = \sum_{i=0}^{2-1} h_i c_i(x, t) - \mu \sum_{i=0}^{1-1} h_i c_i(x, t), \tag{8.12}$$

where for $i = 0$

$$
\begin{aligned}
c_i(x, t) &= \sum_{n=0}^{\infty} \left(\frac{R}{D}\right)^n \frac{\exp[-\lambda t]t^{-1-n}}{\Gamma(-n)n!} x^{n\alpha+i} \times \left(\frac{\partial}{\partial x}\right)^n E_{\alpha-\beta, \beta n+i+1}(\mu x^{\alpha-\beta}) \\
&- \mu \sum_{n=0}^{\infty} \left(\frac{R}{D}\right)^n \frac{\exp[-\lambda t]t^{-1-n}}{\Gamma(-n)n!} x^{n\alpha+i+\alpha-\beta} \times \left(\frac{\partial}{\partial x}\right)^n E_{\alpha-\beta, \beta n+i+1+\alpha-\beta}(\mu x^{\alpha-\beta})
\end{aligned}
\tag{8.13}
$$

and for $i = 1$

$$
c_i(x, t) = \sum_{n=0}^{\infty} \left(\frac{R}{D}\right)^n \frac{\exp[-\lambda t]t^{-1-n}}{\Gamma(-n)n!} x^{n\alpha+i} \times \left(\frac{\partial}{\partial x}\right)^n E_{\alpha-\beta, \beta n+i+1}(\mu x^{\alpha-\beta}). \tag{8.14}
$$

The coefficients h_i are found by using the initial and boundary conditions in $c(x, t)$.

■ Example 2

Our primary concern here is to consider Eq. (8.1) for $\alpha = 2$ and $0 < \beta \leq 2$. Owing to what was discussed earlier, we present the two solutions of space–time derivative of hydrodynamic advection–dispersion equation as

$$
\begin{aligned}
c_1(x, t) &= \sum_{n=0}^{\infty} \left(\frac{R}{D}\right)^n \frac{\exp[-\lambda t]t^{-1-n}}{\Gamma(-n)n!} x^{2n} \times \left(\frac{\partial}{\partial x}\right)^n E_{2-\beta, \beta n+1}(\mu x^{2-\beta}) \\
&- \mu \sum_{n=0}^{\infty} \left(\frac{R}{D}\right)^n \frac{\exp[-\lambda t]t^{-1-n}}{\Gamma(-n)n!} x^{2n+2-\beta} \\
&\quad \times \left(\frac{\partial}{\partial x}\right)^n E_{2-\beta, \beta n+3-\beta}(\mu x^{2-\beta}), \\
c_2(x, t) &= \sum_{n=0}^{\infty} \left(\frac{R}{D}\right)^n \frac{\exp[-\lambda t]t^{-1-n}}{\Gamma(-n)n!} x^{2n+1} \times \left(\frac{\partial}{\partial x}\right)^n E_{2-\beta, \beta n+2}(\mu x^{2-\beta}).
\end{aligned}
\tag{8.15}
$$

These solutions are linearly independent and they result in the fundamental system of solutions to space–time Caputo fractional derivative of hydrodynamic advection–dispersion equation [122]. Some of analytical methods and their recent development that have been used for solving nonlinear fractional partial differential equation can found in [119,121,122,165,123,127,124–126,128–130] and other classical books on analytical and numerical methods.

8.2 Time Fractional Convection–Dispersion Equation With Caputo–Fabrizio Type

In this section, the convection dispersion equation with the Caputo–Fabrizio derivative is analyzed. It is important to note that the Caputo–Fabrizio has exponential decay kernel, which allows it to describe materials in heterogeneous media as was indicated in the Caputo–Fabrizio paper. This model can be used where the model with the power law has elapsed; considerations should be done with much thought on a groundwater system associated with significant heterogeneity, because this phenomenon causes one aquifer system's hydraulic properties to differ significantly from an adjacent aquifer system, and/or even within the same aquifer system. Based on the literature review, this operator was introduced due to the necessity of employing the behavior of classical viscoelastic material; and in the case of this study, the viscoelastic material represents the geological formation through which movement pollution occurs.

8.3 Time Fractional Convection–Dispersion Equation With Atangana–Baleanu Derivative

In this section, we consider the groundwater transport model where the time derivative is converted to the fractional time derivative with the generalized Mittag–Leffler function as kernel. The modified equation is therefore given as

$$\frac{AB(\alpha)}{1-\alpha}\int_0^t \frac{\partial C}{\partial y} E_\alpha\left(-\frac{\alpha}{1-\alpha}(t-y)^\alpha\right) = -v\frac{\partial C}{\partial x} + D\frac{\partial^2 C}{\partial x^2} \tag{8.16}$$

with the following initial condition

$$C(x,0) = C_0. \tag{8.17}$$

The above equation will be analyzed numerically and semi-analytically.

8.3.1 Derivation of Numerical Solution

As presented earlier, we consider the numerical approximation of the Atangana–Baleanu fractional derivative given as follows

$$\frac{AB(\alpha)}{1-\alpha}\int_0^{t_n} \frac{\partial C}{\partial y} E_\alpha\left(-\frac{\alpha}{1-\alpha}(t_n-y)^\alpha\right) = \frac{AB(\alpha)}{1-\alpha}\sum_{j=1}^n \left(\frac{C_i^{j+1}-C_i^j}{\Delta t}\right)\Sigma_{j,i}^\alpha \tag{8.18}$$

where

$$\delta_{j,i}^{\alpha} = \Delta t \left((t_n - t_j) E_{\alpha,2} \left(-\frac{\alpha}{1-\alpha}(t_n - t_j) \right) - (t_n - t_{j+1}) E_{\alpha,2} \left(-\frac{\alpha}{1-\alpha}(t_n - t_{j+1}) \right) \right).$$

(8.19)

Inserting Eq. (8.18) into (8.16), also considering the numerical approximation of first and second order space-derivative, we obtain

$$\frac{AB(\alpha)}{1-\alpha} \sum_{j=1}^{n} \left(\frac{C_i^{j+1} - C_i^{j}}{\Delta t} \right) \delta_{j,i}^{\alpha} = -v \left[\left(\frac{C_{i+1}^{j+1} - C_{i-1}^{j+1}}{2\Delta x} \right) + \left(\frac{C_{i+1}^{j} - C_{i-1}^{j}}{2\Delta x} \right) \right] +$$
$$\frac{D}{2(\Delta x)^2} \left(\left(C_{i+1}^{n+1} - 2C_i^{n+1} + C_{i-1}^{n+1} \right) + \left(C_{i+1}^{n} - 2C_i^{n} + C_{i-1}^{n} \right) \right). \quad (8.20)$$

The above equation can be reformulated to obtain

$$\left(\frac{AB(\alpha)}{(1-\alpha)\Delta t} \delta_{n,i}^{\alpha} + \frac{D}{(\Delta x)^2} \right) C_i^{n+1} = \left(\frac{AB(\alpha)}{(1-\alpha)\Delta t} \delta_{n,i}^{\alpha} - \frac{D}{(\Delta x)^2} \right) C_i^{n} -$$
$$\frac{AB(\alpha)}{1-\alpha} \sum_{j=1}^{n-1} \left(\frac{C_i^{j+1} - C_i^{j}}{\Delta t} \right) \delta_{j,i}^{\alpha} - v \left[\left(\frac{C_{i+1}^{j+1} - C_{i-1}^{j+1}}{2\Delta x} \right) + \left(\frac{C_{i+1}^{j} - C_{i-1}^{j}}{2\Delta x} \right) \right] +$$
$$\frac{D}{2(\Delta x)^2} \left(\left(C_{i+1}^{n+1} + C_{i-1}^{n+1} \right) + \left(C_{i+1}^{n} + C_{i-1}^{n} \right) \right). \quad (8.21)$$

The above recursive formula can be used to generate the following numerical simulation for different values of fractional orders α.

8.4 Limitation of Fractional Derivative for Groundwater Pollution Model

Although the concepts of fractional differentiation and integration appear to be the best mathematical tools to describe the movement of groundwater flow within the geological formation called aquifers, one will note that this concept has its limitations. We shall inform that field and numerical experiments of solute transport by means of heterogeneous porous and fractured aquifers show that the movement of pollutant plumes may not display constant scaling; nevertheless, instead transition diffusive states that is to say, display complex diffusion such as super-diffusion, sub-diffusion and Fickian diffusion, which occur at multiple transport scales. These transitions are therefore perhaps likely associated with physical properties of the geological formations, for instance variability in medium heterogeneity. In the geological formation through which the movement or the transport of solute take place, many possible mechanisms may be accountable for the occurrence of and also more precisely the transition between diffusion states. Let us start with the so-called normal diffusion typically occurring at a particular spatiotemporal scales that may transfer to or perhaps be transferred from any other non-Fickian diffusion linked to the properties of geological formations and tracer. The fractional Brownian motion is assumed for tracer particles in Darcy-scale homogeneous geological formations or also at long travel distance, much longer than the maximum correlation scale of heterogeneity in a stationary geological formation which is caused by the central limit theory. A limitation here is that for a non-stationary geological formation, nevertheless, the approximate Fickian may not be obtained because the characteristics of particle waiting period evolve with travel scale as was studied in [188,189] and such dynamical system is referred to as anomalous diffusion. Another problem is that of super-diffusion that may be considered as space dependent, where the scaling rate may influence the local aquifer properties with the most usual condition for fast transport is a preferential flow path that the contaminant can access. In this case, one would expect the scaling rate or displacement force to either be increased, that is when the flow becomes more particular along the solute, contaminant particle path or also decreased when low-permeable deposit surround and separate the flow, depending of course on the plume position in the matrix rock and geometry of the aquifer. The last problem is the so-called sub-diffusion because of the mass exchange that occurs between high and low permeable depositional zones or layers that may be both time and space dependent. Many of these models can describe the similar transition from non-Fickian to Fickian because of the trapping of the solute particle of pollutant within low-velocity zones. We shall note that the super-diffusion cause by the fast motion and sub-diffusion caused by mass exchange could occur at the same time, if the preferential flow paths and stagnant flow zones co-existed. In this case the physical mathematical model must be able to accurately replicate characterization not only for each super-diffusion or sub-diffusion, but both, which may be very difficult to be handled with fractional constant order derivatives.

CHAPTER 9

Fractional Variable Order Derivatives

This chapter is devoted to the concept of derivatives with variable order. They are differential operators whose order of differentiation is a function of the independent variable. Clearly speaking this means, if for instance the derivative is taken with respect to time, it follows that the order of differentiation is also a function of time. The concept of variable order was introduced to further explain complex behavior of some physical problems where the concept of fractional derivative with constant order failed. Variable-order fractional calculus is the generalization of classical calculus and fractional calculus, which were invented by Newton and Leibnitz hundreds of years ago. Now the study on it becomes a hotpot in recent ten years. It has turned out that many problems in physics, biology, engineering, and finance can be described excellently by models using mathematical tools from variable-order fractional calculus, such as mechanical applications, diffusion process, multi-fractional Gaussian noise, and FIR filters. For more details, see [113–117] and the references therein. As the generalized form of fractional differential equations (see [113,114] for a comprehensive review of fractional calculus and fractional differential equations), the variable-order fractional differential equations mean the differential equations with variable-order fractional derivatives. For the lack of directly viewing physical interpretations, the variable-order fractional calculus developed at a very low pace during the foregone several decades. In 1995, the fractional integration and differentiation of variable-order fractional were studied by Samko in [113–117], which contains mathematical analysis of variable-order fractional differential and integral operators but not variable-order fractional models. The later research results have shown that many complex physical phenomena can be better described by using variable-order fractional differential equations. For instance, an experimental investigation of temperature-dependent variable-order fractional integrator and differentiator is presented in [113–117]. Based on the investigation, it is easier to understand the physical meaning of variable-order fractional calculus and better to know how the variable-order fractional operators are applied in physical processes. As a concrete example, the variable-order fractional calculus through the description of a simple problem in mechanics is presented in [113–117]. A mathematical definition for the variable-order fractional differential operator is given and is suitable for mechanical modeling therein.

9.1 Definition of Existing Variable Order Derivatives

A thorough literature review showed a handful of publications within the scope of variable order calculus. It is noted that a literature review of such a narrow subject is scarce due to the fact that there is no established terminology as in the case of derivatives with fractional order. According to the literature, the concept of variable-order operators goes back to the work done in [113–117] in 1998. In their work, they suggested a generalized linear Riemann–Liouville integration variable-order operator defined as follows

$$_0D_t^{-q(t)}f(t) = \int_0^t \frac{(t-y)^{q_e(t,y)-1}}{\Gamma(q_g(t,y))}dy, q(t) > 0. \tag{9.1}$$

In the above definition, the functions $q_e(t, y)$ and $q_g(t, y)$ have different arguments. For instance, these arguments can be defined as $q_e(t, y) = q(t)$, $q_e(t, y) = q(y)$ and $q_e(t, y) = q(t - y)$. It is important to note that based on the behavior of the operator for different functions $f(t)$, the q-arguments meaning $q_e(t, y)$ and $q_g(t, y)$ were assumed to be equal, therefore $q(t, y) = q(t - y)$ was favored for its adherence to the index rule:

$$_0D_t^{-q(t)}\left(_0D_t^{-\beta(t)}\right)f(t) =_0 D_t^{-\beta(t)}\left(_0D_t^{-q(t)}\right)f(t). \tag{9.2}$$

Eq. (9.2) can be viewed as convolution as follows

$$_0D_t^{-q(t)}f(t) = \left(\frac{t^{-q(t)-1}}{\Gamma(q(t))}\right) * f(t). \tag{9.3}$$

Based upon equation (9.3) the associated Riemann–Liouville variable-order derivative was obtained and defined as:

$$_0D_t^{q(t)}f(t) = \frac{d^n}{dt^n}\int_0^t \frac{(t-y)^{n-q(t-y)-1}}{\Gamma(n-q(t-y))}f(y)dy, n-1 < q(t) < 0. \tag{9.4}$$

A detailed investigation on the three arguments for the variable-order $q(t, y)$ were done in [113–117]. Their investigation showed that the arguments $q(t, y) = q(y)$ and $q(t, y) = q(t - y)$ result in a memory of not only the prior history of the function, but also of the order. A measure of the memory retention of the order was introduced to answer the question of how we obtain a particular variable order for a given real world problem? In addition to this is the introduction of the generalized q-argument $q(t, y) = q(at + by)$ that allows for a tailored memory of the order. Coimbra in 2003 [112] took different approach and independently obtained his derivative with variable order using the Laplace transform of the Caputo fractional derivative. He suggested the following definition or formula:

$$_0D_t^{q(t)}f(t) = \int_0^t \frac{(t-y)^{-q(t)}}{\Gamma(1-q(t))}\frac{df(y)}{dy}dy + \frac{\left(f(0^+ - f(0^-))\right)}{\Gamma(1-q(t))}t^{-q(t)}, 0 \le q(t) < 1. \tag{9.5}$$

It is worth noting that the additional term in (9.5) accounts for the behavior of the physical system when it departs from dynamic equilibrium. Also, the value of $f(0)$ implies a Heaviside distribution. In 2004 Ingman and Suzdalnitsky suggested another formula for variable differ-integral operator [118]:

$$_0 D_t^{q(t)} f(t) = \frac{d^n}{dt^n} \int_a^t \frac{(y-a)^{n-q(y)-1}}{\Gamma(n-q(y))} \left[f(t-y+a) - \sum_j^{n-1} \frac{f^j(a)}{j!} (t-a)^j \right] dy, \quad (9.6)$$

where $q(t) > 0$ and $n - 1 < q(t) < n$. Nevertheless, when $q(t) \geq 0$, the following definition was suggested:

$$_0 D_t^{q(t)} f(t) = \frac{d^n}{dt^n} \int_a^t \frac{(y-a)^{n-q(y)-1}}{\Gamma(n-q(y))} \left[f(t-y+a) \right] dy. \quad (9.7)$$

9.2 Limitations and Advantages

Anomalous diffusion phenomena are extensively observed in physics, chemistry, and biology fields [113–117]. To characterize anomalous diffusion phenomena, constant-order fractional diffusion equations are introduced and have received tremendous success. However, it has been found that the constant order fractional diffusion equations are not capable of characterizing some complex diffusion processes, for instance, diffusion process in inhomogeneous or heterogeneous medium [117]. In addition, when we consider diffusion process in porous medium, if the medium structure or external field changes with time, in this situation, the constant-order fractional diffusion equation model cannot be used to well characterize such phenomenon [114,115]. Still in some biology diffusion processes, the concentration of particles will determine the diffusion pattern [115,117]. To solve the above problems, the variable-order (VO) fractional diffusion equation models have been suggested for use [115–117]. The ground-breaking work of VO operator can be traced to Samko et al. by introducing the variable-order integration and Riemann–Liouville derivative in [115–117]. It has been recognized as a powerful modeling approach in the fields of viscoelasticity [115–117], viscoelastic deformation, viscous fluid, and anomalous diffusion [113–117].

9.3 Atangana–Koca Variable-Order Derivative

The concept of derivative with variable order has been used for complex diffusion problems with great results, however the design of these derivatives was complicated and very difficult to handle analytically. These derivatives could only be evaluated numerically, however there exist a lot of complex problems in real world that request analytical solutions. Recently, Atangana and Koca proposed a new derivative with variable order that is simple to use and also can be used for analytical problems. The definition is given as:

Definition 9.3.1 (Atangana–Koca variable order in Caputo sense). *Let $u(x)$ be a continuous function on a closed interval $[a, b]$. Let $f(t)$ be a continuous function that does not need to be differentiable on an open interval I, and then the derivative of $f(t)$ with variable order $u(x)$ is given by*

$$_0^{AK}D_t^{f(x)}[h(t)] = \int_0^t \frac{dh(y)}{dy}\exp[-f(x)(t-y)]dy. \tag{9.8}$$

As with the discussion about all Caputo fractional derivatives with constant and variable-order type, the above definition required that the function should be differentiable. Therefore, they proposed the following definition.

Definition 9.3.2 (Atangana–Koca derivative with variable order in Riemann–Liouville sense). *Let $u(x)$ be a continuous function on a closed interval $[a, b]$. Let $f(t)$ be a continuous function that does not need to be differentiable on an open interval I, and then the derivative of $f(t)$ with variable order $u(x)$ is given by*

$$_0^{A}D_t^{u(x)}[f(t)] = \frac{d}{dt}\int_0^t f(y)\exp[-u(x)(t-y)]dy. \tag{9.9}$$

The above derivative does not have any singularity as the existing one in the literature.

Remark 1. The above formula is equivalent to

$$_0^{A}D_t^{u(x)}[f(t)] = f(t) - u(x)\int_0^t f(y)\exp[-u(x)(t-y)]dy.$$

Theorem 9.3.3. *Existence: If the function is continuous in a closed interval, say $[a, b]$, and $u(t)$ be a positive function on a closed interval I, then the derivative of $f(x)$ with variable order $u(t)$ exists; more importantly, it is of exponential order.*

Proof. We call $C[a, b]$ the space of continuous functions f on $[a, b]$ awarded with the norm

$$\|f(t)\| = \max_{t\in[a,b]} |f(t)| \tag{9.10}$$

By definition, we have the following

$$\left\| {_0^A D_t^{u(x)}}[f(t)] \right\| = \left\| \frac{d}{dt}\left\{ \int_0^t f(y)\exp[-u(x)(t-y)]dy \right\} \right\|$$

$$= \left\| f(t) - u(x)\int_0^t f(y)\exp[-u(x)(t-y)]dy \right\|$$

$$\leq \|f(t)\| + u(x)\left\| \int_0^t f(y)\exp[-u(x)(t-y)]dy \right\|$$

$$\leq M + u(x)\int_0^t \|f(y)\|\exp[-u(x)(t-y)]dy$$

$$\leq M + Mu(x)\int_0^t \exp[-u(x)(t-y)]dy$$

$$= M(1+u(x))\left[\frac{1}{u(x)} - \frac{1}{u(x)}\exp[-u(x)t] \right]$$

Therefore

$$\left\| {_0^A D_t^{u(x)}}[f(t)] \right\| \leq M(1+u(x))\left[\frac{1}{u(x)} - \frac{1}{u(x)}\exp[-u(x)t] \right]. \tag{9.11}$$

$\square$

Theorem 9.3.4. *If the function $f(t)$ is infinitely differentiable at the origin and $u(x)$ be a positive function on a closed interval I, then the derivative of $f(t)$ with variable order $u(x)$ is given by*

$$_0^A D_t^{u(x)}[f(t)] = \sum_{k=0}^n \frac{f^{(k)}(0)}{k!} t^{n-k}$$

$$+ \sum_{k=0}^n \frac{f^{(k)}(0)}{k!}\exp[-tu(x)]t^{n-k+1}$$

$$\cdot \begin{pmatrix} \Gamma(n-k+1,-tu(x)) \\ -\Gamma(n-k+1) \end{pmatrix}(-tu(x))^{n-k+1} + R_n. \tag{9.12}$$

Proof. If the function is infinitely differentiable, then we have the following

$$f(t) = \sum_{k=0}^{n} \frac{f^{(k)}(0)}{k!} t^{n-k} + \frac{1}{n!} \int_{0}^{t} (t-y)^n f^{(n+1)}(y) dy. \tag{9.13}$$

Therefore applying the derivative of variable order to both sides of Eq. (9.13) we obtain

$$
\begin{aligned}
{}_{0}^{A}D_t^{u(x)}[f(t)] &= {}_{0}^{A}D_t^{u(x)} \left\{ \sum_{k=0}^{n} \frac{f^{(k)}(0)}{k!} t^{n-k} + \frac{1}{n!} \int_{0}^{t} (t-y)^n f^{(n+1)}(y) dy \right\} \\
&= \frac{d}{dt} \int_{0}^{t} \left\{ \sum_{k=0}^{n} \frac{f^{(k)}(0)}{k!} \tau^{n-k} + \frac{1}{n!} \int_{0}^{\tau} (\tau-y)^n f^{(n+1)}(y) dy \right\} \\
&\quad . \exp[-u(x)(t-\tau)] d\tau \\
&= \sum_{k=0}^{n} \frac{f^{(k)}(0)}{k!} t^{n-k} + \frac{1}{n!} \int_{0}^{t} (t-y)^n f^{(n+1)}(y) dy - \\
&\quad \frac{1}{u(x)} \int_{0}^{t} \left\{ \sum_{k=0}^{n} \frac{f^{(k)}(0)}{k!} \tau^{n-k} + \frac{1}{n!} \int_{0}^{\tau} (\tau-y)^n f^{(n+1)}(y) dy \right\} \\
&\quad . \exp[-u(x)(t-\tau)] d\tau \\
&= \sum_{k=0}^{n} \frac{f^{(k)}(0)}{k!} t^{n-k} - u(x) \int_{0}^{t} \left\{ \sum_{k=0}^{n} \frac{f^{(k)}(0)}{k!} \tau^{n-k} \right\} \\
&\quad . \exp[-u(x)(t-\tau)] d\tau \\
&\quad + u(x) \int_{0}^{t} \left\{ \frac{1}{n!} \int_{\tau}^{0} (\tau-y)^n f^{(n+1)}(y) dy \right\} . \exp[-u(x)(t-\tau)] d\tau.
\end{aligned}
$$

However,

$$u(x) \int_{0}^{t} \left\{ \sum_{k=0}^{n} \frac{f^{(k)}(0)}{k!} \tau^{n-k} \right\} \exp[-u(x)(t-\tau)] d\tau \tag{9.15}$$

$$
\begin{aligned}
&= u(x) \sum_{k=0}^{n} \frac{f^{(k)}(0)}{k!} \exp[-tu(x)] t^{n-k+1} \\
&\quad . (\Gamma(n-k+1, -tu(x)) - \Gamma(n-k+1)) (-tu(x))^{n-k+1}.
\end{aligned}
$$

Now putting Eq. (9.15) into (9.15), we obtain

$$
{}_0^A D_t^{u(x)}[f(t)] = \sum_{k=0}^n \frac{f^{(k)}(0)}{k!} t^{n-k}
$$

$$
+ \sum_{k=0}^n \frac{f^{(k)}(0)}{k!} \exp[-tu(x)] t^{n-k+1}
$$

$$
\cdot \left(\begin{array}{c} \Gamma(n-k+1, -tu(x)) \\ -\Gamma(n-k+1) \end{array} \right) (-tu(x))^{n-k+1} + R_n
$$

with

$$
R_n = u(x) \int_0^t \left\{ \frac{1}{n!} \int_\tau^0 (\tau - y)^n f^{(n+1)}(y) dy \right\} \cdot \exp[-u(x)(t - \tau)] d\tau. \tag{9.16}
$$

This completes the proof. □

Theorem 9.3.5. *If the function $f(t)$ is infinitely differentiable at the origin and $u(x)$ be a positive function on a closed interval I, then the derivative of $f(t)$ with variable order $u(x)$ of the version proposed by Atangana and Koca is as follows*

$$
{}_0^{AK} D_t^{u(x)}[f(t)] = \sum_{k=0}^n \frac{f^{(k)}(0)}{k!} \exp[-tu(x)] t^{n-k}
$$

$$
((n-k)\Gamma(n-k, -tu(x)) + \Gamma(n-k+1))
$$

$$
\cdot (-tu(x))^{n-k} + R_n.
$$

The proof is the same as previously.

Remark 2. The new derivative with variable order has Lipschitz condition

$$
\left\| {}_0^A D_t^{u(x)}[f(t)] - {}_0^A D_t^{u(x)}[h(t)] \right\| \tag{9.17}
$$

$$
= \left\| \frac{d}{dt} \int_0^t \{f(y) - h(y)\} \exp[-u(x)(t - y)] dy \right\|
$$

$$
\leq \Phi_1 \left\| \int_0^t \{f(y) - h(y)\} \exp[-u(x)(t - y)] dy \right\|
$$

$$\le \ \Phi_1 \left\| \int_0^t \|f(y) - h(y)\| \exp[-u(x)(t-y)]dy \right\|$$

$$\le \ \Phi_1 \|f(t) - h(t)\| \int_0^t \exp[-u(x)(t-y)]dy$$

$$\le \ \Phi_1 \|f(t) - h(t)\| \left(\frac{1 - \exp[-u(x)t]}{u(x)} \right)$$

$$= \ L \|f(t) - h(t)\|.$$

Remark 3. When the variable order is zero, we recover the initial function

$$_0^A D_t^0[f(t)] = \frac{d}{dt} \int_0^t f(y)dy = f(t). \tag{9.18}$$

Theorem 9.3.6. *The relation between the two derivatives with variable order is given by*

$$_0^A D_t^{f(x)}[h(t)] = _0^{AK} D_t^{f(x)}[h(t)] + h(0)\exp[-f(x)t]. \tag{9.19}$$

And then if the function h is zero at the origin then both derivatives coincide.

Proof. Integrating by part the formula of Eq. (9.9) we obtain the following

$$_0^A D_t^{f(x)}[h(t)] \ = \ \frac{d}{dt} \int_0^t h(y)\exp[-f(x)(t-y)]dy \tag{9.20}$$

$$= \ h(t) - f(x) \int_0^t h(y)\exp[-f(x)(t-y)]dy.$$

But by integration by part the below formula produces

$$\int_0^t h(y)\exp[-f(x)(t-y)]dy \ = \ \frac{h(t)}{f(x)} - \frac{h(0)}{f(x)}\exp[-f(x)t]$$

$$- \frac{1}{f(x)} \int_0^t \frac{dh(y)}{dy}\exp[-f(x)(t-y)]dy.$$

Replacing the above into (9.20) we obtain

$$_0^A D_t^{f(x)}[h(t)] =_0^{AK} D_t^{f(x)}[h(t)] + h(0)\exp[-f(x)t].$$

This completes the proof. □

Theorem 9.3.7. *The new derivative with variable-order stationary points for the function*

$$f(t) = \exp[-u(x)t]$$

is given as

$$t = \frac{1}{u(x)}. \tag{9.21}$$

Proof. In general the derivative of a function exp[at] of variable order $u(x)$ is given as

$$\begin{aligned}_0^A D_t^{u(x)}[f(t)] &= \frac{d}{dt}\int_0^t \exp[ay]\exp[-u(x)(t-y)]dy \tag{9.22}\\ &= \frac{a\exp[at] + \exp[-u(x)t]u(x)}{a+u(x)}.\end{aligned}$$

Therefore if the function exp[$-u(x)t$] using Eq. (9.22), we obtain

$$_0^A D_t^{u(x)}[f(t)] = \exp[-u(x)t](1 - tu(x)). \tag{9.23}$$

Now equating the above equals to zero we obtain the desired result. □

9.4 Relation With Some Integral Transform

In this section, we present some useful relation of the variable-order derivative with some existing integral transform operator, starting with the Laplace transform. Note that these relations are established because they are used to solve some differential equations with the new derivative with variable order. The Laplace transform of the new derivative with variable order is given as

$$L\left[_0^A D_t^{f(x)}[h(t)]\right](s) = \frac{sL[h(t)](s)}{s+f(x)}. \tag{9.24}$$

Proof. By definition

$$L\left[{}_0^A D_t^{f(x)}[h(t)]\right](s) = L\left\{\frac{d}{dt}\int_0^t h(y)\exp[-f(x)(t-y)]dy\right\}(s)$$

Using the properties of Laplace transform, we obtain the following

$$L\left\{\frac{d}{dt}\int_0^t h(y)\exp[-f(x)(t-y)]dy\right\}(s) \qquad (9.25)$$

$$= sL\{h(t)\}(s)L\{\exp[-f(x)t]\}(s) = \frac{sL\{h(t)\}(s)}{s+f(x)}.$$

This completes the proof. The Fourier transform of the new derivative with variable order is given as

$$F\left\{{}_0^A D_t^{u(x)}[f(t)]\right\}(w) = F\left\{\frac{d}{dt}\int_0^t f(y)\exp[-u(x)(t-y)]dy\right\}(w) \qquad (9.26)$$

$$= wF\{f(t)\}(w)F\{\exp[-u(x)t]\}(w)$$

$$= wF\{f(t)\}(w)\frac{2u(x)}{u^2(x)+w^2}.$$

The relation with Mellin Transform

$$M\left\{{}_0^A D_t^{u(x)}[f(t)]\right\}(s) = M\left\{\frac{d}{dt}\int_0^t f(y)\exp[-u(x)(t-y)]dy\right\}(s) \qquad (9.27)$$

$$= -(s-1)M\left\{\int_0^t f(y)\exp[-u(x)(t-y)]dy\right\}(s)$$

$$= -(s-1)M\{f(t)\}(s)M\{\exp[-u(x)t]\}(s)$$

$$= -(s-1)M\{f(t)\}(s)u(x)^s\Gamma(s)$$

Relation with Sumudu Transform

$$S\left\{{}_0^A D_t^{u(x)}[f(t)]\right\}(u) = S\left\{\frac{d}{dt}\int_0^t f(y)\exp[-u(x)(t-y)]dy\right\}(u) \qquad (9.28)$$

$$= \frac{1}{u} S \left\{ \int_0^t f(y) \exp[-u(x)(t-y)] dy \right\}(u)$$

$$= \frac{1}{u} S\{f(t)\}(u) S\{\exp[-u(x)t]\}(u)$$

$$= \frac{1}{u} S\{f(t)\}(u) \frac{1}{1+u(x)u} \qquad \square$$

Theorem 9.4.1. *The variable-order integral associate to the new derivative is given as*

$$_0^A I_t^{u(x)}\,(f(t)) = f(t) + u(x) \int_0^t f(y) dy \qquad (9.29)$$

Proof. To obtain the above result, we shall solve the following equation (9.29) using one of the above integral transforms for instance the Laplace transform

$$_0^A D_t^{u(x)}\,(f(t)) = h(t). \qquad (9.30)$$

Therefore applying the Laplace transform, we obtain

$$\frac{sF(s)}{s+u(x)} = H(s) \Rightarrow F(s) = H(s) + u(x)\frac{H(s)}{s} \qquad (9.31)$$

Applying the inverse Laplace, we obtain

$$f(t) = h(t) + u(x) \int_0^t h(y) dy \qquad (9.32)$$

This completes the proof. $\qquad \square$

Remark 4. If the function f and the variable order u are bounded then the variable-order integral is bounded.

Proof. By definition,

$$_0^A I_t^{u(y)}\left[f(x,t)\right] = f(x,t) + u(y)\int_0^t f(x,l) dl$$

$$\left\| _0^A I_t^{u(y)}\left[f(x,t)\right] \right\|_{C[I\times J]} \leq \|f(x,t)\|_{C[I\times J]} + \|u(y)\|_{C[a,b]} T \|f(x,t)\|_{C[I\times J]}$$

$$\leq M + aTM. \qquad \square$$

Corollary 9.4.2. *The following equation* ${}_0^A D_t^{u(x)}(f(t)) = K$ *has the following solution*

$$f(t) = K + u(x)t. \tag{9.33}$$

Proof. The proof is achieved by means of the use of Laplace transform as follows

$$
\begin{aligned}
Ł({}_0^A D_t^{u(x)}(f(t)))(s) &= L(K)(s) \Leftrightarrow \\
\frac{sF(s)}{s + u(x)} &= \frac{K}{s} \\
F(s) &= \frac{K}{s} + \frac{u(x)}{s^2} \Leftrightarrow \\
f(t) &= K + u(x)t. \qquad \square
\end{aligned}
$$

Corollary 9.4.3. *The following equation* ${}_0^A D_t^{u(x)}(f(t)) = f(t) + K$ *has the below solution*

$$f(t) = -K\left(1 + \frac{\delta(t)}{u(x)}\right). \tag{9.34}$$

Proof. Again using the Laplace transform, we obtain

$$
\begin{aligned}
Ł({}_0^A D_t^{u(x)}(f(t)))(s) &= L(f(t) + K)(s) \Leftrightarrow \\
\frac{sF(s)}{s + u(x)} &= \frac{K}{s} + F(s) \\
F(s) &= -\frac{K}{s} - \frac{K}{u(x)} \Leftrightarrow \\
f(t) &= -K - \frac{K}{u(x)}\delta(t). \qquad \square
\end{aligned}
$$

9.5 Partial Derivative With Variable Order

Many physical problems involve time and space components, therefore, we extend the concept of variable-order derivative to partial variable-order derivative. We provide the definition as follows.

Definition 9.5.1. *Let* $f(x, t)$ *be a continuous function defined in* $[a_1, b_1][t_1, t_2]$. *Let* $u(y)$ *be a non-constant continuous function defined in an open interval* I, *then the derivative of* $f(x, t)$ *with variable order* $u(y)$ *is defined as follows*

$$ {}_0^A D_x^{u(y)}\left[f(x, t)\right] = \frac{\partial}{\partial x} \int_0^x f(l, t) \exp[-u(y)(x - l)] dl. \tag{9.35}$$

Theorem 9.5.2. *Let $f(x,t)$ be a continuous function defined in $[a_1, b_1][t_1, t_2]$. Let $u(y)$ and $v(y)$ be non-constant continuous functions defined in an open interval I, then the following relation holds*

$$
{}_0^A D_x^{u(y)} \left[{}_0^A D_t^{v(y)} f(x,t) \right] = {}_0^A D_t^{v(y)} \left[{}_0^A D_x^{u(y)} f(x,t) \right]. \tag{9.36}
$$

Proof. By definition the left-hand side of Eq. (9.36) can be formulated as follows

$$
{}_0^A D_x^{u(y)} \left[{}_0^A D_t^{v(y)} f(x,t) \right] = \frac{\partial}{\partial x} \int_0^x {}_0^A D_t^{v(y)} f(k,t) \exp[-u(y)(x-k)] dk \tag{9.37}
$$

$$
= \frac{\partial}{\partial x} \int_0^x \frac{\partial}{\partial t} \int_0^t f(k,l) \exp[-v(y)(t-l)] dl
$$
$$
. \exp[-u(y)(x-k)] dk
$$

$$
= \frac{\partial^2}{\partial x \partial t} \int_0^x \int_0^t f(k,l) \exp[-v(y)(t-l)]
$$
$$
. \exp[-u(y)(x-k)] dk dl.
$$

For simplicity, let us put

$$
F(x,t) = \int_0^x \int_0^t f(k,l) \exp[-v(y)(t-l)] \exp[-u(y)(x-k)] dk dl. \tag{9.38}
$$

The above function is continuous in both directions and the first partial derivatives exist and are continuous, therefore using the Clauraut's theorem, we obtain the following

$$
\frac{\partial^2}{\partial x \partial t} F(x,t) = \frac{\partial^2}{\partial t \partial x} F(x,t) \Rightarrow {}_0^A D_x^{u(y)} \left[{}_0^A D_t^{v(y)} f(x,t) \right] \tag{9.39}
$$

$$
= \frac{\partial^2}{\partial t \partial x} \int_0^x \int_0^t f(k,l) \exp[-v(y)(t-l)]
$$
$$
. \exp[-u(y)(x-k)] dl dk
$$

$$
= \frac{\partial}{\partial t} \int_0^t \left[\frac{\partial}{\partial x} \int_0^x f(k,l) \exp[-u(y)(x-k)] dk \right]
$$
$$
. \exp[-v(y)(t-l)] dl
$$

$$= \frac{\partial}{\partial t} \int_0^t \left[{}_0^A D_t^{v(y)} \left(f(x,l) \right) \right] \exp[-v(y)(t-l)] dl$$

$$= {}_0^A D_t^{v(y)} \left[{}_0^A D_x^{u(y)} \left(f(x,t) \right) \right].$$

This completes the proof.   □

Theorem 9.5.3. *Let $f(x,t) \in C^1[x_1, x_2][t_1, t_2]$ and $0 < u(y) \in I$ and non-constant, then the following equation*

$$ {}_0^A D_t^{u(y)} \Theta(x,t) = \Theta(x,t) + f(x,t) \tag{9.40}$$

has as exact solution

$$\Theta(x,t) = -\frac{1}{u(y)}[f'(x,t) + \delta(t)f(x,0) + f(x,t)]. \tag{9.41}$$

Proof. To obtain the above result, we apply on both sides of Eq. (9.41) the Laplace transform, then we obtain

$$\frac{s\Theta(x,s)}{s+u(y)} = \Theta(x,s) + F(x,s) \Rightarrow \tag{9.42}$$

$$\Theta(x,s) \left(\frac{s}{s+u(y)} - 1 \right) = F(x,s)$$

$$\Theta(x,s) = -\frac{s+u(y)}{u(y)} F(x,s). \tag{9.43}$$

Now employing the inverse Laplace on both sides of Eq. (9.43), we obtain

$$\Theta(x,t) = -\frac{1}{u(y)}[f'(x,t) + \delta(t)f(x,0) + f(x,t)].$$

This completes the proof.   □

9.6 Atangana–Koca Derivative With Variable Order in Caputo Sense

The first variable-order derivative was based on the idea introduced by Riemann–Liouville. This version has its advantages and disadvantages and can be complemented with the version based on the Caputo concept. In this part of the book, we shall present the Caputo version of Atangana–Koca variable-order derivative. This version is very useful when dealing with the partial differential equation or ordinary differential equations with initial conditions, in particular when dealing with the solvability by means of Green-function which involves the Laplace transform operator. We shall start here with the definition

Definition 9.6.1. *Let there be a positive function, $l(x) \in C^1[a, b]$, let $a(t)$ be a differentiable function in an open interval I, then the derivative of fractional variable order $l(x)$ of $a(t)$ is given as follows*

$$_0^{AK} D_t^{l(x)}[a(t)] = \int_0^t \frac{da(y)}{dn} \exp[-l(x)(t-y)]dn. \tag{9.44}$$

As discussed earlier, one can find out that the above definition is user-friendlier than the existing Caputo variable-order derivative. The above definition is more easy to handle analytically than the existing in the literature derivative with fractional variable order. With the above one can model those physical problems involving initial conditions. We shall note that, if the variable order of the derivative is zero, we recover the initial function, however if it is one then we recover the first derivative. In addition to this, if the function $a(t)$ is a constant, we obtain zero at the right-hand side of Eq. (9.44). It is very important to notice that, Eq. (9.44) can be seen as the convolution. We shall now present some interesting relationships of the proposed derivative with existing integral operators. With this information in mind, we present the relationship of this new definition with existing integral transforms operators.

9.6.1 Relation With Integral Transforms

The relation of the new derivative with existing integral transform operators including Laplace transform, Sumudu transform, Fourier transform and Mellin transform are very important as they could be used to solve analytically those ordinary and partial differential equations with non-linear terms. Also they can be used to construct the Green-function of such function. We shall start with the following theorem.

Theorem 9.6.2. *The Laplace transform of Eq. (9.44) produces*

$$L\left(_0^{AK} D_t^{l(x)}[a(t)]\right) = \frac{sL(a(t)) - a(0)}{s + l(x)}. \tag{9.45}$$

Proof. To prove the above theorem, we make use of the definition of the variable-order derivative in Caputo sense provided in Eq. (9.44):

$$L\left(_0^{AK} D_t^{l(x)}[a(t)]\right) = \int_0^\infty \left(\int_0^z \frac{da(y)}{dy} \exp[-l(x)(t-y)]dy\right) \exp[-sz]dz. \tag{9.46}$$

To continue with our proof, we have to employ the convolution theorem for Laplace transform to obtain

$$L\left(_0^{AK} D_t^{l(x)}[a(t)]\right) = L\left(\frac{da(y)}{dy}\right) L(\exp[-l(x)(t)]) = \frac{sL(a(t)) - a(0)}{s + l(x)}. \tag{9.47}$$

This completes the proof. □

The Sumudu transform operator is also commonly used to solve both non-linear and linear ordinary and partial differential equations. In non-linear case, some iterative methods have been developed using the Sumudu and the decomposition approach. In this part therefore, we present its relation with the new derivative.

Theorem 9.6.3. *The Sumudu transform of Eq. (9.44) produces*

$$S\left({}_{0}^{AK}D_{t}^{l(x)}[a(t)]\right) = \frac{S(a(t)) - a(0)}{s\,(1 + sl(x))}. \tag{9.48}$$

Proof. Using the definition of the new derivative and that of the Sumudu transform, we obtain

$$S\left({}_{0}^{AK}D_{t}^{l(x)}[a(t)]\right) = \int_{-\infty}^{\infty}\left(\int_{0}^{z}\frac{da(y)}{dy}\exp[-l(x)(t-y)]dy\right)\frac{\exp[-\frac{z}{s}]}{s}dz. \tag{9.49}$$

Using the convolution theorem for Sumudu transform that is similar to that of Laplace transform, we obtain the following

$$S\left({}_{0}^{AK}D_{t}^{l(x)}[a(t)]\right) = S\left(\frac{da(y)}{dy}\right)S\left(\exp[-l(x)(t)]\right) = \frac{S(a(t)) - a(0)}{s\,(1 + sl(x))}. \tag{9.50}$$

In many cases in the literature one will realize that, when dealing with spatio-differential equation for which boundaries conditions are infinity, the Laplace transform is not suitable. Also the Sumudu transform, in this case the Fourier transform, is appropriate. In this part therefore, we present the relation of the Fourier transform with the new derivative and this is done in the following theorem.

Theorem 9.6.4. *Let $a(t)$ be a function for which ${}_{0}^{AK}D_{t}^{l(x)}[a(t)]$ exists. The Fourier transform of the derivative of fractional variable order $l(x)$ of $a(t)$ is given as:*

$$F\left({}_{0}^{AK}D_{t}^{l(x)}[a(t)]\right) = 2\pi ikA(k)\sqrt{2\pi}\delta(k + il(x)).$$

Proof. By definition of Fourier transform, we have the following

$$F\left({}_{0}^{AK}D_{t}^{l(x)}[a(t)]\right) = \int_{-\infty}^{\infty}\left(\int_{0}^{t}\frac{da(y)}{dy}\exp[-l(x)(t-y)]dy\right)\exp(-2\pi ikt)dt \tag{9.51}$$

Using the convolution theorem for Fourier transform, we obtain

$$F\left({}_{0}^{AK}D_{t}^{l(x)}[a(t)]\right) = F\left(\frac{da(y)}{dy}\right)F\left(\exp[-l(x)(t)]\right) \tag{9.52}$$

thus the following result is obtained

$$F\left({}_{0}^{AK}D_{t}^{l(x)}[a(t)]\right) = 2\pi ikA(k)\sqrt{2\pi}\delta(k + il(x)).$$

This completes the proof.   □

Partial and ordinary differential equations with singularities with power are a very important class of differential equations that are used to model real-world problems. These class
of equations cannot be solved using Sumudu, Laplace, and Fourier transform. In this case
one must rely on the Mellin transform. In this part therefore the relation of Mellin with new
derivative will be presented and this is done in the following theorem.

Theorem 9.6.5. *Let $a(t)$ be a function for which $_0^{AK}D_t^{l(x)}[a(t)]$ exists. The Mellin transform
of the derivative of fractional variable order $l(x)$ of $a(t)$ is given as:*

$$M\left(_0^{AK}D_t^{l(x)}[a(t)]\right)(s) = \frac{\Gamma(s)^2}{\Gamma(s-1)}l(x)^{-s}.\qquad\square$$

Proof. By definition of Mellin transform, we have the following result

$$\varphi(s) = M\left(_0^{AK}D_t^{l(x)}[a(t)]\right)(s) = \int_0^\infty t^{s-1}\left(\int_0^t \frac{da(y)}{dy}\exp[-l(x)(t-y)]dy\right)dt.\quad(9.53)$$

Using the convolution theorem for Mellin transform, we obtain

$$M\left(_0^{AK}D_t^{l(x)}[a(t)]\right)(s) = M\left(\frac{da(y)}{dy}\right)(s)M\left(\exp[-l(x)(t)]\right)(s).\qquad(9.54)$$

Then we obtain following

$$M\left(_0^{AK}D_t^{l(x)}[a(t)]\right)(s) = \frac{\Gamma(s)^2}{\Gamma(s-1)}l(x)^{-s}.$$

This completes the proof. $\qquad\square$

Theorem 9.6.6. *Let us consider a positive function, $l(x) \in C^1[a,b]$, let $a(t)$ be a differentiable function in an open interval J, then*

$$_0^{AK}D_t^{l(x)}[a(t)] = r(t)$$

and we have the following

$$a(t) = a(0) + r(t) + l(x)\int_0^t r(y)dy.$$

Proof. Let us consider taking the Laplace transform of $_0^{AK}D_t^{l(x)}[a(t)] = r(t)$. Then we obtain

$$A(s) = \frac{a(0)}{s} + \left(1 + \frac{l(x)}{s}\right)R(s).\qquad(9.55)$$

Employing the Inverse Laplace operator on both sides we obtain, using the linearity,

$$a(t) = a(0) + r(t) + L^{-1}\left(\frac{l(x)}{s}R(s)\right). \tag{9.56}$$

Nevertheless, we have

$$L^{-1}\left(\frac{l(x)}{s}R(s)\right) = l(x)\int_0^t r(y)dy. \tag{9.57}$$

Thus

$$a(t) = a(0) + r(t) + l(x)\int_0^t r(y)dy. \tag{9.58}$$

$\square$

Theorem 9.6.7. *Let us consider a positive function, $l(x) \in C^1[a, b]$, let $a(t)$ be a differentiable function in an open interval I, then*

$$_0^{AK}D_t^{l(x)}[a(t)] = a(t)$$

and we have the following

$$a(t) = -\frac{a(0)}{a(x)}\delta(t).$$

Proof. Let us consider taking the Laplace transform of $_0^{AK}D_t^{l(x)}[a(t)] = a(t)$. Thus the following is obtained

$$A(s) = (1 + \frac{l(x)}{s})A(s) + \frac{a(0)}{s}. \tag{9.59}$$

Therefore we have

$$A(s) = -\frac{a(0)}{l(x)}. \tag{9.60}$$

Employing on both sides the above equation the Inverse Laplace operator, we obtain below what completes the proof:

$$a(t) = -\frac{a(0)}{l(x)}\delta(t). \tag{9.61}$$

$\square$

Theorem 9.6.8. *Let us consider a positive function, $l(x) \in C^1[a,b]$, let $a(t)$ be a differentiable within an open interval J, then*

$$\,_0^{AK} D_t^{l(x)}[a(t)] = c_1$$

and we have following

$$a(t) = a(0) + c_1 + c_1 l(x) t.$$

Proof. Let us employ on both sides of our equation Laplace transform $\,_0^{AK} D_t^{l(x)}[a(t)] = c_1$. Then taking the results obtained earlier, we obtain

$$A(s) = \frac{a(0)}{s} + \frac{c_1}{s} + \frac{c_1 l(x)}{s^2}. \tag{9.62}$$

Now taking the Inverse Laplace operator on both sides and using the Laplace transforms linearity, we obtain below what completes our proof:

$$a(t) = a(0) + c_1 + c_1 l(x) t. \tag{9.63}$$

$\square$

Theorem 9.6.9. *Let us consider a positive function, $l(x) \in C^1[a,b]$, let $a(t)$ be a differentiable function within an open interval J, then the derivative of fractional variable order $l(x)$ of $a(t)$, $\,_0^{AK} D_t^{l(x)}[a(t)]$ is satisfying the Lipschitz condition.*

Proof.

$$\left\| \,_0^{AK} D_t^{l(x)}(a(t)) - \,_0^{AK} D_t^{l(x)}(r(t)) \right\| = \left\| \int_0^t \frac{d}{dy}(a(y) - r(y)) \exp[-l(x)(t-y)] dy \right\| \tag{9.64}$$

Using the triangular inequality for integral operator, we obtain

$$\left\| \,_0^{AK} D_t^{l(x)}(a(t)) - \,_0^{AK} D_t^{l(x)}(r(t)) \right\| < \int_0^t \left\| \frac{d}{dy}(a(y) - r(y)) \exp[-f(x)(t-y)] dy \right\| \tag{9.65}$$

Nevertheless, we should remember that the function exponential is always positive; this produces

$$\left\| \,_0^{AK} D_t^{l(x)}(a(t)) - \,_0^{AK} D_t^{l(x)}(r(t)) \right\| < \int_0^t \left\| \frac{d}{dy}(a(y) - r(y)) \right\| \exp[-l(x)(t-y)] dy \tag{9.66}$$

Using the Lipschitz condition of the derivative, we can find a positive constant such that,

$$\left\| \frac{d}{dy}(a(y) - r(y)) \right\| < K \|a - r\|. \tag{9.67}$$

Putting everything together, we obtain

$$\left\| {}_0^{AK}D_t^{l(x)}(a(t)) - {}_0^{AK}D_t^{l(x)}(r(t)) \right\| < K\|a-r\| \int_0^t \exp[-l(x)(t-y)]dy. \tag{9.68}$$

Nevertheless,

$$\int_0^t \exp[-l(x)(t-y)]dy = \frac{1}{l(x)}\left[1-\exp[-l(x)t]\right] > 0. \tag{9.69}$$

Therefore by letting

$$L = K\frac{1}{l(x)}\left[1-\exp[-l(x)t]\right] \tag{9.70}$$

we obtain

$$\left\| {}_0^{AK}D_t^{l(x)}(a(t)) - {}_0^{AK}D_t^{l(x)}(r(t)) \right\| < L\|a-r\|. \tag{9.71}$$

$\square$

9.6.2 Numerical Approximation of Atangana–Koca Fractional Variable Order

It was observed in many situations that models of complex real-world problem could only be solved by means of numerical methods. To use these numerical methods, one needs to have numerical representations of the used derivatives. We therefore devote this section to the discussion underpinning the numerical approximation of the proposed fractional variable-order derivative.

$$ {}_0^{AK}D_t^{l(x)}[a(t)] = \int_0^t \frac{da(y)}{dy}\exp[-l(x)(t-y)]dy \tag{9.72}$$

We consider a particular point $t = t_n$ $(n > 1)$; we have the following

$$
\begin{aligned}
{}_0^{AK}D_t^{l(x)}[a(t_n)] &= \int_0^{t_n} \frac{da(y)}{dy}\exp[-l(x)(t_n-y)]dy \\
&= \sum_{j=1}^n \int_{t_{j-1}}^{t_j} \frac{da(y)}{dy}\exp[-l(x)(t_n-y)]dy \\
&= \sum_{j=1}^n \int_{t_{j-1}}^{t_j} \frac{a^{j+1}-a^j}{\Delta t}\exp[-l(x)(t_n-y)]dy
\end{aligned}
\tag{9.73}
$$

$$= \sum_{j=1}^{n} \frac{a^{j+1} - a^j}{\Delta t} \int_{t_{j-1}}^{t_j} \exp[-l(x)(t_n - y)]dy$$

$$= \sum_{j=1}^{n} \frac{a^{j+1} - a^j}{\Delta t} \frac{1}{l(x)} \left[\exp[-l(x)(t_n - t_j)] - \exp[-l(x)(t_n - t_{j+1})] \right].$$

Then for $0 < l(x) < 1$, the approximate numerical representation of the Atangana–Koca derivative is given as

$$\substack{AK \\ 0} D_t^{l(x)} (a(t_n)) = \sum_{j=1}^{n} \frac{a^{j+1} - a^j}{\Delta t} \delta_{(n,j)}^{l(x)}.$$

Theorem 9.6.10. *Let $a(t) \in C^1[a, b]$ and assume that the fractional variable order $0 < l(x) < 1$ exists, then the numerical approximation of the new variable-order derivative is*

$$\substack{AK \\ 0} D_t^{l(x)} [a(t_n)] = \sum_{j=1}^{n} \frac{a^{j+1} - a^j}{\Delta t} \delta_{(n,j)}^{l(x)} + O(\Delta t). \tag{9.74}$$

Proof. Let assume that the conditions of the theorem are satisfied, then

$$
\begin{aligned}
\substack{AK \\ 0} D_t^{l(x)} [a(t_n)] &= \int_0^{t_n} \frac{da(y)}{dy} \exp[-l(x)(t_n - y)]dy \tag{9.75} \\
&= \sum_{j=1}^{n} \int_{t_{j-1}}^{t_j} \frac{da(y)}{dy} \exp[-l(x)(t_n - y)]dy \\
&= \sum_{j=1}^{n} \int_{t_{j-1}}^{t_j} \left(\frac{a^{j+1} - a^j}{\Delta t} + O(\Delta t) \right) \exp[-l(x)(t_n - y)]dy \\
&= \sum_{j=1}^{n} \frac{a^{j+1} - a^j}{\Delta t} \int_{t_{j-1}}^{t_j} \exp[-l(x)(t_n - y)]dy \\
&\quad + O(\Delta t) \sum_{j=1}^{n} \int_{t_{j-1}}^{t_j} \exp[-l(x)(t_n - y)]dy \\
&= \sum_{j=1}^{n} \frac{a^{j+1} - a^j}{\Delta t} \delta_{(n,j)}^{l(x)} + O(\Delta t) \sum_{j=1}^{n} \delta_{(n,j)}^{l(x)} \\
&= \sum_{j=1}^{n} \frac{a^{j+1} - a^j}{\Delta t} \delta_{(n,j)}^{l(x)} + O(\Delta t).
\end{aligned}
$$

This completes the proof. □

9.6.3 Second Order Variable-Order Approximation of Atangana–Koca Derivative

In this section, we present the numerical approximation of the space fractional variable-order derivative. However, for the second order variable-order derivative which we defined, we proposed the following definition.

Definition 9.6.11. *Let us consider a positive function and twice differentiable a and a continuous function $1 < l(x) < 2 \in C^2[0, 1]$. If $a(t)$ has as domain $[a, b]$, then the derivative of fractional variable-order $l(x)$ of $a(t)$ is provided as*

$$^{AK}_{0} D_t^{l(x)}[a(t)] = \int_0^t \frac{d^2 a(y)}{dy^2} \exp[-[l(x)(t-y)]^2] dy. \qquad (9.76)$$

If we consider the derivative at the point $t = t_n$ $(n \geq 1)$, we have the following expressions

$$^{AK}_{0} D_t^{l(x)}[a(t_n)] = \int_0^{t_n} \frac{d^2 a(y)}{dy^2} \exp[-[l(x)(t_n-y)]^2] dy \qquad (9.77)$$

$$^{AK}_{0} D_t^{l(x)}[a(t_n)] = \sum_{j=1}^{n} \int_{t_j}^{t_{j+1}} \frac{d^2 a(y)}{dy^2} \exp[-[l(x)(t_n-y)]^2] dy \qquad (9.78)$$

$$^{AK}_{0} D_t^{l(x)}[a(t_n)] = \sum_{j=1}^{n} \int_{t_j}^{t_{j+1}} \left(\frac{a^{j+1} - 2a^j + a^{j-1}}{(\Delta t)^2} \right) \exp[-[l(x)(t_n-y)]^2] dy \qquad (9.79)$$

$$^{AK}_{0} D_t^{l(x)}[a(t_n)] = \sum_{j=1}^{n} \left(\frac{a^{j+1} - 2a^j + a^{j-1}}{(\Delta t)^2} \right) \int_{t_j}^{t_{j+1}} \exp[-[l(x)(t_n-y)]^2] dy \qquad (9.80)$$

$$^{AK}_{0} D_t^{l(x)}[a(t_n)] = \sum_{j=1}^{n} \left(\frac{a^{j+1} - 2a^j + a^{j-1}}{(\Delta t)^2} \right)$$
$$\times \frac{\sqrt{\pi} \left(Erf[(t_{j+1} - t_n)l[x]] + Erf[(-t_j + t_n)l[x]] \right)}{2l[x]} \qquad (9.81)$$

$$^{AK}_{0} D_t^{l(x)}[a(t_n)] = \sum_{j=1}^{n} \left(\frac{a^{j+1} - 2a^j + a^{j-1}}{(\Delta t)^2} \right) \omega_{(n,j)}^{l(x)} \qquad (9.82)$$

Following the derivation presented above, we present the following theorem.

Theorem 9.6.12. *Let us consider a function $a(t)$ to be twice differentiable, then the second order approximation of the Atangana–Koca derivative with variable-order $1 < l(x) < 2$ is given as*

$$^{AK}_{0} D_t^{l(x)}[a(t_n)] = \sum_{j=1}^{n} \left(\frac{a^{j+1} - 2a^j + a^{j-1}}{(\Delta t)^2} \right) \omega_{(n,j)}^{l(x)} + O\left((\Delta t)^2\right). \qquad (9.83)$$

Proof. Employing the definition and considering the step of discretization, we have the following relations

$$\,_0^{AK} D_t^{l(x)} [a(t_n)] = \int_0^t \frac{d^2 a(y)}{dy^2} \exp[-[l(x)(t_n - y)]^2] dy \tag{9.84}$$

$$\,_0^{AK} D_t^{l(x)} [a(t_n)] = \sum_{j=1}^{n} \int_{t_j}^{t_{j+1}} \frac{d^2 a(y)}{dy^2} \exp[-[l(x)(t_n - y)]^2] dy \tag{9.85}$$

$$\,_0^{AK} D_t^{l(x)} [a(t_n)] = \sum_{j=1}^{n} \int_{t_j}^{t_{j+1}} \left(\frac{a^{j+1} - 2a^j + a^{j-1}}{(\Delta t)^2} + O\left((\Delta t)^2\right) \right) \tag{9.86}$$

$$\times \exp[-[l(x)(t_n - y)]^2] dy$$

$$\,_0^{AK} D_t^{l(x)} [a(t_n)] = \sum_{j=1}^{n} \int_{t_j}^{t_{j+1}} \left(\frac{a^{j+1} - 2a^j + a^{j-1}}{(\Delta t)^2} \right) \exp[-[l(x)(t_n - y)]^2] dy \tag{9.87}$$

$$+ O\left((\Delta t)^2\right) \sum_{j=1}^{n} \int_{t_j}^{t_{j+1}} \exp[-[l(x)(t_n - y)]^2] dy$$

$$\,_0^{AK} D_t^{f(x)} [a(t_n)] = \sum_{j=1}^{n} \int_{t_j}^{t_{j+1}} \left(\frac{a^{j+1} - 2a^j + a^{j-1}}{(\Delta t)^2} \right) \exp[-[l(x)(t_n - y)]^2] dy$$

$$+ O\left((\Delta t)^2\right) \sum_{j=1}^{n} \omega_{(n,j)}^{l(x)}$$

$$\,_0^{AK} D_t^{l(x)} [a(t_n)] = \sum_{j=1}^{n} \int_{t_j}^{t_{j+1}} \left(\frac{a^{j+1} - 2a^j + a^{j-1}}{(\Delta t)^2} \right) \exp[-[l(x)(t_n - y)]^2] dy \tag{9.88}$$

$$+ O\left((\Delta t)^2\right).$$

This completes the proof. ☐

Groundwater Flow Model in Self-similar Aquifer With Atangana–Baleanu Fractional Operators

10.1 Groundwater Flow Model in a Self-similar Aquifer

The fractal flow model has attracted many researchers across the globe. Through the work of Barker [222], we have good understanding of aquifer geometry where the author introduced the generalized flow equation and gave the interpretation of a flow dimension.

$$S_0 \partial_t h(r,t) = \frac{K}{r^{d-1}} \partial_r [r^{d-1} \partial_r h(r,t)].$$ (10.1)

Eq. (10.1) holds for the specific initial and boundary conditions as well as a constant-rate condition, in the form

$$
\begin{aligned}
h(r,0) &= 0, \\
\lim_{r \to \infty} h(r,t) &= 0, \\
Q &= \frac{2\pi^{\frac{d}{2}}}{\Gamma\left(\frac{d}{2}\right)} r w^{d-1} K b^{3-d} \partial_r h(r_w,t).
\end{aligned}
$$

In 2001, Roberts and Beauhein modified the model developed by Barker to justify the flow dimension function $n(r)$ with a no-flow case boundaries, say from -1 to 2, and that the flow dimension may be considered a function of radius r in the form

$$n(r) = \frac{d \log\left(\frac{K}{A}\right)}{d \log(r)} + 1.$$ (10.2)

The approach used by Barker as well as the approach of the resulting variable flow dimension have considerable advantages over numerous other traditional approaches, as they usually generate an estimate of a lumped flow dimension parameter representing conductance which may solely have qualitative character because there is no physical meaning associated with it. Furthermore, assuming theories of diffusion slowdown in disordered systems as given by O'Shaughnessy and Procaccia [223], Havlin and Ben-Avraham [224], an analytical model

Fractional Operators with Constant and Variable Order
with Application to Geo-Hydrology
DOI: 10.1016/B978-0-12-809670-3.00010-2

which consists of a fractal dimension "D" was given consideration by Chang and Yortsos [225]. Later on, consideration was also given to a solution for a model as such, by making use of a simplified Green function technique. Thereafter, using a slightly modified version of the model as given by Chang and Yortsos [226], the analysis of the model was presented. To add, the last version of the model was researched by many other scientific researchers. As a result, a considerable amount of research articles have been published in a number of hydrology journals. Certainly this mathematical model could be used to generate representations of an observed real-world problem. However, when making comparisons between observations and mathematical equations, an adequate agreement is not achieved. Ning et al. [227] addressed this issue by proposing a new model which is governed by fractional differentiation, whereby the fractional Caputo derivative replaces the time derivative. Despite that models based on fractional differentiation appear as more suitable, they still have associated difficulty when applied to a few real-world problems. First of all, the fractional derivative related to the aforementioned model is based on the power law which is associated with a singular function. Consequently, unrealistic predictions regarding fractures are made. On the other hand, it appears impractical describing fractal effects by solely introducing r^{d-1} and what is known as the fractal dimension. In recent times, literature brings up the concept of fractional differentiation which was made by using several other real-world problems for modeling fractal effects in a dynamic system. This differentiation, however, is not made use of in this work. Furthermore, Atangana and Baleanu [196] recently established a new approach to differentiation, whereby the fractional derivative is based on a non-singular and non-local kernel called the generalized Mittag–Leffler. The generalized Mittag–Leffler function was brought up in literature to address difficulty associated with the use of the power law. To add, this function incorporates the effect of memory which is fundamental to investigations regarding groundwater flow. The memory effect is imperative to account for because it is essential for a water molecule to "remember" its path of flow within a fractured network. As a result, for an accurate representation of what is observed as a means of a mathematical model, the following equation is proposed:

$$S_{00}{}^{ABC}D_t^\alpha h(r,t) = \frac{K}{r^{d-1-\theta}}\partial_r[r^{d-1-\theta}\partial_r h(r,t)],$$

$$_0^{ABC}D_t^\alpha h(r,t) = \frac{AB(\alpha)}{1-\alpha}\int_0^t \partial_r[h(r,x)]E_\alpha\left[\frac{-\alpha}{1-\alpha}(t-x)^\alpha\right]dx. \tag{10.3}$$

$_0^{ABC}D_t^\alpha h(r,t)$ is known as a fading memory, and d is the fractal dimension. In order to accommodate readers that are familiar with the new findings, we use the previous initial conditions. The most important aspect of modeling is the ability to show that the model has a positive solution. The study of existence of positive solutions of a given nonlinear differential equation has generated a lot of attention. In this section, we use the concept of fixed-point theorem to present the analysis of the new model, and to achieve this we start by presenting some useful preliminaries. It was demonstrated by Atangana and Koca [104] that if y is the solution of

$$\substack{ABC\\0}D_t^\alpha y(t) = f(t, y(t)), \tag{10.4}$$

then y is also a solution of the integral equation

$$y(t) - y(0) = \frac{1-\alpha}{AB(\alpha)} f(t, y(t)) + \frac{\alpha}{AB(\alpha)\Gamma(\alpha)} \int_0^t (t-x)^{\alpha-1} f(x, y(x))dx. \tag{10.5}$$

Also, we consider the following function:

$$f(r, t, h(r, t)) = \frac{K}{r^{d-1}} \partial_r \left[r^{d-1-\theta} \partial_r(r, t) \right]. \tag{10.6}$$

We assume $M > 0$ such that $\| f(r, t, h(r, t))\| < M$. Considering the physical problem under study, one can find a constant $H > 0$ such that $\forall (x, t) \in [a, b] * [0, T], |h(r, t)|$.

Lemma 10.1.1. *The mapping $G : H_1 \to H_1$ defined as:*

$$Th(r, t) = \frac{1-\alpha}{AB(\alpha)} f(r, t, h(r, t)) + \frac{\alpha}{AB(\alpha)\Gamma(\alpha)} \int_0^t (t-x)^{\alpha-1} f(r, x, h(r, x))dx \tag{10.7}$$

is completely continuous.

Lemma 10.1.2. *Let NCH_1 be bounded. Assume that there exists $l > 0$ such that*

$$|h(r, t_1) - h(r, t_2)| < l|t_1 - t_2|, \forall h \in N,$$

then, $T(N)$ is compact.

Proof. Let $f = \max\left\{ \frac{1-\alpha}{AB(\alpha)} f(r, t, h(r, t)) \right\}$ for $0 \le h \le M$ for $h \in N$. □

We can evaluate the following inequality: The mapping $G : H_1 \to H_1$ defined as:

$$\|Th(r, t)\| \le \frac{1-\alpha}{AB(\alpha)}\| f(r, t, h(r, t))\| + \frac{\alpha}{AB(\alpha)\Gamma(\alpha)} \int_0^t (t-x)^{\alpha-1}\| f(r, x, h(r, x))\|dx$$

$$\le \frac{1-\alpha}{AB(\alpha)}M + \frac{\alpha Mt^\alpha}{AB(\alpha)} < \infty. \tag{10.8}$$

Consider $h \in N, t_1 < t_2$. Then for any given $\epsilon > 0$ if $|t_1 < t_2| < \Phi$, then

$$\|Th(r, t_2) - Th(r, t_1)\| \leq \frac{1-\alpha}{AB(\alpha)} \|f(r, t_2, h(r, t_2)) - f(r, t_1, h(r, t_1))$$

$$+ \left\| \frac{\alpha}{AB(\alpha)\Gamma(\alpha)} \int_0^t (t_1 - x)^{\alpha-1} f(r_1, x_1, h(r, x)) dx \right. \quad (10.9)$$

$$- \frac{\alpha}{AB(\alpha)\Gamma(\alpha)} \int_0^t (t_1 - x)^{\alpha-1} f(r, x, h(r, x)) dx \bigg\| .$$

We shall evaluate the above step-by-step:

$$\|f(r, t_2, h(r, t_2)) - f(r, t_1, h(r, t_1))\| = \frac{K}{r^{d-1}} \partial_r \left\| r^{d-1-\theta} \partial_r h(r, t_2) - \partial_r \left[r^{d-1-\theta} \partial_r h(r, t_1) \right] \right\|$$

$$\leq \frac{K}{r^{d-1}} \theta_1 r^{d-1-\theta} \|\partial_r h(r, t_2) - \partial_r h(r, t_1)\| \quad (10.10)$$

$$\leq K\theta_1\theta_2 \|h(r, t_2) - h(r, t_1)\| b \leq K\theta_1\theta_2 l |t_2 - t_1|.$$

Next, we evaluate

$$\left\| \int_0^{t_2} (t_2 - x)^{\alpha-1} f(r, x, h(r, x)) dx - \int_0^{t_1} (t_1 - x)^{\alpha-1} f(r, x, h(r, x)) dx \right\|$$

$$\leq \frac{\alpha}{AB(\alpha)\Gamma(\alpha)} \left\| \int_0^{t_2} (t_2 - x)^{\alpha-1} dx - \int_0^{t_1} (t_1 - x)^{\alpha-1} dx \right\| . \quad (10.11)$$

But

$$\int_0^{t_2} (t_2 - x)^{\alpha-1} dx - \int_0^{t_1} (t_1 - x)^{\alpha-1} dx = \int_{t_1}^{t_2} (t_2 - x)^{\alpha-1} dx$$

$$+ \int_0^{t_1} (t_1 - x)^{\alpha-1} - (t_2 - x)^{\alpha-1} dx$$

$$= \frac{2}{\Gamma(\alpha+1)} (t_1 - t_2)^{\alpha}, \quad (10.12)$$

choose

$$\Phi = \frac{\epsilon}{K\theta_1\theta_2 l + \frac{2\alpha}{AB(\alpha)\Gamma(\alpha+1)}} .$$

We obtain the requested result $T(N)$ is equicontinuous and based on the known Anzela–Asule theorem, $\overline{T(N)}$ is compact.

Theorem 10.1.3. *Let* $T : [a_1, b_1][a_2, b_2] \times [0, \infty) \to [0, \infty)$ *be a continuous function and* $T(r, t)$ *for each* $t \in [a, b]$. *Assume there exist* h_1 *and* h_2 *satisfying*

$$G(D)h_1 \leq f(r, x, h_1(r, x)), \quad G(D)h_1 \leq f(r, x, h_2(r, x)), \quad a < t < b.$$

10.2 Positive Solution of the Modified Groundwater Flow in Self-similar Aquifer With Atangana–Baleanu Fractional Operators

We proceed here by presenting the analysis and conditions within which the solution is unique. We establish these conditions by considering the following:

$$\|Th(r, t_2) - Th(r, t_1)\| \tag{10.13}$$

$$\leq \frac{1-\alpha}{AB(\alpha)} \| f(r, t, h_2(r, t)) - f(r, t, h_1(r, t))$$

$$+ \frac{\alpha}{AB(\alpha)\Gamma(\alpha)} \int_0^t (t - x)^{\alpha-1} \| f(r, x, h_2(r, x)) - f(r, x, h_1(r, x))\| dx$$

$$\leq \frac{1-\alpha}{AB(\alpha)} \delta \|h_2(r, t) - h_1(r, t)\| + \frac{\alpha}{AB(\alpha)\Gamma(\alpha+1)} \delta b^\alpha \|h_2 - h_1\|$$

$$\leq \left(\frac{1-\alpha}{AB(\alpha)} + \frac{\alpha}{AB(\alpha)\Gamma(\alpha+1)} \right) \|h_2 - h_1\| \leq K \|h_2 - h_1\|.$$

Thus $K < 1$, then T is in contraction. This indicates that there is a fixed point, which is regarded as the positive solution of our equation. In this section, we present the derivation of a solution to the new model. We adopt the mixture of integral transform with perturbation method here. The methodology of the used method is first presented for the general fractional differential equation based on the fractional derivative with Mittag–Leffler function. Let the following equation be a general nonlinear partial differential equation with the new fractional derivative based on the Mittag–Leffler function

$$_{0}^{ABC}D_t^\alpha U(x, t) = LU(x, t) + NU(x, t) + f(x, t). \tag{10.14}$$

To solve Eq. (10.14), we apply the fractional integral transform on both sides of this equation to have

$$U(x, t) = U(x, t) + \frac{1-\alpha}{AB(\alpha)} \{LU(x, t) + NU(x, t)\} \tag{10.15}$$

$$+ \frac{\alpha}{AB(\alpha)\Gamma(\alpha)} \int_0^t (t - y)^{1-\alpha} \{LU(x, t) + NU(x, t) + f(x, y)\} dy,$$

where $U_0(x, t) = U(x, 0)$, L is called the linear operator, N is known as a nonlinear operator that represents the function $f(x, t)$, and $U(x, 0)$ is the given initial condition. In what follows, we present the stability of the method.

Theorem 10.2.1. *Assume operator N is Lipschitz, then the method is stable if*

$$\frac{1-\alpha}{AB\alpha} K_1 + \frac{\alpha T^\alpha}{AB(\alpha)\Gamma(\alpha+1)} K_2 < 1.$$

Proof. Let $n, m \in \mathbb{N}$ and assume that the nonlinear operator N has Lipschitz condition

$$
\begin{aligned}
\|U_{n+1} - U_{m+1}\| &= \left\| \frac{1-\alpha}{AB(\alpha)}\{L(U_n - U_m) + NU_n - NU_m\} \right. \\
&\quad + \left. \frac{\alpha}{AB(\alpha)\Gamma(\alpha)} \int_0^t (t-y)^{\alpha-1}\{L(U_n - U_m) + NU_n + NU_m\}dy \right\|.
\end{aligned}
$$

Now, using the properties of norm, we have

$$
\begin{aligned}
\|U_{n+1}U_{m+1}\| &= \left\| \frac{1-\alpha}{AB(\alpha)}\{\|U_n - U_m\|K_1 + \|U_n - U_m\|K_2\} \right. \\
&\quad + \left. \frac{\alpha}{AB(\alpha)\Gamma(\alpha)}\{\|U_n - U_m\|K_1 + \|U_n - U_m\|K_2\} \int_0^t (t-y)^{\alpha-1}dy \right\| \\
&\leq \left\{ \frac{1-\alpha}{AB(\alpha)}K_1 + K_2 + \frac{\alpha}{AB(\alpha)\Gamma(\alpha)}T^\alpha \right\} \|U_n - U_m\|. \tag{10.16}
\end{aligned}
$$

Repeating this process (n, m) times, we obtain

$$
\|U_{n+1} - U_{m+1}\| \leq \left\{ \frac{1-\alpha}{AB(\alpha)}K_1 + K_2 + \frac{\alpha T^\alpha}{AB(\alpha)\Gamma(\alpha+1)} \right\}_1^2 \|U_{n-1} - U_{m-2}\|. \tag{10.17}
$$

By choosing $m = n - 1$, we get the following:

$$
\|U_{n+1} - U_{m+1}\| \leq \left\{ \frac{1-\alpha}{AB(\alpha)}K_1 + K_2 + \frac{\alpha T^\alpha}{AB(\alpha)\Gamma(\alpha+1)} \right\}^n U(x, 0). \tag{10.18}
$$

If $n \to \infty$, the entire right-hand side of (10.18) equals zero which means that the sequence is converged since

$$
\frac{1-\alpha}{AB(\alpha)}K_1 + K_2 + \frac{\alpha T^\alpha}{AB(\alpha)\Gamma(\alpha+1)} < 1. \qquad \square
$$

10.3 Semi-Analytical and Numerical Solutions

Applying the Atangana–Baleanu fractional integral on both sides of Eq. (10.3), we obtain

$$
\begin{aligned}
S_0[h(r, t) - h(r, 0)] &= \frac{1-\alpha}{AB(\alpha)}\frac{K}{r^{d-1}}\partial_r\left[r^{d-1-\theta}\partial_r h(r, t)\right] \tag{10.19} \\
&\quad + \frac{\alpha}{AB(\alpha)\Gamma(\alpha)}\int_0^t (t-y)^{\alpha-1}\frac{K}{r^{d-1-\theta}}\partial_r\left[r^{d-1-\theta}\partial_r h(r, y)\right]dy.
\end{aligned}
$$

Next, we consider the recursive formula

$$h_{n+1}(r,t) = \frac{1-\alpha}{S_0 A B(\alpha)} \frac{K}{r^{d-1}} \partial_r \left[r^{d-1-\theta} \partial_r h(r,t) \right] \tag{10.20}$$

$$+ \frac{\alpha}{S_0 A B(\alpha) \Gamma(\alpha)} \int_0^t (t-y)^{\alpha-1} \frac{K}{r^{d-1}} \partial_r \left[r^{d-1-\theta} \partial_r h_n(r,y) \right] dy.$$

10.3.1 Numerical Analysis With Fractional Integral

In this section, the numerical solution of the new equation is presented, based on four different schemes. The first is based on the Volterra version of our new equation given below as

$$h_{n+1}(r,t) = \frac{1-\alpha}{S_0 A B(\alpha)} \frac{K}{r^{d-1}} \partial_r \left[r^{d-1-\theta} \partial_r h(r,t) \right] \tag{10.21}$$

$$+ \frac{\alpha}{S_0 A B(\alpha) \Gamma(\alpha)} \int_0^t (t-y)^{\alpha-1} \frac{K}{r^{d-1-\theta}} \partial_r \left[r^{d-1-\theta} \partial_r h(r,y) \right] dy.$$

The above equation will be solved numerically using a new numerical scheme. We begin with the numerical scheme and its application.

10.3.2 Numerical Approximation of Fractional Integral

In this section we suggest an approximation of fractional integral. Let us consider a function f defined in a closed interval I. By the definition of fractional integral, we have

$$^{RL}I_t^\alpha f(t) = \frac{1}{\Gamma(\alpha)} \int_0^t (t-y)^{\alpha-1} f(y) dy.$$

For any given $n > 1$, the above can be formulated as

$$^{RL}I_t^\alpha f(t_n) = \frac{1}{\Gamma(\alpha)} \int_0^{t_n} (t_n - y)^{\alpha-1} f(y) dy, \tag{10.22}$$

$$= \frac{1}{\Gamma(\alpha)} \sum_{k=0}^n \int_{t_k}^{t_{k+1}} (t_n - y)^{\alpha-1} \frac{f(t_{k+1}) + f(t_k)}{2} dy$$

$$+ \frac{1}{\Gamma(\alpha)} \sum_{k=0}^n (t_n - y)^{\alpha-1} (f(y) - f(t_{k+1})) dy,$$

$$= \frac{1}{\Gamma(\alpha)} \sum_{k=0}^n \frac{f(t_{k+1}) + f(t_k)}{2} \int_{t_k}^{t_{k+1}} (t_n - y)^{\alpha-1} dy$$

$$+ \frac{1}{\Gamma(\alpha)} \sum_{k=0}^n (t_n - y)^{\alpha-1} \frac{(f(y) - f(t_{k+1})) (y - t_{k+1})}{y - tk + 1} dy. \tag{10.23}$$

We shall develop both components of the above equation; we start with the first component

$$\frac{1}{\Gamma(\alpha)} \sum_{k=0}^{n} \frac{f(t_{k+1}) + f(t_k)}{2} \int_{t_k}^{t_{k+1}} (t_n - y)^{\alpha-1} dy =$$

$$\frac{(\Delta t)^\alpha}{\Gamma(\alpha+1)} \sum_{k=0}^{n} \frac{f(t_{k+1}) + f(t_k)}{2} \{(n-k)^\alpha - (n-k-1)^\alpha\}, \qquad (10.24)$$

$$= \frac{(\Delta t)^\alpha}{\Gamma(\alpha+1)} \sum_{k=0}^{n} \frac{f(t_{k+1}) + f(t_k)}{2} \delta_{n,k}^\alpha.$$

The remaining second component can be evaluated as:

$$
\begin{aligned}
R_{n,k}^\alpha(t) &= \frac{1}{\Gamma(\alpha)} \sum_{k=0}^{n} \int_{t_k}^{t_{k+1}} \left\{ (t_n - y)^{\alpha-1} \frac{(f(y) - f(t_{k+1}))(y - t_{k+1})}{y - tk + 1} dy \right\}, \\
&= \frac{1}{\Gamma(\alpha)} \sum_{k=0}^{n} \int_{t_k}^{t_{k+1}} (t_n - y)^{\alpha-1} f^{(')}(\lambda)(y - t_{k+1}) dy, \quad y \le \lambda \le t_n, \\
&\le \frac{\Delta t}{\Gamma(\alpha)} \sum_{k=0}^{n} \int_{t_k}^{t_{k+1}} (t_n - y)^{\alpha-1} f^{(')} dy, \quad y \le \lambda \le t_n, \qquad (10.25) \\
&\le \frac{\Delta t}{\Gamma(\alpha)} \max_{0 \le t \le t_n} \left\{ f^{(')(t)} \right\} \sum_{k=0}^{n} \int_{t_k}^{t_{k+1}} (t_n - y)^{\alpha-1} dy, \\
&\le \frac{(\Delta t)^{\alpha+1}}{\Gamma(\alpha)} \max_{0 \le t \le t_n} \left\{ f^{(')(t)} \right\} \sum_{k=0}^{n} \{(n-k)^\alpha - (n-k-1)^\alpha\} dy.
\end{aligned}
$$

Theorem 10.3.1. *Let f be defined in an open interval I, then the fractional integral of f can be approximated as*

$$^{RL} I_t^\alpha f(t_n) = \frac{(\Delta t)^\alpha}{\Gamma(\alpha+1)} \sum_{k=0}^{n} \frac{f(t_{k+1}) + f(t_k)}{2} \{(n-k)^\alpha - (n-k-1)^\alpha\} + R_{n,k}^\alpha(t),$$

$$\|R_{n,k}^\alpha(t)\| < \frac{(\Delta t)^{\alpha+1}}{\Gamma(\alpha+1)} \max_{0 \le t \le t_n} \left\{ f^{(')(t)} \right\} \sum_{k=0}^{n} \{(n-k)^\alpha - (n-k-1)^\alpha\} dy.$$

$$(10.26)$$

10.3.3 Application to New Modified Model

This section is devoted to construction of numerical solution of the new fractal flow model developed with the concept of fractional differentiation based on the generalized Mittag–Leffler law. To do this, we start by reformulating Eq. (10.3) in the form:

$$
\begin{aligned}
h(r,t) &= \frac{1-\alpha}{S_0 A B(\alpha)} \left\{ \frac{K(d-1-\theta)}{r^{1+\theta}} \partial_r h(r,t) + \frac{K}{r^\theta} \partial_{rr}^2 h(r,t) \right\} \\
&+ \frac{\alpha}{S_0 A B(\alpha) \Gamma(\alpha)} \int_0^t (t-y)^{\alpha-1} \left\{ \frac{K(d-1-\theta)}{r^{1+\theta}} \partial_r h(r,t) + \frac{K}{r^\theta} \partial_{rr}^2 h(r,t) \right\} dy.
\end{aligned}
\tag{10.27}
$$

We now start the discretization of Eq. (10.27). We recall the first and second approximations for the local derivative. Different numerical methods such as implicit, explicit, and Crank–Nicholson schemes are employed for local derivative. Under the framework of implicit scheme, we have the following approximation:

$$
\frac{\partial h(r,t)}{\partial t} = \frac{h_i^{j+1} - h_i^j}{\Delta t}, \quad \frac{\partial h(r,t)}{\partial r} = \frac{h_{i+1}^{j+1} - h_{i-1}^{j+1}}{(2\Delta x)}, \quad \frac{\partial^2 h(r,t)}{\partial r^2} = \frac{h_{i-1}^{j+1} - 2h_i^{j+1} + h_{i+1}^{j+1}}{(\Delta r)^2}.
\tag{10.28}
$$

Using the above with the integral approximation as suggested earlier, one can conveniently reformulate Eq. (10.27) as

$$
\begin{aligned}
h_i^j &= \frac{1-\alpha}{S_0 A B(\alpha)} \left\{ \frac{K(d-\theta-1)}{r_i^{1+\theta}} \frac{h_{i+1}^{j+1} - h_{i-1}^{j+1}}{2\Delta r} + \frac{K}{r^\theta} \frac{h_{i-1}^{j+1} - 2h_i^{j+1} + h_{i+1}^{j+1}}{(\Delta r)^2} \right\} + \\
&\frac{(\Delta t)^{\alpha+1}}{S_0 A B(\alpha) \Gamma(\alpha)} \sum_{k=0}^{j} \left\{ \frac{K(d-\theta-1)}{r_i^{1+\theta}} \frac{h_{i+1}^{k+1} - h_{i-1}^{k+1}}{2\Delta r} + \frac{K}{r_i^\theta} \frac{h_{i-1}^{k+1} - 2h_i^{k+1} + h_{i+1}^{k+1}}{(\Delta r)^2} \right\} b_{n,k}^\alpha, \\
b_{n,k}^\alpha &= (n-k)^\alpha - (n-k-1)^\alpha.
\end{aligned}
\tag{10.29}
$$

From the above, with index j equal 0, we obtain

$$
h_i^0 = \frac{1-\alpha}{S_0 A B(\alpha)} \left\{ \frac{K}{r^\theta} \frac{2h_i^1}{(\Delta r)^2} \right\} + \frac{(\Delta t)^{\alpha+1}}{S_0 A B(\alpha) \Gamma(\alpha)} \frac{K}{r_i^\theta} \frac{2h_i^1}{(\Delta r)^2}.
\tag{10.30}
$$

For any $j > 0$, we let

$$
\begin{aligned}
a &= \frac{1-\alpha}{2\Delta r S_0 A B(\alpha)} K(d-\theta-1), \quad b = \frac{(1-\alpha)K}{\Delta r S_0 A B(\alpha)}, \\
c &= \frac{K(d-\theta-1)(\Delta t)^{\alpha+1}}{2\Delta r S_0 A B(\alpha) \Gamma(\alpha)}, \quad e = \frac{K(\Delta t)^{\alpha+1}}{(\Delta r)^2 S_0 A B(\alpha) \Gamma(\alpha)}.
\end{aligned}
\tag{10.31}
$$

Then, the numerical solution of the new model using the implicit scheme is recursively presented as

$$h_i^j = \left(\frac{-2b}{r_i^\theta} - \frac{-2e}{r_i^\theta} b_{n,k}^\alpha\right) h_i^{j+1} + a\frac{h_{i+1}^{j+1} - h_{i-1}^{j+1}}{r_i^{1+\theta}} + b\frac{h_{i-1}^{j+1} + h_{i+1}^{j+1}}{r_i^\theta} +$$

$$\sum_{k=0}^{j-1}\left\{c\frac{h_{i+1}^{k+1} - h_{i-1}^{k+1}}{r_i^{1+\theta}} + e\frac{h_{i-1}^{k+1} + h_{i+1}^{k+1}}{r_i^{1+\theta}}\right\}b_{j,k}^\alpha, \tag{10.32}$$

$$b_{j,k}^\alpha = (j-k)^\alpha - (j-k-1)^\alpha.$$

Within the framework of explicit difference scheme, we have the following numerical approximation

$$\frac{\partial h(r,t)}{\partial t} = \frac{h_i^{j+1} - h_i^j}{\Delta t}, \frac{\partial h(r,t)}{\partial t} = \frac{h_{i+1}^j - h_i^j}{2\Delta x}, \frac{\partial^2 h(r,t)}{\partial r^2} = \frac{h_{i-1}^j - 2h_i^j + h_{i+1}^j}{(\Delta r)^2}. \tag{10.33}$$

Using the above with the integral approximation mentioned earlier, we reformulate (10.27) as:

$$h_i^{j+1} = \frac{1-\alpha}{S_0 AB(\alpha)}\left\{K(d-\theta-1)\frac{h_{i+1}^j - h_i^j}{r_i^{1+\theta}}\frac{1}{2\Delta r} + \frac{K}{r^\theta}\frac{h_{i-1}^j - 2h_i^j + h_{i+1}^j}{(\Delta r)^2}\right\} +$$

$$\frac{(\Delta t)^{\alpha+1}}{S_0 AB(\alpha)\Gamma(\alpha)}\sum_{k=0}^{j}\left\{K(d-\theta-1)\frac{h_{i+1}^k - h_i^k}{r_i^{1+\theta}}\frac{1}{2\Delta r} + \frac{K}{r_i^\theta}\frac{h_{i-1}^k - 2h_i^k + h_{i+1}^k}{(\Delta r)^2}\right\}b_{j,k}^\alpha, \tag{10.34}$$

$$b_{j,k}^\alpha = (j-k)^\alpha - (j-k-1)^\alpha.$$

For $j = 0$, we have

$$h_j^1 = \frac{1-\alpha}{S_0 AB(\alpha)}\left\{-\frac{K(d-\theta-1)}{r_i^{1+\theta}}\frac{h_i^0}{2\Delta r} - \frac{K}{r^\theta}\frac{2h_i^0}{(\Delta r)^2}\right\}. \tag{10.35}$$

Similarly, for an index $j > 0$ we have the following recursive formula

$$h_i^{j+1} = h_i^j\left(-b_{n,j}^\alpha\frac{(\Delta t)^{\alpha+1}}{2r^{1+\theta}\Delta r S_0 AB(\alpha)\Gamma(\alpha)} - \frac{K(d-\theta-1)}{r_i^{1+\theta}}\frac{1-\alpha}{2\Delta r S_0 AB(\alpha)} - \frac{K}{(\Delta r)^2 r_i^\theta}\right) +$$

$$\frac{1-\alpha}{S_0 AB(\alpha)}\left\{\frac{K(d-\theta-1)}{r_i^{1+\theta}}\frac{h_{i+1}^j}{2\Delta r} + \frac{K}{r_i^\theta}\frac{h_{i-1}^j + h_{i+1}^j}{(\Delta r)^2}\right\} + \tag{10.36}$$

$$\frac{(\Delta)^{\alpha+1}}{S_0 AB(\alpha)\Gamma(\alpha)}\sum_{k=0}^{j-1}\left\{\frac{K(d-\theta-1)}{r_i^{1+\theta}}\frac{h_{i+1}^k - k_i^k}{2\Delta r} + \frac{K}{r_i^\theta}\frac{h_{i-1}^k - 2h_i^k + h_{i+1}^k}{(\Delta r)^2}\right\}b_{j,k}^\alpha +$$

$$\left\{\frac{K(d-\theta-1)}{r_i^{1+\theta}}\frac{h_{i+1}^j - h_i^j}{2\Delta r} + \frac{K}{r_i^\theta}\frac{h_{i-1}^j + h_{i+1}^j}{(\Delta r)^2}\right\}\frac{(\Delta t)^{\alpha+1}}{S_0 AB(\alpha)\Gamma(\alpha)}.$$

The formula above can be used to obtain the numerical simulations. By using the Crank–Nicholson formula, we get the following numerical approximations:

$$
\begin{aligned}
\frac{\partial h(r,t)}{\partial t} &= \frac{h_i^{j+1} - h_i^j}{\Delta t}, \\[2mm]
\frac{\partial h(r,t)}{\partial r} &= \frac{\left(h_{i+1}^{j+1} - h_{i-1}^{j+1}\right) + \left(h_{i+1}^{j-1} - h_{i-1}^{j-1}\right)}{2\Delta r}, \\[2mm]
\frac{\partial^2 h(r,t)}{\partial r^2} &= \frac{h_{i-1}^{j+1} - 2h_i^{j+1} + h_{i+1}^{j+1}}{(\Delta)^2} + \frac{h_{i-1}^j - 2h_i^j + h_{i+1}^j}{(\Delta)^2}, \\[2mm]
h(r,t) &= \frac{h_i^{j+1} + h_i^j}{2}.
\end{aligned}
\tag{10.37}
$$

By replacing the above expressions in the equation, we get

$$
\begin{aligned}
\frac{h_i^{j+1} + h_i^j}{2} &= \frac{1-\alpha}{S_0 A B(\alpha)}\left\{ \frac{K(d-\theta-1)}{r_i^{1+\theta}} \frac{\left(h_{i+1}^{j+1} - h_{i-1}^{j+1}\right) + \left(h_{i+1}^{j-1} - h_{i-1}^{j-1}\right)}{2\Delta r} \right.\\
&\quad \left. + \frac{K}{r^\theta}\left\{ \frac{h_{i-1}^{j+1} - 2h_i^{j+1} + h_{i+1}^{j+1}}{(\Delta)^2} + \frac{h_{i-1}^j - 2h_i^j + h_{i+1}^j}{(\Delta)^2}, \right\}\right\} \\
&\quad + \frac{(\Delta)^{\alpha+1}}{S_0 A B(\alpha)\Gamma(\alpha)} \sum_{k=0}^{j} \left\{ \frac{K(d-\theta-1)}{r_i^{1+\theta}} \left\{ \frac{\left(h_{i+1}^{j+1} - h_{i-1}^{j+1}\right) + \left(h_{i+1}^{j-1} - h_{i-1}^{j-1}\right)}{2\Delta r} \right\} \right. \\
&\quad \left. + \frac{K}{r^\theta}\left\{ \frac{h_{i-1}^{k+1} - 2h_i^{k+1} + h_{i+1}^{k+1}}{(\Delta)^2} + \frac{h_{i-1}^k - 2h_i^k + h_{i+1}^k}{(\Delta)^2}, \right\}\right\} b_{j,k}^\alpha, \\[2mm]
b_{j,k}^\alpha &= (j-k)^\alpha - (j-k-1)^\alpha.
\end{aligned}
\tag{10.38}
$$

A recursive formula for any index greater than zero is therefore given as

$$
\begin{aligned}
&h_i^{j+1}\left\{ \frac{1}{2} + \frac{1-\alpha}{S_0 A B(\alpha)} \frac{2K}{(\Delta r)^2 r_i^\theta} + \frac{2(\Delta t)^{\alpha+1}}{(\Delta r)^2 S_0 A B(\alpha)\Gamma(\alpha)} \frac{K}{r_i^\theta} b_{j,j}^\alpha \right\} = \\
&h_i^{j+1}\left\{ \frac{1}{2} + \frac{1-\alpha}{S_0 A B(\alpha)} \frac{2K}{(\Delta r)^2 r_i^\theta} + \frac{2(\Delta t)^{\alpha+1}}{(\Delta r)^2 S_0 A B(\alpha)\Gamma(\alpha)} \frac{K}{r_i^\theta} b_{j,j}^\alpha \right\} \\
&\quad + \frac{1-\alpha}{S_0 A B(\alpha)}\left\{ \frac{K(d-\theta-1)}{r_i^{1+\theta}} \frac{\left(h_{i+1}^{j+1} - h_{i-1}^{j+1}\right) + \left(h_{i+1}^{j-1} - h_{i-1}^{j-1}\right)}{2\Delta r} \right.\\
&\quad \left. + \frac{K}{r_i^\theta}\left\{ \frac{h_{i-1}^{j+1} + h_{i+1}^{j+1}}{(\Delta r)^2} + \frac{h_{i-1}^j + h_{i+1}^j}{(\Delta r)^2} \right\}\right\} \\
&\quad + \frac{(\Delta)^{\alpha+1}}{S_0 A B(\alpha)\Gamma(\alpha)} \sum_{k=0}^{j-1} \left\{ \frac{K(d-\theta-1)}{r_i^{1+\theta}} \frac{\left(h_{i+1}^{k+1} - h_{i-1}^{jk+1}\right) + \left(h_{i+1}^{k-1} - h_{i-1}^{k-1}\right)}{2\Delta r} \right.\\
&\quad \left. + \frac{K}{r_i^\theta}\left\{ \frac{h_{i-1}^{k+1} - 2h_i^{k+1} + h_{i+1}^{k+1}}{(\Delta r)^2} + \frac{h_{i-1}^k - 2h_i^k + h_{i+1}^k}{(\Delta r)^2} \right\}\right\} b_{j,k}^\alpha +
\end{aligned}
\tag{10.39}
$$

$$\frac{(\Delta)^{\alpha+1}}{S_0 A B(\alpha)\Gamma(\alpha)}\left\{ \begin{array}{l} \frac{K(d-\theta-1)}{r_i^{1+\theta}}\left\{ \frac{\left(h_{i+1}^{j+1}-h_{i-1}^{j+1}\right)+\left(h_{i+1}^{j-1}-h_{i-1}^{j-1}\right)}{2\Delta r}\right\} \\ +\frac{K}{r_i^{\theta}}\left\{ \frac{h_{i-1}^{j+1}+h_{i+1}^{j+1}}{(\Delta r)^2}+\frac{h_{i-1}^{j}+h_{i+1}^{j}}{(\Delta r)^2}\right\} \end{array}\right\}$$

$$b_{j,k}^{\alpha}=(j-k)^{\alpha}-(j-k-1)^{\alpha}.$$

10.3.4 Numerical Solution With Fractional Derivative

In this section, the new partial fractional differential equation is solved numerically. We first present the numerical approximation of the fractional derivative based on the Mittag–Leffler function. The numerical method of approximation was earlier in the paper proposed with the title "Solutions of Cattaneo–Hristov model of elastic heat diffusion with Caputo–Fabrizio and Atangana–Baleanu fractional derivatives." The following derivation was made:

$$\begin{aligned}
{}_0^{ABC}D_t^{\alpha}\{f(t_j)\} &= \frac{AB(\alpha)}{1-\alpha}\int_0^{t_j}E_{\alpha}\left\{-\frac{\alpha}{1-\alpha}(t_j-y)^{\alpha}\right\}\frac{df(y)}{dy}dy, \\
&= \frac{AB(\alpha)}{1-\alpha}\sum_{k=0}^{j}\int_{t_k}^{t_{k+1}}E_{\alpha}\left\{-\frac{\alpha}{1-\alpha}(t_j-y)^{\alpha}\right\}\frac{f^{k+1}-f^k}{\Delta t}dy, \quad (10.40) \\
&= \frac{AB(\alpha)}{1-\alpha}\sum_{k=0}^{j}\frac{f^{k+1}-f^k}{\Delta t}\int_{t_k}^{t_{k+1}}E_{\alpha}\left\{-\frac{\alpha}{1-\alpha}(t_j-y)^{\alpha}\right\}dy, \\
&= \frac{AB(\alpha)}{1-\alpha}\sum_{k=0}^{j}\frac{f^{k+1}-f^k}{\Delta t}\delta_{j,k}^{\alpha},
\end{aligned}$$

$$\begin{aligned}
\delta_{j,k}^{\alpha} &= (t_j-t_{k+1})E_{\alpha}\left\{-\frac{\alpha}{1-\alpha}(t_j-t_{k+1})\right\} \\
&\quad -(t_j-t_{k+1})E_{\alpha}\left\{-\frac{\alpha}{1-\alpha}(t_j-t_{k+1})\right\}.
\end{aligned}$$

Using the above discretization with the explicit numerical scheme, the numerical solution of the fractal flow with the novel differentiation is given by

$$S_0\frac{AB(\alpha)}{\alpha}\sum_{k=0}^{j}\frac{h_i^{k+1}-h_i^k}{\Delta t}\delta_{j,k}^{\alpha}=\frac{K(d-\theta-1)}{r_i^{\theta+1}}\left\{\frac{h_{i+1}^j-h_i^j}{2\Delta r}\right\}+\frac{K}{r_i^{\theta}}\left\{\frac{h_{i-1}^j-2h_i^j+h_{i+1}^j}{(\Delta r)^2}\right\},$$

$$(10.41)$$

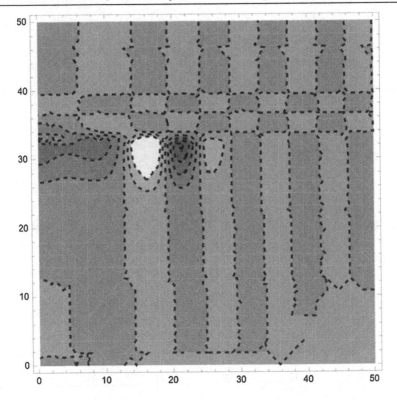

Figure 10.1: **Numerical simulation for** $\alpha = 0.95$.

which is reformulated as

$$
\begin{aligned}
h_i^{j+1} \frac{AB(\alpha)}{\alpha \Delta t} S_0 = {} & h_i^j \left\{ \frac{AB(\alpha)}{\alpha \Delta t} S_0 - \frac{K(d - \theta - 1)}{2 r_i^{\theta+1} \Delta r} - \frac{2K}{r_i^{\theta} (\Delta r)^2} \right\} \\
& - S_0 \frac{AB(\alpha)}{\alpha} \sum_{k=0}^{j} \frac{h_i^{k+1} - h_i^k}{\Delta t} \delta_{j,k}^{\alpha} \\
& + \frac{K(d - \theta - 1)}{r_i^{\theta+1}} \left\{ \frac{h_{i+1}^j}{2 \Delta r} \right\} + \frac{K}{r_i^{\theta}} \left\{ \frac{h_{i-1}^j + h_{i+1}^j}{(\Delta r)^2} \right\}.
\end{aligned} \tag{10.42}
$$

When $j = 0$, the recursive formula is given as

$$
h_i^1 \frac{AB(\alpha)}{\alpha \Delta t} S_0 = h_i^0 \left\{ \frac{AB(\alpha)}{\alpha \Delta t} S_0 - \frac{K(d - \theta - 1)}{2 r_i^{\theta+1} \Delta r} - \frac{2K}{r_i^{\theta} (\Delta r)^2} \right\}. \tag{10.43}
$$

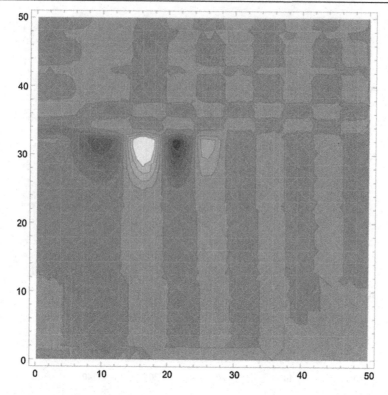

Figure 10.2: Numerical simulation for $\alpha = 0.75$.

Also, the following inequality is observed

$$\left\| \frac{h_i^1}{h_i^0} \right\| = \left\| \frac{\left\{ \frac{AB(\alpha)}{\alpha \Delta t} S_0 - \frac{K(d-\theta-1)}{2r_i^{\theta+1}\Delta r} - \frac{2K}{r_i^{\theta}(\Delta r)^2} \right\}}{\frac{AB(\alpha)}{\alpha \Delta t} S_0} \right\| \leq 1. \qquad (10.44)$$

With the implicit scheme the numerical solution of the fractal flow with the novel differentiation is given by

$$S_0 \frac{AB(\alpha)}{\alpha} \sum_{k=0}^{j} \frac{h_i^{k+1} - h_i^k}{\Delta t} \delta_{j,k}^{\alpha} = \frac{K(d-\theta-1)}{r_i^{\theta+1}} \left\{ \frac{h_{i+1}^{j+1} - h_i^{j+1}}{2\Delta r} \right\}$$

$$+ \frac{K}{r_i^{\theta}} \left\{ \frac{h_{i-1}^{j+1} - 2h_i^{j+1} + h_{i+1}^{j+1}}{(\Delta r)^2} \right\}, \qquad (10.45)$$

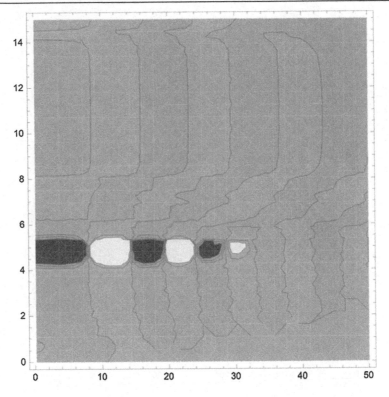

Figure 10.3: Numerical simulation for $\alpha = 0.65$.

which is also reformulated as

$$h_i^{j+1}\left\{\frac{AB(\alpha)}{\alpha\,\Delta t}S_0 - \frac{K(d-\theta-1)}{2r_i^{\theta+1}\Delta r} - \frac{2K}{r_i^\theta(\Delta r)^2}\right\} =$$
$$h_i^j\left\{\frac{AB(\alpha)}{\alpha\,\Delta t}S_0\right\} - S_0\frac{AB(\alpha)}{\alpha}\sum_{k=0}^{j-1}\frac{h_i^{k+1}-h_i^k}{\Delta t}\delta_{j,k}^\alpha +$$
$$\frac{K(d-\theta-1)}{r_i^{\theta+1}}\left\{\frac{h_{i+1}^{j+1}}{2\Delta r}\right\} + \frac{K}{r_i^\theta}\left\{\frac{h_{i-1}^{j+1}+h_{i+1}^{j+1}}{(\Delta r)^2}\right\}. \tag{10.46}$$

At $j=0$, the recursive formula is presented as

$$h_i^0\frac{AB(\alpha)}{\alpha\,\Delta t}S_0 = h_i^1\left\{\frac{AB(\alpha)}{\alpha\,\Delta t}S_0 - \frac{K(d-\theta-1)}{2r_i^{\theta+1}\Delta r} - \frac{2K}{r_i^\theta(\Delta r)^2}\right\}. \tag{10.47}$$

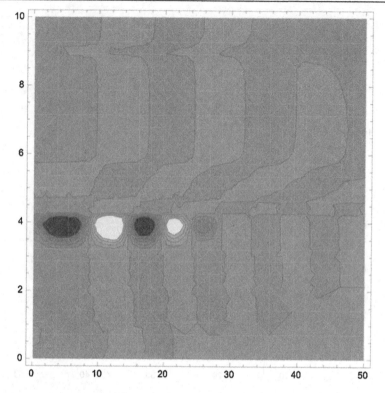

Figure 10.4: Numerical simulation for $\alpha = 0.3$.

Also, the following inequality is obtained

$$\left| \frac{h_i^1}{h_i^0} \right| = \left| \frac{\frac{AB(\alpha)}{\alpha \Delta t} S_0}{\frac{AB(\alpha)}{\alpha \Delta t} S_0 - \frac{K(d-\theta-1)}{2r_i^{\theta+1}\Delta r} - \frac{2K}{r_i^{\theta}(\Delta r)^2}} \right| \leq 1. \qquad (10.48)$$

By using the Crank–Nicholson numerical scheme, the numerical solution of the fractal flow with the novel differentiation is given as

$$S_0 \frac{AB(\alpha)}{\alpha} \sum_{k=0}^{j} \frac{h_i^{k+1} - h_i^k}{\Delta t} \delta_{j,k}^{\alpha} = \frac{K(d-\theta-1)}{r_i^{\theta+1}} \left\{ \frac{\left(h_{i+1}^{j+1} - h_{i-1}^{j+1} \right) + \left(h_{i+1}^{j} - h_{i-1}^{j} \right)}{2\Delta r} \right\} +$$
$$\frac{K}{r_i^{\theta}} \left\{ \frac{h_{i-1}^{j+1} - 2h_i^{j+1} + h_{i+1}^{j+1}}{(\Delta r)^2} + \frac{h_{i-1}^{j} - 2h_i^{j} + h_{i+1}^{j}}{(\Delta r)^2} \right\}. \qquad (10.49)$$

The above equation is also reformulated as follows:

$$
h_i^{j+1} \left\{ \frac{AB(\alpha)}{\alpha \Delta t} S_0 + \frac{2K}{r_i^\theta (\Delta r)^2} \right\} =
$$

$$
h_i^j \left\{ \frac{AB(\alpha)}{\alpha \Delta t} S_0 - \frac{2K}{r_i^\theta (\Delta r)^2} \right\} - S_0 \frac{AB(\alpha)}{\alpha} \sum_{k=0}^{j-1} \frac{h_i^{k+1} - h_i^k}{\Delta t} \delta_{j,k}^\alpha + \tag{10.50}
$$

$$
\frac{K(d - \theta - 1)}{r_i^{\theta+1}} \left\{ \frac{h_{i+1}^{j+1} + h_{i-1}^{j+1}}{2\Delta r} + \frac{h_{i+1}^j + h_{i-1}^j}{2\Delta r} \right\} + \frac{K}{r_i^\theta} \left\{ \frac{h_{i-1}^j + h_{i+1}^j}{(\Delta r)^2} + \frac{h_{i+1}^j + h_{i-1}^j}{2\Delta r} \right\}.
$$

At $j = 0$, the recursive formula is given as

$$
h_i^0 \left\{ \frac{AB(\alpha)}{\alpha \Delta t} S_0 - \frac{2K}{r_i^\theta (\Delta r)^2} \right\} = h_i^1 \left\{ \frac{AB(\alpha)}{\alpha \Delta t} S_0 + \frac{2K}{r_i^\theta (\Delta r)^2} \right\}. \tag{10.51}
$$

The numerical simulations are depicted in Figs. 10.1, 10.2, 10.3, 10.4 for different values of fractional orders α.

CHAPTER 11

Groundwater Flow Within a Fracture, Matrix Rock and Leaky Aquifers: Fractal Geometry

In this chapter, we shall start by providing to reader a clear understanding of a dual media or porosity aquifers. We will note that, in dual porosity aquifers, the main storage of water occurs in the inter-granular pore spaces, and the main groundwater transmission system is the fracture system in the rock. These types of aquifers are commonly referred to as dual porosity aquifers, and their permeability reflects the frequency, openness, and inter-connectivity of the fractures. These aquifers are often the sedimentary rocks.

11.1 Model of Groundwater Flow in a Fractal Dual Media Accounting for Elasticity

In this section, we start of by recalling that, in a dual media the water flows within the fracture network and also within the matrix soils [219]. Within the matrix rock, the medium of flow is non-viscous and homogeneous with memory effect. However, the water flowing within the matrix rock in a geological formation with elastic property cannot be described adequately with the time classical derivative but can efficiently be described with the non-local operator with a power law kernel known as Riemann–Liouville or Caputo fractional derivative. Thus, in order to formulate the mathematical equations that describe the effect of elasticity of the matrix rock, the time local derivative will be replaced by the Caputo fractional derivative to obtain:

$$
\begin{aligned}
S_{S_j} {}_0^C D_t^\alpha \left(h_f(r,t)\right) &= \nabla.\left(k_f.\nabla h_f(r,t)\right) + \eta(h_m(r,t) - h_f(r,t)) + q_f, \\
S_{S_m} {}_0^C D_t^\alpha \left(h_m(r,t)\right) &= \eta(h_m(r,t) - h_f(r,t)).
\end{aligned}
\tag{11.1}
$$

The above formulation was suggested in the work of Atangana and Pacome [219]. We shall first present the existence and uniqueness of the above system, which is very important aspect to ascert that the exact solution exists and is unique under some conditions that will be derived here.

Fractional Operators with Constant and Variable Order
with Application to Geo-Hydrology
DOI: 10.1016/B978-0-12-809670-3.00011-4

11.1.1 Existence of System Solutions by Means of Fixed-Point Theorem Method

The existence of a positive solution for a given fractional differential equation is a big challenge to mathematician because sometimes there exist some complex differential systems that cannot be solved analytically, but the proof of existence helps us know that there exists a solution under some conditions within a well-constructed Sobolev space. In this chapter, we consider the following Sobolev space [219]:

$$H^1(0, T) = \{u \in L^2(a, b)/u' \in L^2(a, b)\}.$$

Likewise, we consider the Hilbert space where

$$H \in \left\{ u, v / \int_0^t (t - y)^\alpha vu\,dy < \infty \right\}.$$

To prove the existence of Eq. (11.1), we express the change of hydraulic head within the matrix soil in terms of the change of hydraulic head within the fracture. To achieve this, we employ the Laplace transform in time to have

$$
\begin{aligned}
S_{S_m} P^\alpha h_m(r, p) &= \eta \left(h_m(r, p) - h_f(r, p) \right), \\
\left(S_{S_m} p^\alpha - \eta \right) h_m(r, p) &= -\eta h_f(r, p), \\
h_m(r, p) &= \frac{-\eta h_f(r, p)}{S_{S_m} \left(p^\alpha - \frac{\eta}{S_{S_m}} \right)}, \\
h_m(r, t) &= L^{-1} \left(\frac{-\eta h_f(r, p)}{S_{S_m} \left(p^\alpha - \frac{\eta}{S_{S_m}} \right)} \right), \\
h_m(r, t) &= -\frac{\eta}{S_{S_m}} \int_0^t h_f(r, y) E_\alpha \left(\frac{\eta}{S_{S_m}} (t - y) \right) dy.
\end{aligned}
\tag{11.2}
$$

By replacing (11.2) in system (11.1) we obtain

$$
\begin{aligned}
S_{S_j\,0}^{\ C} D_t^\alpha \left(h_f(r, t) \right) = {}& \nabla. \left(k_f. \nabla h_f(r, t) \right) \\
&+ \eta \left(-\frac{\eta}{S_{S_m}} \int_0^t h_f(r, y) E_\alpha \left(\frac{\eta}{S_{S_m}} (t - y) \right) dy - h_f(r, t) \right) + q_f.
\end{aligned}
$$

Let

$$
\begin{aligned}
\Gamma_1 &: H^1(0, T) \to H^1(0, T) \\
h &\to \Gamma_1 h = {}_0^{RL} I_t^\alpha \Bigg(\nabla. \left(k_j. \nabla h_j(r, t) \right) \\
&\quad + \eta \left\{ \frac{-\eta}{S_{S_m}} \int_0^t h_f(r, y) E_\alpha \left(\frac{\eta}{S_{S_m}} (t - y) dy \right) - h_j(r, t) \right\} + q_f \Bigg)
\end{aligned}
\tag{11.3}
$$

where

$$
{}_{0}^{RL}I_{t}^{\alpha} = \frac{1}{\Gamma(\alpha)} \int_{0}^{t} (t - \tau)^{\alpha-1} f(\tau) d\tau.
$$

We aim to prove that it possesses Lipschitz condition. Let

$$
K(r, t, h) = \nabla.(k_{j}.\nabla h_{j}(r, t)) - \frac{-\eta^{2}}{SS_{m}} \int_{0}^{t} h_{f}(r, y) E_{\alpha}\left(\frac{\eta}{SS_{m}}(t - y)\right) dy. \qquad (11.4)
$$

Let $h_{1}h_{2} \in H^{1}(0, T)$, then

$$
\|K(r, t, h_{1}) - K(r, t, h_{2})\|_{H^{1}(0,T)} \le \|\nabla.(K_{j}.\nabla h_{j}(r, t)) - \nabla(k_{f}\nabla h_{2}(r, t))\| +
$$

$$
\left\|\frac{\eta^{2}}{SS_{m}}\right\| \int_{0}^{t} \|h_{1}(r, y) - h_{2}(r, y)\| E_{\alpha}\left(\frac{\eta}{SS_{m}}(t - y)\right) dy
$$

$$
\le \theta_{1}\theta_{2}\|k_{f}\|\|h_{1} - h_{2}\| + \left\|\frac{\eta^{2}}{SS_{m}}\right\| \|h_{1} - h_{2}\|_{H^{1}(0,T)} M + \|\eta\|\|h_{1} - h_{2}\| \qquad (11.5)
$$

$$
\le \left(\theta_{1}\theta_{2}\|k_{f}\| + \left\|\frac{\eta^{2}}{SS_{m}}\right\| M + \|\eta\|\right) \|h_{1} - h_{2}\|_{H^{1}(0,T)}
$$

$$
\le l\|h_{1} - h_{2}\|.
$$

With the definition of K, we define the following function:

$$
\Gamma h = \frac{1}{\Gamma(\alpha)} \int_{0}^{t} K(r, \tau, h)^{\alpha-1} d\tau.
$$

Let h_{1} and h_{2} be elements of $H^{1}(0, T)$, then

$$
\|\Gamma_{1}h_{1} - \Gamma_{1}h_{2}\|_{H^{1}(0,T)} = \left\|\frac{1}{\Gamma(\alpha)} \int_{0}^{t} (K(r, \tau, h_{1}) - K(r, \tau, h_{2}))(t - \tau)^{\alpha-1} d\tau\right\|_{H^{1}(0,T)}
$$

$$
\le \frac{1}{\Gamma(\alpha)} \int_{0}^{t} \|K(r, \tau, h_{1}) - K(r, \tau, h_{2})\|(t - \tau)^{\alpha-1} d\tau
$$

$$
\le \frac{1}{\Gamma(\alpha)} \int_{0}^{t} l\|h_{1} - h_{2}\|(t - \alpha)^{\alpha-1} d\tau \qquad (11.6)
$$

$$
\le \frac{l}{\Gamma(\alpha)}\|h_{1} - h_{2}\|_{H^{1}(0,T)} \frac{T^{\alpha-1}}{\alpha}
$$

$$
\le \frac{lT^{\alpha}}{\Gamma(\alpha+1)}\|h_{1} - h_{2}\|_{H^{1}(0,T)}
$$

$$
\le \beta\|h_{1} - h_{2}\|_{H^{1}(0,T)}.
$$

Let us consider the following recursive formula:

$$h_j^{n+1}(r,t) = \Gamma_1 h_f^m = h_f(r,0) + \frac{1}{\Gamma(\alpha)} \int_0^t K(h_f^n, r, \tau)$$

$$\left\| \Gamma_1 h_f^n - \Gamma_1 h_f^{n-1} \right\|_{H^1(0,T)} = \left\| h_f^{n+1} - h_f^n \right\|_{H^1(0,T)} \tag{11.7}$$

$$= \frac{1}{\Gamma(\alpha)} \left\| \int_0^t \left\{ K(h_f^n, r, \tau) - K(h_f^{n-1}, r, \tau) \right\} (t-\tau)^{\alpha-1} d\tau \right\|$$

$$\leq \frac{l}{\Gamma(\alpha)} \left\| h_f^{n-1} - h_f^n \right\|_{H^1(0,T)} \frac{T^\alpha}{\alpha} = \frac{lT^\alpha}{\Gamma(\alpha+1)} \left\| h_f^{n-1} - h_f^n \right\|_{H^1(0,T)}.$$

Recursively on n, we get

$$\left\| \Gamma_1 h_f^n - \Gamma_1 h_f^{n-1} \right\| \leq \left(\frac{lT^\alpha}{\Gamma(\alpha+1)^n} \right) \| h_f^1(r,t) \|. \tag{11.8}$$

We choose $\frac{lT^\alpha}{\Gamma(\alpha+1)^n}$ such that $\frac{lT^\alpha}{\Gamma(\alpha+1)^n} < 1$, for $n \to \infty$: $\left\| \Gamma_1 h_f^n - \Gamma_1 h_f^{n-1} \right\| \to 0$, thus $(h_f^n)_{n \in N}$ to a Cauchy sequence in a Banach space therefore converges toward h_f. Taking the limit on both sides, we have

$$\lim_{x \to \infty} h_f^{n+1} \Leftrightarrow h_f = \Gamma h_f$$

which shows that Γ has a solution and is unique.

11.2 Derivation of the Numerical Solution

In this section, we argue the fact that the storativity coefficients within the aquifer follow the power decay law as suggested in Eq. (11.1). Here we suggest that the storativity coefficient may follow the exponential law with an upper boundary. With this in mind, we present the numerical solution of the system of equations using the upwind numerical scheme in space and the Crank–Nicolson in space. We first present the numerical approximation for time derivative for each nonlocal operator [219]. We present first the numerical approximation of Caputo fractional derivative in time. Let $t_{n+1} - t_n = \Delta t$, $u(t_n, x_i) = u_i^n$, then

$${}_0^C D_t^\alpha u(x_i, t_n) = \frac{1}{\Gamma(1-\alpha)} \int_0^{t_n} \frac{\partial u(x_i, \tau)}{\partial t} (t_n - \tau)^{-\alpha} d\tau$$

$$= \frac{1}{\Gamma(1-\alpha)} \sum_{j=0}^n \int_{t_j}^{t_{j+1}} \frac{u_i^{j+1} - u_i^j}{\Delta t} (t_n - \tau)^{-\alpha} d\tau$$

$$= \frac{1}{\Gamma(1-\alpha)} \sum_{j=0}^{n} \frac{u_i^{j+1} - u_i^j}{\Delta t} \int_{t_j}^{t_{j+1}} (t_n - \tau)^{-\alpha} d\tau$$

$$= \frac{1}{\Gamma(1-\alpha)} \sum_{j=0}^{n} \frac{u_i^{j+1} - u_i^j}{\Delta t} \left(-\frac{Y^{-\alpha+1}}{1-\alpha} \Big|_{t_n - t_j}^{t_n - t_{j+1}} \right) \tag{11.9}$$

$$= \frac{1}{\Gamma(2-\alpha)} \sum_{j=0}^{n} \frac{u_i^{j+1} - u_i^j}{\Delta t} \left\{ ((n-j)\Delta t)^{1-\alpha} - ((n-j-1)\Delta t)^{1-\alpha} \right\}$$

$$= \frac{1}{\Gamma(2-\alpha)} \sum_{j=0}^{n} (u_i^{j+1} - u_i^j) \left\{ ((n-j)\Delta t)^{1-\alpha} - ((n-j-1)\Delta t)^{1-\alpha} \right\}.$$

Next, we consider the second-order upwind scheme for the first-order space derivative:

$$u_x^- = \frac{3u_i^n - 4u_{i-1}^n + 3u_{i-2}^n}{2(\Delta x)}, \qquad u_x^+ = \frac{-u_{i+2}^n - 4u_{i+1}^n + 3u_i^n}{2(\Delta x)}. \tag{11.10}$$

This numerical scheme is known as the linear upwind differencing scheme which has been recognized as a powerful mathematical scheme which is less diffusive compared to the classical first-order accurate scheme. However one can have the third-order for the upwind numerical scheme, we have the following discretization:

$$u_x^- = \frac{2u_i^n - 3u_{i-1}^n - 6u_{i-2}^n + u_{i-3}^n}{6(\Delta x)}, \qquad u_x^+ = \frac{-2u_{i+3}^n + 6u_{i+2}^n - 3u_{i+1}^n - u_i^n}{6(\Delta x)}. \tag{11.11}$$

It is mentioned that this numerical scheme (11.11) is less diffusive compared to the second-order scheme (11.10). We should also mention that the scheme is known to have introduced slight dispersive errors in the region where the gradient is elevated. Thus, with second-order upwind scheme, we have the following numerical formulas:

$$\sum_{j=0}^{n} \frac{h_{fi}^{j+1} - h_{fi}^j}{(\Delta t)^\alpha \Gamma(2-\alpha)} \left\{ (n-j)^{1-\alpha} - (n-j-1)^{1-\alpha} \right\} S_{s_f}(r_i)$$

$$= \frac{k_f^{j+1} - k_f^j}{(\Delta r)} \bullet \frac{3h_{fi}^n 4h_{f(i-1)}^n + 3h_{f(i-2)}^n}{2(\Delta r)} +$$

$$k_f^i \left\{ \frac{h_{fi(i+1)}^{n+1} - 2h_{fi(i)}^{n+1} + h_{fi(i-1)}^{n+1}}{2(\Delta r)^2} + \frac{h_{fi(i+1)}^n - 2h_{fi(i)}^n + h_{fi(i-1)}^n}{2(\Delta r)^2} \right\} + \eta \left\{ h_{mi}^n - h_{fi}^n \right\}, \tag{11.12}$$

$$\sum_{j=0}^{n} \frac{h_{mi}^{j+1} - h_{mi}^j}{(\Delta t)^\alpha \Gamma(2-\alpha)} \left\{ (n-j)^{1-\alpha} - (n-j-1)^{1-\alpha} \right\} S_{s_m}(r_i) = \eta \left(h_{mi}^n - h_{fi}^n \right),$$

$$\text{if } \frac{k_f^{j+1} - k_f^j}{\Delta r} > 0$$

and

$$\sum_{j=0}^{n} \frac{h_{fi}^{j+1} - h_{fi}^j}{(\Delta t)^\alpha \Gamma(2-\alpha)} \left\{ (n-j)^{1-\alpha} - (n-j-1)^{1-\alpha} \right\} S_{s_f}(r_i)$$

$$= \frac{k_f^{j+1} - k_f^j}{(\Delta r)} \bullet \frac{-h_{f(i+2)}^n - 4h_{f(i+1)}^n + 3h_{f(i)}^n}{2(\Delta r)} +$$

$$k_f^i \left\{ \frac{h_{f_1(i+1)}^{n+1} - 2h_{f_1(i)}^{n+1} + h_{f_1(i-1)}^{n+1}}{2(\Delta r)^2} + \frac{h_{f_1(i+1)}^n - 2h_{f_1(i)}^n + h_{f_1(i-1)}^n}{2(\Delta r)^2} \right\} + \eta \left\{ h_{mi}^n - h_{fi}^n \right\},$$

$$\tag{11.13}$$

$$\sum_{j=0}^{n} \frac{h_{mi}^{j+1} - h_{mi}^j}{(\Delta t)^\alpha \Gamma(2-\alpha)} \left\{ (n-j)^{1-\alpha} - (n-j-1)^{1-\alpha} \right\} S_{s_m}(r_i) = \eta \left(h_{mi}^n - h_{fi}^n \right),$$

$$\text{if } \frac{k_f^{j+1} - k_f^j}{\Delta r} < 0.$$

Thus with the third-order upwind scheme, we obtain

$$\sum_{j=0}^{n} \frac{h_{fi}^{j+1} - h_{fi}^j}{(\Delta t)^\alpha \Gamma(2-\alpha)} \left\{ (n-j)^{1-\alpha} - (n-j-1)^{1-\alpha} \right\} S_{s_f}(r_i)$$

$$= \frac{k_f^{j+1} - k_f^j}{(\Delta r)} \bullet \frac{2h_{fi}^n + 3h_{f(i-1)}^n - 6h_{f(i-2)}^n + h_{f(i-3)}^n}{6(\Delta r)} +$$

$$k_f^i \left\{ \frac{h_{f_1(i+1)}^{n+1} - 2h_{f_1(i)}^{n+1} + h_{f_1(i-1)}^{n+1}}{2(\Delta r)^2} + \frac{h_{f_1(i+1)}^n - 2h_{f_1(i)}^n + h_{f_1(i-1)}^n}{2(\Delta r)^2} \right\} + \eta \left\{ h_{mi}^n - h_{fi}^n \right\},$$

$$\tag{11.14}$$

$$\sum_{j=0}^{n} \frac{h_{mi}^{j+1} - h_{mi}^j}{(\Delta t)^\alpha \Gamma(2-\alpha)} \left\{ (n-j)^{1-\alpha} - (n-j-1)^{1-\alpha} \right\} S_{s_m}(r_i) = \eta \left(h_{mi}^n - h_{fi}^n \right),$$

$$\text{if } \frac{k_f^{j+1} - k_f^j}{\Delta r} > 0$$

and

$$\sum_{j=0}^{n} \frac{h_{fi}^{j+1} - h_{fi}^{j}}{(\Delta t)^{\alpha} \Gamma(2-\alpha)} \left\{ (n-j)^{1-\alpha} - (n-j-1)^{1-\alpha} \right\} S_{S_f}(r_i)$$

$$= \frac{k_f^{j+1} - k_f^{j}}{(\Delta r)} \bullet \frac{2h_{f(i+3)}^{n} + 6h_{f(i+2)}^{n} - 3h_{f(i+1)}^{n} - 2h_{fi}^{n}}{6(\Delta r)} +$$

$$k_f^{i} \left\{ \frac{h_{f_1(i+1)}^{n+1} - 2h_{f_1(i)}^{n+1} + h_{f_1(i-1)}^{n+1}}{2(\Delta r)^2} + \frac{h_{f_1(i+1)}^{n} - 2h_{f_1(i)}^{n} + h_{f_1(i-1)}^{n}}{2(\Delta r)^2} \right\} + \eta \left\{ h_{mi}^{n} - h_{fi}^{n} \right\},$$

$$(11.15)$$

$$\sum_{j=0}^{n} \frac{h_{mi}^{j+1} - h_{mi}^{j}}{(\Delta t)^{\alpha} \Gamma(2-\alpha)} \left\{ (n-j)^{1-\alpha} - (n-j-1)^{1-\alpha} \right\} S_{S_m}(r_i) = \eta \left(h_{mi}^{n} - h_{fi}^{n} \right),$$

if $\dfrac{k_f^{j+1} - k_f^{j}}{\Delta r} < 0.$

11.3 Model of Groundwater Flow in Fractal Dual Media With Viscoelasticity Effect

As indicated earlier, a double media has water flow through both the fractured network and the matrix rock which has different characteristics. In this section we consider matrix rock having properties of viscoelasticity. It is noteworthy that a suitable or realistic depiction of the subsurface may be obtained by associating mechanical properties of elastic material and properties of viscous fluids. The stress of the resultant media or material is dependent on both strain and strain rate. Additionally, it is dependent on higher time derivatives of the strain. Subsequently, these geological formations having both solid-like and liquid-like behavior are known to be viscoelastic [219]. Furthermore, this section concentrates on fractal dual flow through a common geological formations showing properties of heterogeneity and inelasticity, within the framework of linear viscoelasticity. Herein, it is assumed that water flows through the matrix rock having a property of viscoelasticity. It is known that this scenario cannot be adequately described using the time classical derivative. On the other hand, the Atangana–Baleanu derivative in the Caputo and Riemann–Liouville sense, which is a nonlocal operator with a Mittag–Leffler kernel, can be used to describe the above-mentioned scenario. Accordingly, in order to account for matrix rock having properties of elasticity, the Atangana–Baleanu fractional derivative replaces the time local derivative to obtain the following:

$$S_{S_j 0}^{ABC} D_t^{\alpha} \left(h_f(r,t) \right) = \nabla \cdot \left(k_f \cdot \nabla h_f(r,t) \right) + \eta(h_m(r,t) - h_f(r,t)) + q_f,$$

$$S_{S_m 0}^{ABC} D_t^{\alpha} \left(h_m(r,t) \right) = \eta(h_m(r,t) - h_f(r,t)). \tag{11.16}$$

In what follows, we shall first present the existence and uniqueness of the above system.

11.3.1 Existence of System Solutions

In this section, the well-constructed Sobolev space is considered. In this paper, we consider the following Sobolev space:

$$H^1(0, T) = \{u \in L^2(a, b)/u' \in L^2(a, b)\}.$$

We also consider the following Hilbert space where

$$H \in \left\{ u, v / \int_0^t E_\alpha \left[-\frac{\alpha}{1-\alpha}(t-y)^\alpha \right] vu \, dy \infty \right\}.$$

To proof the existence of Eq. (11.16), we express the change of hydraulic head within the matrix soil in terms of the change of hydraulic head within the fracture as presented earlier in the case of power law. To achieve this, we employ the Laplace transform in time to obtain:

$$\frac{S_{S_m}(r, \alpha) p^\alpha h_m(r, p)}{p^\alpha + \frac{\alpha}{1-\alpha}} = \eta(h_m(r, p) - h_f(r, p)),$$

$$\left(\frac{S_{S_m}(r, \alpha) p^\alpha}{p^\alpha + \frac{\alpha}{1-\alpha}} - \eta \right) h_m(r, p) = -\eta h_f(r, p),$$

$$h_m(r, p) = \frac{-\eta h_f(r, p)}{\left(\frac{S_{S_m}(r,\alpha) p^\alpha}{p^\alpha + \frac{\alpha}{1-\alpha}} - \eta \right)},$$

$$h_m(r, t) = L^{-1} \left[\frac{-\eta h_f(r, p)}{\left(\frac{S_{S_m}(r,\alpha) p^\alpha}{p^\alpha + \frac{\alpha}{1-\alpha}} - \eta \right)} \right], \tag{11.17}$$

$$h_m(r, t) = \frac{\eta}{\eta - S_{S_m}(r, \alpha)} \int_0^t h_f(r, y) E_\alpha \left[\frac{\alpha\eta}{(\eta - S_{S_m}(r, \alpha))(1 - \alpha)} (t-y)^\alpha \right] dy +$$

$$\frac{\eta\alpha}{(1-\alpha)(\eta - S_{S_m}(r, \alpha))} \int_0^t h_f(r, y) E_\alpha \left[\frac{\alpha\eta}{(\eta - S_{S_m}(r, \alpha))(1 - \alpha)} (t-y)^\alpha \right] dy.$$

Eq. (11.17) can now be replaced in system (11.16) to obtain

$$S_{S_f} {}_0^{ABC} D_t^\alpha (h_f(r, t)) = \nabla.(k_f.\nabla h_f(r, t)) +$$

$$\eta \left[\frac{\eta}{\eta - S_{S_m}(r, \alpha)} \int_0^t h_f(r, y) E_\alpha \left[\frac{\alpha\eta}{(\eta - S_{S_m}(r, \alpha))(1 - \alpha)} (t-y)^\alpha \right] dy + \right. \tag{11.18}$$

$$\left. \frac{\eta\alpha}{(1-\alpha)(\eta - S_{S_m}(r, \alpha))} \int_0^t h_f(r, y) E_\alpha \left[\frac{\alpha\eta}{(\eta - S_{S_m}(r, \alpha))(1 - \alpha)} (t-y)^\alpha \right] dy - h_f(r, t) \right] + q_f,$$

Let us consider the following function:

$$T_1 : H^1(0, T) \to H^1(0, T)$$

$$v \to T_1 v = {}^{AB}_0 I^\alpha_t \Big\{ \nabla.(k_f.\nabla v(r, t))$$

$$+ \eta \Bigg[\frac{\eta}{\eta - S_{S_m}(r, \alpha)} \int_0^t h_f(r, y) E_\alpha \left(\frac{\alpha \eta}{(\eta - S_{S_m}(r, \alpha))(1 - \alpha)} (t - y)^\alpha \right) dy + \qquad (11.19)$$

$$\frac{\eta \alpha}{(1 - \alpha)(\eta - S_{S_m}(r, \alpha))} \int_0^t h_f(r, y) E_\alpha \left(\frac{\alpha \eta}{(\eta - S_{S_m}(r, \alpha))(1 - \alpha)} (t - y)^\alpha \right) dy - h_f(r, t) \Bigg]$$

$$+ q_f \Big\}.$$

The fractional integral used here is known as Atangana–Baleanu fractional integral and is given as:

$$^{AB}_0 I^\alpha_t f(t) = \frac{1 - \alpha}{AB(\alpha)} f(t) + \frac{\alpha}{AB(\alpha)\Gamma(\alpha)} \int_0^t (t - \tau)^{\alpha-1} f(\tau) d\tau. \qquad (11.20)$$

We aim to prove that T_i possesses Lipschitz condition.

Let us consider the following operator:

$$B(r, t, v) = \nabla.(k_f.\nabla v(r, t)) +$$

$$\eta \Bigg[\frac{\eta}{\eta - S_{S_m}(r, \alpha)} \int_0^t v(r, y) E_\alpha \left[\frac{\alpha \eta}{(\eta - S_{S_m}(r, \alpha))(1 - \alpha)} (t - y)^\alpha \right] dy + \qquad (11.21)$$

$$\frac{\eta \alpha}{(1 - \alpha)(\eta - S_{S_m}(r, \alpha))} \int_0^t v(r, y) E_\alpha \left[\frac{\alpha \eta}{(\eta - S_{S_m}(r, \alpha))(1 - \alpha)} (t - y)^\alpha \right] dy - v(r, t) \Bigg].$$

Let $h_1, h_2 \in H^1(0, T)$, then

$$\|B(r, t, h_1) - B(r, t, h_2)\|_{H^1(0,T)} \le \|\nabla.(k_f.\nabla h_1(r, t)) - \nabla(k_f \nabla h_2(r, t))\| +$$

$$\left\| \frac{\eta}{\eta - S_{S_m}(r, \alpha)} \right\| \left\| \int_0^t \|h_1(r, y) - h_2(r, y)\| E_\alpha \left[\frac{\alpha \eta}{(\eta - S_{S_m}(r, \alpha))(1 - \alpha)} (t - y)^\alpha \right] dy + \right.$$

$$\left\| \frac{\eta}{\eta - S_{S_m}(r, \alpha)(1 - \alpha)} \right\| \left\| \int_0^t \|h_1(r, y) - h_2(r, y)\| E_\alpha \left[\frac{\alpha \eta}{(\eta - S_{S_m}(r, \alpha))(1 - \alpha)} (t - y)^\alpha \right] dy$$

$$\qquad (11.22)$$

$$\le \theta_1 \theta_2 \|k_f\| \|h_1 - h_2\| + \left\| \frac{\eta M_2}{\eta - S_{S_m}(r, \alpha)} + \frac{\eta \alpha M_1}{\eta - S_{S_m}(r, \alpha)(1 - \alpha)} \right\| \|h_1 - h_2\|_{H^1(0,T)}$$

$$+ \|\eta\| \|h_1 - h_2\|$$

$$\leq \left[\theta_1 \theta_2 \|k_f\| + \left\| \frac{\eta M_2}{\eta - S_{S_m}(r,\alpha)} + \frac{\eta \alpha M_1}{\eta - S_{S_m}(r,\alpha)(1-\alpha)} \right\| + \|\eta\| \right] \|h_1 - h_2\|_{H^1(0,T)}$$

$$\leq l_1 \|h_1 - h_2\|.$$

Using the definition of B presented previously, we consider the following operator:

$$Ph = \frac{1-\alpha}{AB(\alpha)} B(r,\tau,h) + \frac{\alpha}{AB(\alpha)\Gamma(\alpha)} \int_0^t K(r,\tau,h)(t-\tau)^{1-\alpha} d\tau. \tag{11.23}$$

Let h_1 and h_2 be elements of $H^1(0,T)$, then

$$\|Ph_1 - Ph_2\|_{H^1(0,T)} = \left\| \begin{array}{c} \frac{1-\alpha}{AB(\alpha)}\{(B(r,\tau,h_1)-B(r,\tau,h_2))\}+ \\ \frac{\alpha}{AB(\alpha)\Gamma(\alpha)} \int_0^t (B(r,\tau,h_1)-B(r,\tau,h_2))(t-\tau)^{\alpha-1} d\tau \end{array} \right\|_{H^1(0,T)}$$

$$\leq \frac{1-\alpha}{AB(\alpha)} \|\{(B(r,\tau,h_1) - B(r,\tau,h_2))\}\|_{H^1(0,T)} +$$

$$\frac{\alpha}{AB(\alpha)\Gamma(\alpha)} \int_0^t \|\{(B(r,\tau,h_1) - B(r,\tau,h_2))\}\| (t-\tau)^{\alpha-1} d\tau$$

$$\leq \frac{(1-\alpha)l_1}{AB(\alpha)} + \frac{\alpha}{AB(\alpha)\Gamma^2(\alpha)} \int_0^t l_1 \|h_1 - h_2\| (t-\tau)^{\alpha-1} d\tau \tag{11.24}$$

$$\leq \left\{ \frac{(1-\alpha)l_1}{AB(\alpha)} + \frac{l_1 T^\alpha}{AB(\alpha)\Gamma(\alpha)} \right\} \|h_1 - h_2\|_{H^1(0,T)}$$

$$\leq \left\{ \frac{(1-\alpha)l_1}{AB(\alpha)} + \frac{l_1 T^\alpha}{AB(\alpha)\Gamma^2(\alpha)} \right\} \|h_1 - h_2\|_{H^1(0,T)}$$

$$\leq \beta_1 \|h_1 - h_2\|_{H^1(0,T)}.$$

In this section, using the Picard iterative method, we aim to establish the existence of the system solutions. Thus, we consider the following Volterra equation subject to Atangana–Baleanu fractional integral:

$$h_f^{n+1}(r,t) = Ph_f^n = \frac{1-\alpha}{AB(\alpha)} B(h_f^n, r, \tau) + \frac{\alpha}{AB(\alpha)\Gamma(\alpha)} \int_0^t B(h_f^n, r, \tau)(t-\tau)^{\alpha-1} d\tau. \tag{11.25}$$

Thus,

$$\left\| Ph_f^n - Ph_f^{n-1} \right\|_{H^1(0,T)} = \left\| h_f^{n+1} - h_f^n \right\|_{H^1(0,T)}$$

$$= \frac{1-\alpha}{AB(\alpha)} \left(B\left(h_f^n, r, \tau\right) - B\left(h_f^{n-1}, r, \tau\right) \right) +$$

$$\frac{\alpha}{AB(\alpha)\Gamma(\alpha)} \left\| \int_0^t \left\{ B\left(h_f^n, r, \tau\right) - B\left(h_f^{n-1}, r, \tau\right) \right\}$$

$$\times (t-\tau)^{\alpha-1} d\tau \bigg\|$$

$$\leq \left\{ \frac{(1-\alpha)l_1}{AB(\alpha)} + \frac{l_1 T^\alpha}{AB(\alpha)\Gamma(\alpha)} \right\} \left\| h_f^{n-1} - h_f^n \right\|_{H^1(0,T)}. \quad (11.26)$$

Recursively on n, we have

$$\left\| Ph_f^n - Ph_f^{n-1} \right\| \leq \left\{ \frac{(1-\alpha)l_1}{AB(\alpha)} + \frac{l_1 T^\alpha}{AB(\alpha)\Gamma(\alpha)} \right\}^n \left\| h_f^1 \right\|_{H^1(0,T)}. \quad (11.27)$$

We choose l_1 such that, for a very large n,

$$\left\{ \frac{(1-\alpha)l_1}{AB(\alpha)} + \frac{l_1 T^\alpha}{AB(\alpha)\Gamma(\alpha)} \right\}^n \rightarrow 0. \quad (11.28)$$

$\left\| Ph_f^n - Ph_f^{n-1} \right\| \rightarrow 0$, thus $(h_f^n)_{n\in N}$ is a Cauchy sequence in a Banach space therefore converges towards h_f. Taking limit of both sides, we have

$$\lim_{x\to\infty} h_f^{n+1} = Ph_f^n \Leftrightarrow h_f = Ph_f$$

which shows that B has a solution, and it is unique. The unique solution is the solution of Eq. (11.16).

11.3.2 Derivation of the Numerical Solution With Mittag–Leffler Law

We present the numerical solution of the system of equations using the upwind scheme in space and the Crank–Nicolson in time. We present first the numerical approximation of the Atangana–Baleanu fractional derivative in Caputo sense. Let $t_{n+1} - t_n = \Delta t$, $u(t_n, x_i) = u_i^n$.

Then the Atangana–Baleanu fractional derivative in Caputo sense is approximated as [219]:

$$
\begin{aligned}
{}_0^{ABC} D_t^\alpha u(x_i, t_n) &= \frac{AB(\alpha)}{(1-\alpha)} \int_0^t \frac{\partial u(x_i, \tau)}{\partial t} E_\alpha \left\{ -\frac{\alpha}{1-\alpha}(t_n-\tau)^\alpha \right\} d\tau \\
&= \frac{AB(\alpha)}{(1-\alpha)} \sum_{j=0}^n \int_{t_j}^{t_{j+1}} \frac{u_i^{j+1} - u_i^j}{\Delta t} E_\alpha \left\{ -\frac{\alpha}{1-\alpha}(t_n-\tau)^\alpha \right\} d\tau \\
&= \frac{AB(\alpha)}{(1-\alpha)} \sum_{j=0}^n \frac{u_i^{j+1} - u_i^j}{\Delta t} \int_{t_j}^{t_{j+1}} E_\alpha \left\{ -\frac{\alpha}{1-\alpha}(t_n-\tau)^\alpha \right\} d\tau \\
&= \frac{AB(\alpha)}{(1-\alpha)} \sum_{j=0}^n \frac{u_i^{j+1} - u_i^j}{\Delta t} \left\{ -(t_n-t_{j+1})E_{\alpha,2}\left(-\frac{\alpha}{1-\alpha}(t_n-t_{j+1}) \right) + \right.
\end{aligned}
$$

$$(t_n - t_j) E_{\alpha,2} \left(-\frac{\alpha}{1-\alpha}(t_n - t_j) \right) \Bigg\}$$ (11.29)

$$= \frac{AB(\alpha)}{(1-\alpha)} \sum_{j=0}^{n} \frac{u_i^{j+1} - u_i^j}{\Delta t} \{ -\Delta t (n - j - 1)$$

$$\times E_{\alpha,2} \left(-\frac{\alpha \Delta t}{1-\alpha}(n - j - 1) \right) +$$

$$\Delta t (n - j) E_{\alpha,2} \left(-\frac{\alpha \Delta t}{1-\alpha}(n - j) \right) \Bigg\}$$

$$= \frac{AB(\alpha)}{(1-\alpha)} \sum_{j=0}^{n} (u_i^{j+1} - u_i^j) \left\{ (n - j) E_{\alpha,2} \left(-\frac{\alpha \Delta t}{1-\alpha}(n - j) \right) - \right.$$

$$\left. (n - j - 1) E_{\alpha,2} \left(-\frac{\alpha \Delta t}{1-\alpha}(n - j - 1) \right) \right\}.$$

Using the above numerical approximation and the upwind second-order in space, we obtain the numerical formulas [219]:

$$\frac{AB(\alpha)}{(1-\alpha)} \sum_{j=0}^{n} \left(h_{fi}^{j+1} - h_{fi}^j \right) \left\{ \begin{array}{l} (n-j) E_{\alpha,2}\left(-\frac{\alpha \Delta t}{1-\alpha}(n-j) - \right) \\ (n-j-1) E_{\alpha,2}\left(-\frac{\alpha \Delta t}{1-\alpha}(n-j-1) \right) \end{array} \right\} S_{S_f}(r_i)$$

$$= \frac{k_f^{j+1} - k_f^j}{(\Delta r)} \bullet \frac{3h_{f(i)}^n - 4h_{f(i-1)}^n + 3h_{f(i-1)}^n}{2(\Delta r)} +$$

$$k_f^i \left\{ \frac{h_{f_1(i+1)}^{n+1} - 2h_{f_1(i)}^{n+1} + h_{f_1(i-1)}^{n+1}}{2(\Delta r)^2} + \frac{h_{f_1(i+1)}^n - 2h_{f_1(i)}^n + h_{f_1(i-1)}^n}{2(\Delta r)^2} \right\} + \eta \left\{ h_{mi}^n - h_{fi}^n \right\},$$

$$\frac{AB(\alpha)}{(1-\alpha)} \sum_{j=0}^{n} \left(h_{fi}^{j+1} - h_{fi}^j \right) \left\{ \begin{array}{l} (n-j) E_{\alpha,2}\left(-\frac{\alpha \Delta t}{1-\alpha}(n-j) - \right) \\ (n-j-1) E_{\alpha,2}\left(-\frac{\alpha \Delta t}{1-\alpha}(n-j-1) \right) \end{array} \right\} S_{S_m}(r_i) = \eta \left(h_{mi}^n - h_{fi}^n \right)$$

$$\text{if } \frac{k_f^{j+1} - k_f^j}{(\Delta r)} > 0$$

and

$$\frac{AB(\alpha)}{(1-\alpha)} \sum_{j=0}^{n} \left(h_{fi}^{j+1} - h_{fi}^j \right) \left\{ \begin{array}{l} (n-j) E_{\alpha,2}\left(-\frac{\alpha \Delta t}{1-\alpha}(n-j) - \right) \\ (n-j-1) E_{\alpha,2}\left(-\frac{\alpha \Delta t}{1-\alpha}(n-j-1) \right) \end{array} \right\} S_{S_f}(r_i)$$

$$= \frac{k_f^{j+1} - k_f^j}{(\Delta r)} \bullet \frac{-h_{f(i+2)}^n - 4h_{f(i+1)}^n + 3h_{f(i)}^n}{2(\Delta r)} +$$

$$k_f^i \left\{ \frac{h_{f1(i+1)}^{n+1} - 2h_{f1(i)}^{n+1} + h_{f1(i-1)}^{n+1}}{2(\Delta r)^2} + \frac{h_{f1(i+1)}^{n} - 2h_{f1(i)}^{n} + h_{f1(i-1)}^{n}}{2(\Delta r)^2} \right\} + \eta \left\{ h_{mi}^n - h_{fi}^n \right\},$$

$$\frac{AB(\alpha)}{(1-\alpha)} \sum_{j=0}^{n} \left(h_{fi}^{j+1} - h_{fi}^{j} \right) \left\{ \begin{array}{l} (n-j)E_{\alpha,2}\left(-\frac{\alpha\Delta t}{1-\alpha}(n-j)-\right) \\ (n-j-1)E_{\alpha,2}\left(-\frac{\alpha\Delta t}{1-\alpha}(n-j-1)\right) \end{array} \right\} S_{S_m}(r_i) = \eta \left(h_{mi}^n - h_{fi}^n \right)$$

if $\dfrac{k_f^{j+1} - k_f^{j}}{(\Delta r)} > 0.$

Thus with the third-order upwind numerical scheme, we obtain [219]:

$$\frac{AB(\alpha)}{(1-\alpha)} \sum_{j=0}^{n} \left(h_{fi}^{j+1} - h_{fi}^{j} \right) \left\{ \begin{array}{l} (n-j)E_{\alpha,2}\left(-\frac{\alpha\Delta t}{1-\alpha}(n-j)-\right) \\ (n-j-1)E_{\alpha,2}\left(-\frac{\alpha\Delta t}{1-\alpha}(n-j-1)\right) \end{array} \right\} S_{S_f}(r_i)$$

$$= \frac{k_f^{j+1} - k_f^{j}}{(\Delta r)} \cdot \frac{2h_{f(i)}^{n} + 3h_{f(i-1)}^{n} - 6h_{f(i-2)}^{n} + h_{f(i-3)}^{n}}{2(\Delta r)} +$$

$$k_f^i \left\{ \frac{h_{f1(i+1)}^{n+1} - 2h_{f1(i)}^{n+1} + h_{f1(i-1)}^{n+1}}{2(\Delta r)^2} + \frac{h_{f1(i+1)}^{n} - 2h_{f1(i)}^{n} + h_{f1(i-1)}^{n}}{2(\Delta r)^2} \right\} + \eta \left\{ h_{mi}^n - h_{fi}^n \right\},$$

$$\frac{AB(\alpha)}{(1-\alpha)} \sum_{j=0}^{n} \left(h_{fi}^{j+1} - h_{fi}^{j} \right) \left\{ \begin{array}{l} (n-j)E_{\alpha,2}\left(-\frac{\alpha\Delta t}{1-\alpha}(n-j)-\right) \\ (n-j-1)E_{\alpha,2}\left(-\frac{\alpha\Delta t}{1-\alpha}(n-j-1)\right) \end{array} \right\} S_{S_m}(r_i) = \eta \left(h_{mi}^n - h_{fi}^n \right)$$

if $\dfrac{k_f^{j+1} - k_f^{j}}{(\Delta r)} > 0$

and

$$\frac{AB(\alpha)}{(1-\alpha)} \sum_{j=0}^{n} \left(h_{fi}^{j+1} - h_{fi}^{j} \right) \left\{ \begin{array}{l} (n-j)E_{\alpha,2}\left(-\frac{\alpha\Delta t}{1-\alpha}(n-j)-\right) \\ (n-j-1)E_{\alpha,2}\left(-\frac{\alpha\Delta t}{1-\alpha}(n-j-1)\right) \end{array} \right\} S_{S_f}(r_i)$$

$$= \frac{k_f^{j+1} - k_f^{j}}{(\Delta r)} \cdot \frac{-2h_{f(i+3)}^{n} + 6h_{f(i+2)}^{n} - 3h_{f(i+1)}^{n} - 2h_{fi}^{n}}{2(\Delta r)} +$$

$$k_f^i \left\{ \frac{h_{f1(i+1)}^{n+1} - 2h_{f1(i)}^{n+1} + h_{f1(i-1)}^{n+1}}{2(\Delta r)^2} + \frac{h_{f1(i+1)}^{n} - 2h_{f1(i)}^{n} + h_{f1(i-1)}^{n}}{2(\Delta r)^2} \right\} + \eta \left\{ h_{mi}^n - h_{fi}^n \right\},$$

$$\frac{AB(\alpha)}{(1-\alpha)} \sum_{j=0}^{n} \left(h_{fi}^{j+1} - h_{fi}^{j} \right) \left\{ \begin{array}{l} (n-j)E_{\alpha,2}\left(-\frac{\alpha\Delta t}{1-\alpha}(n-j)-\right) \\ (n-j-1)E_{\alpha,2}\left(-\frac{\alpha\Delta t}{1-\alpha}(n-j-1)\right) \end{array} \right\} S_{S_m}(r_i) = \eta \left(h_{mi}^n - h_{fi}^n \right)$$

if $\dfrac{k_f^{j+1} - k_f^{j}}{(\Delta r)} > 0.$

11.4 Model of Groundwater Flow in a Fractal Dual Media With Heterogeneity and Viscoelasticity Properties

In this section, we consider the model with a more complex nonlocal operator as presented in the work [219]. The considered operator here is a convolution of power-Mittag–Leffler with the unknown function. This nonlocal operator was recently proposed by Atangana in the paper "Derivative with two fractional orders: A new avenue of investigation toward revolution in fractional calculus". Therefore, using the new established nonlocal fractional operator suggested by Atangana, the modified model is given as follows:

$$\begin{aligned} S_{S_f0}{}^{AC}D_t^\alpha \left(h_f(r,t)\right) &= \nabla.\left(k_f.\nabla h_f(r,t)\right)+\eta(h_m(r,t)-h_f(r,t))+q_f, \\ S_{S_m0}{}^{AC}D_t^\alpha \left(h_m(r,t)\right) &= \eta(h_m(r,t)-h_f(r,t)). \end{aligned} \tag{11.30}$$

Here, we will not discuss the analysis of existence and uniqueness of exact solution. Only numerical method of solving the model will be shown. To do this, we present first the numerical approximation of the Atangana fractional derivative with two fractional orders.

$$\begin{aligned} {}_0^{AC}D_t^{\alpha,\beta}(f(x,t)) &= \frac{A(\beta)}{1-\beta}\frac{1}{\Gamma(1-\alpha)}\int_0^t \partial_t f(x,y)(t-y)^{-\alpha}E_\beta\left(-\frac{\beta}{1-\beta}(t-y)^{\alpha+\beta}\right)dy, \\ &= \frac{A(\beta)}{1-\beta}\frac{1}{\Gamma(1-\alpha)}\int_0^t \frac{f(x,y+\Delta y)-f(x,y)}{\Delta t}(t-y)^{-\alpha}E_\beta\left(-\frac{\beta}{1-\beta}(t-y)^{\alpha+\beta}\right)dy, \\ &= \frac{A(\beta)}{1-\beta}\frac{1}{\Gamma(1-\alpha)}\sum_{j=0}^n\int_{t_j}^{t_{j+1}} \frac{f_i^{j+1}-f_i^j}{\Delta t}(t_n-y)^{-\alpha}E_\beta\left(-\frac{\beta}{1-\beta}(t-y)^{\alpha+\beta}\right)dy, \quad (11.31) \\ &= \frac{A(\beta)}{1-\beta}\frac{1}{\Gamma(1-\alpha)}\sum_{j=0}^n\left(\frac{f_i^{j+1}-f_i^j}{\Delta t}\right)\int_{t_j}^{t_{j+1}} (t_n-y)^{-\alpha}E_\beta\left(-\frac{\beta}{1-\beta}(t-y)^{\alpha+\beta}\right)dy. \end{aligned}$$

In the above expression, the integral is given as follows:

$$\begin{aligned} &\int_{t_j}^{t_{j+1}} (t_n-y)^{-\alpha}E_\beta\left(-\frac{\beta}{1-\beta}(t-y)^{\alpha+\beta}\right)dy \\ &= (t_n-t_{j+1})^{1-\alpha}E_{\beta,2-\alpha}\left(-\frac{\beta}{1-\beta}(t_n-t_{j+1})^{\alpha+\beta}\right) \\ &+ (t_n-t_{j+1})^{1-\alpha}E_{\beta,2-\alpha}\left(-\frac{\beta}{1-\beta}(t_n-t_j)^{\alpha+\beta}\right). \end{aligned} \tag{11.32}$$

Replacing (11.31) in (11.32) we obtain the following numerical approximation:

$${}^{AC}_{0}D^{\alpha,\beta}_t f(x_i, t_n)$$

$$= \frac{A(\beta)}{(1-\beta)}\frac{1}{\Gamma(1-\alpha)}\sum_{j=1}^{n}\left(\frac{f_i^{j+1} - f_i^{j}}{\Delta t}\right)\left\{\begin{array}{l}(t_n-t_{j+1})^{1-\alpha}E_{\beta,2-\alpha}\left(-\frac{\beta}{1-\beta}(t_n-t_{j+1})^{\alpha+\beta}\right)+ \\ (t_n-t_j)^{1-\alpha}E_{\beta,2-\alpha}\left(-\frac{\beta}{1-\beta}(t_n-t_j)^{\alpha+\beta}\right)\end{array}\right\}. \quad (11.33)$$

Coupling the upwind for second-order with the above, the numerical solution of Eq. (11.30) is given as:

$$\frac{A(\beta)}{(1-\beta)\Gamma(1-\alpha)}\sum_{j=1}^{n}\left(\frac{f_i^{j+1} - f_i^{j}}{\Delta t}\right)\left\{\begin{array}{l}(t_n-t_{j+1})^{1-\alpha}E_{\beta,2-\alpha}\left(-\frac{\beta}{1-\beta}(t_n-t_{j+1})^{\alpha+\beta}\right)+ \\ (t_n-t_j)^{1-\alpha}E_{\beta,2-\alpha}\left(-\frac{\beta}{1-\beta}(t_n-t_j)^{\alpha+\beta}\right)\end{array}\right\}S_{S_j}$$

$$= \frac{k_f^{j+1} - k_f^{j}}{(\Delta r)}\frac{3h^n_{f(i)} - 4h^n_{f(i-1)} + 3h^n_{f(i-2)}}{2(\Delta r)} +$$

$$k_f^i\left\{\frac{h^{n+1}_{f_1(i+1)} - 2h^{n+1}_{f_1(i)} + h^{n+1}_{f_1(i-1)}}{2(\Delta r)^2} + \frac{h^n_{f_1(i+1)} - 2h^n_{f_1(i)} + h^n_{f_1(i-1)}}{2(\Delta r)^2}\right\} + \eta\left(h^n_{fi} - h^n_{mi}\right),$$

$$\frac{A(\beta)}{(1-\beta)\Gamma(1-\alpha)}\sum_{j=1}^{n}\left(\frac{h^{j+1}_{mi} - h^{j}_{mi}}{\Delta t}\right)\left\{\begin{array}{l}(t_n-t_{j+1})^{1-\alpha}E_{\beta,2-\alpha}\left(-\frac{\beta}{1-\beta}(t_n-t_{j+1})^{\alpha+\beta}\right)+ \\ (t_n-t_j)^{1-\alpha}E_{\beta,2-\alpha}\left(-\frac{\beta}{1-\beta}(t_n-t_j)^{\alpha+\beta}\right)\end{array}\right\}S_{S_j}$$

$$= \eta\left(h^n_{fi} - h^n_{mi}\right)$$

if $\dfrac{k_f^{i+1} - k_f^{i}}{\Delta r} > 0,$ \quad (11.34)

otherwise we obtain the following:

$$\frac{A(\beta)}{(1-\beta)\Gamma(1-\alpha)}\sum_{j=1}^{n}\left(\frac{f_i^{j+1} - f_i^{j}}{\Delta t}\right)\left\{\begin{array}{l}(t_n-t_{j+1})^{1-\alpha}E_{\beta,2-\alpha}\left(-\frac{\beta}{1-\beta}(t_n-t_{j+1})^{\alpha+\beta}\right)+ \\ (t_n-t_j)^{1-\alpha}E_{\beta,2-\alpha}\left(-\frac{\beta}{1-\beta}(t_n-t_j)^{\alpha+\beta}\right)\end{array}\right\}S_{S_j}$$

$$= \frac{k_f^{j+1} - k_f^{j}}{(\Delta r)}\frac{-h^n_{f(i+2)} - 4h^n_{f(i+1)} + 3h^n_{f(i)}}{2(\Delta r)} +$$

$$k_f^i\left\{\frac{h^{n+1}_{f_1(i+1)} - 2h^{n+1}_{f_1(i)} + h^{n+1}_{f_1(i-1)}}{2(\Delta r)^2} + \frac{h^n_{f_1(i+1)} - 2h^n_{f_1(i)} + h^n_{f_1(i-1)}}{2(\Delta r)^2}\right\} + \eta\left(h^n_{fi} - h^n_{mi}\right),$$

$$\frac{A(\beta)}{(1-\beta)\Gamma(1-\alpha)}\sum_{j=1}^{n}\left(\frac{h^{j+1}_{mi} - h^{j}_{mi}}{\Delta t}\right)\left\{\begin{array}{l}(t_n-t_{j+1})^{1-\alpha}E_{\beta,2-\alpha}\left(-\frac{\beta}{1-\beta}(t_n-t_{j+1})^{\alpha+\beta}\right)+ \\ (t_n-t_j)^{1-\alpha}E_{\beta,2-\alpha}\left(-\frac{\beta}{1-\beta}(t_n-t_j)^{\alpha+\beta}\right)\end{array}\right\}S_{S_j}$$

$$= \eta\left(h^n_{fi} - h^n_{mi}\right)$$

if $\dfrac{k_f^{i+1} - k_f^{i}}{\Delta r} < 0.$ \quad (11.35)

Coupling the derived numerical approximation with the upwind for third-order, the numerical solution of Eq. (11.31) is given as

$$
\frac{A(\beta)}{(1-\beta)\Gamma(1-\alpha)} \sum_{j=1}^{n} \left(\frac{f_i^{j+1} - f_i^j}{\Delta t} \right) \left\{ \begin{array}{l} (t_n - t_{j+1})^{1-\alpha} E_{\beta,2-\alpha}\left(-\frac{\beta}{1-\beta}(t_n - t_{j+1})^{\alpha+\beta}\right) + \\ (t_n - t_j)^{1-\alpha} E_{\beta,2-\alpha}\left(-\frac{\beta}{1-\beta}(t_n - t_j)^{\alpha+\beta}\right) \end{array} \right\} S_{S_j}
$$

$$
= \frac{k_f^{j+1} - k_f^j}{(\Delta r)} \frac{2h_{f(i)}^n + 6h_{f(i-1)}^n - 6h_{f(i-2)}^n + 6h_{f(i-3)}^n}{6(\Delta r)} +
$$

$$
k_f^i \left\{ \frac{h_{f_1(i+1)}^{n+1} - 2h_{f_1(i)}^{n+1} + h_{f_1(i-1)}^{n+1}}{2(\Delta r)^2} + \frac{h_{f_1(i+1)}^n - 2h_{f_1(i)}^n + h_{f_1(i-1)}^n}{2(\Delta r)^2} \right\} + \eta \left(h_{fi}^n - h_{mi}^n \right),
$$

$$
\frac{A(\beta)}{(1-\beta)\Gamma(1-\alpha)} \sum_{j=1}^{n} \left(\frac{h_{mi}^{j+1} - h_{mi}^j}{\Delta t} \right) \left\{ \begin{array}{l} (t_n - t_{j+1})^{1-\alpha} E_{\beta,2-\alpha}\left(-\frac{\beta}{1-\beta}(t_n - t_{j+1})^{\alpha+\beta}\right) + \\ (t_n - t_j)^{1-\alpha} E_{\beta,2-\alpha}\left(-\frac{\beta}{1-\beta}(t_n - t_j)^{\alpha+\beta}\right) \end{array} \right\} S_{S_j}
$$

$$
= \eta \left(h_{fi}^n - h_{mi}^n \right)
$$

if $\dfrac{k_f^{i+1} - k_f^i}{\Delta r} > 0.$

(11.36)

Otherwise we have the following:

$$
\frac{A(\beta)}{(1-\beta)\Gamma(1-\alpha)} \sum_{j=1}^{n} \left(\frac{f_i^{j+1} - f_i^j}{\Delta t} \right) \left\{ \begin{array}{l} (t_n - t_{j+1})^{1-\alpha} E_{\beta,2-\alpha}\left(-\frac{\beta}{1-\beta}(t_n - t_{j+1})^{\alpha+\beta}\right) + \\ (t_n - t_j)^{1-\alpha} E_{\beta,2-\alpha}\left(-\frac{\beta}{1-\beta}(t_n - t_j)^{\alpha+\beta}\right) \end{array} \right\} S_{S_j}
$$

$$
= \frac{k_f^{j+1} - k_f^j}{(\Delta r)} \frac{-h_{f(i+3)}^n + 6h_{f(i+2)}^n - 3h_{f(i+1)}^n - 2h_{f(i)}^n}{2(\Delta r)} +
$$

$$
k_f^i \left\{ \frac{h_{f_1(i+1)}^{n+1} - 2h_{f_1(i)}^{n+1} + h_{f_1(i-1)}^{n+1}}{2(\Delta r)^2} + \frac{h_{f_1(i+1)}^n - 2h_{f_1(i)}^n + h_{f_1(i-1)}^n}{2(\Delta r)^2} \right\} + \eta \left(h_{fi}^n - h_{mi}^n \right),
$$

$$
\frac{A(\beta)}{(1-\beta)\Gamma(1-\alpha)} \sum_{j=1}^{n} \left(\frac{h_{mi}^{j+1} - h_{mi}^j}{\Delta t} \right) \left\{ \begin{array}{l} (t_n - t_{j+1})^{1-\alpha} E_{\beta,2-\alpha}\left(-\frac{\beta}{1-\beta}(t_n - t_{j+1})^{\alpha+\beta}\right) + \\ (t_n - t_j)^{1-\alpha} E_{\beta,2-\alpha}\left(-\frac{\beta}{1-\beta}(t_n - t_j)^{\alpha+\beta}\right) \end{array} \right\} S_{S_j}
$$

$$
= \eta \left(h_{fi}^n - h_{mi}^n \right)
$$

if $\dfrac{k_f^{i+1} - k_f^i}{\Delta r} < 0.$

(11.37)

11.4.1 Numerical Simulation for Different Values of Fractional Order

In this section, we present the numerical simulations of the modified groundwater fractal flow in dual media at different instances of fractional order. We display the contour plot of the solutions for readers to see the solution in space and time for a given value of fractional powers α and β. The aquifer parameter values used here are theoretical, not measured from the field; this section is primarily designed to show readers more scenarios that can be described using the concept of fractional differentiation. In the numerical simulations, we suggest that, parameters for fractal dual media aquifers are not only following the power law behavior but do follow other laws that will be suggested. The numerical simulations are shown in figures below. The numerical graphics show some interesting real-world observations. The numerical simulations are generated based on the nonlocal operator with Mittag–Leffler kernel. Numerical simulation that depicts the behavior of the flow within the matrix rock and fractures at different instants of parameters θ and d are shown in Figs. 11.1 to 11.18.

11.5 Modeling Groundwater Flow With Variable Order Derivatives in a Leaky Aquifer

The investigation of water movement within a geological formations called leaky aquifers has been a focus of interest of many researchers, also since the movement of subsurface water is taking place within a given geological formation, the characterization of aquifers referred to as hydrogeology. Related nomenclature: aquitard, which is recognized as a bed of low permeability along an aquifer (see more information in [174]); an aquiclude also known as aquifuge, which is a rock-hard and water-resistant neighborhood underlying or overlying an aquifer. However, we shall note that the resistant neighborhood overlies the aquifer; pressure could affect it to becoming a restrained aquifer. We note in the literature two end-members in the

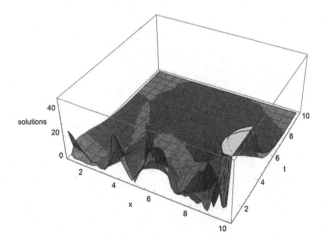

Figure 11.1: Numerical simulation for the flow in the matrix rock and fractures obtained with parameters $\theta = 1$, $d = 2$.

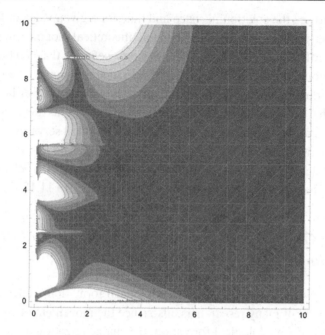

Figure 11.2: Numerical simulation for the flow in the fractures with parameters $\theta = 1$, $d = 2$.

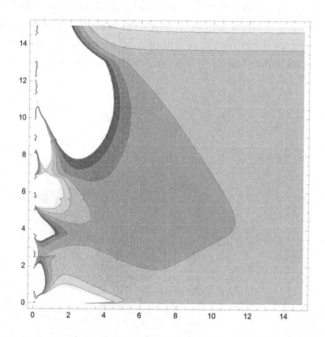

Figure 11.3: Numerical simulation for the flow in the matrix rock with parameters $\theta = 1$, $d = 2$.

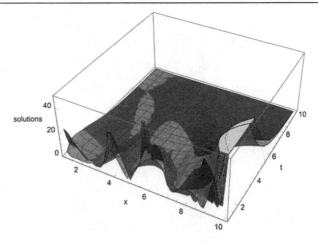

Figure 11.4: **Numerical simulation for the flow in the matrix rock and fractures obtained with parameters** $\theta = 0.95$, $d = 1.95$.

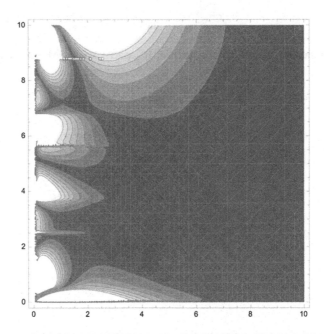

Figure 11.5: **Numerical simulation for the flow in the fractures with parameters** $\theta = 0.95$, $d = 1.95$.

range of kinds of aquifers: we have, on one hand, the confined and on the other, the unconfined, with of course a semi-confined being in-between. Characteristically but not at all times, the shallowest aquifer at a specified setting is unconfined, implying that it does not at all have

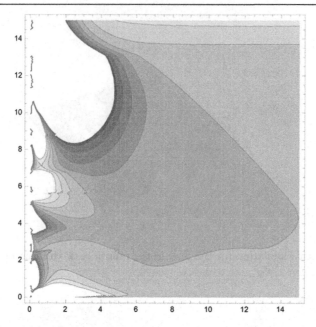

Figure 11.6: Numerical simulation for the flow in the matrix rock with parameters $\theta = 0.95$, $d = 1.95$.

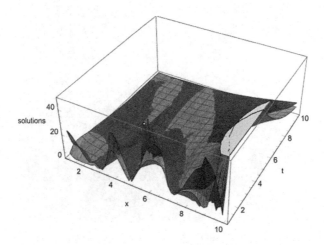

Figure 11.7: Numerical simulation for the flow in the matrix rock and fractures obtained with parameters $\theta = 0.55$, $d = 1.55$.

a confined layer which is an aquitard or an aquiclude, sandwiched between it and the outside. At time when a permeable aquifer is pumped, the piezometric intensity of the aquifer within the borehole is lessening. This lowering spreads thoroughly away from the borehole as pump-

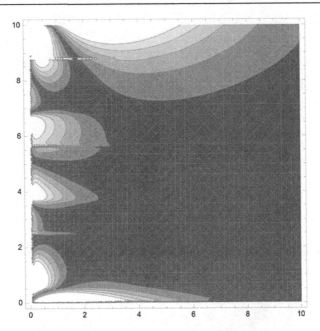

Figure 11.8: Numerical simulation for the flow in the fractures with parameters $\theta = 0.55$, $d = 1.55$.

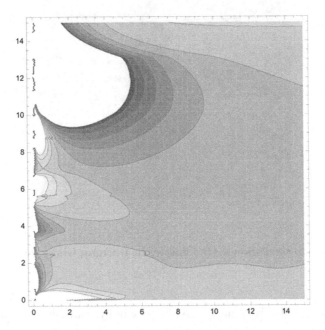

Figure 11.9: Numerical simulation for the flow in the matrix rock with parameters $\theta = 0.55$, $d = 1.55$.

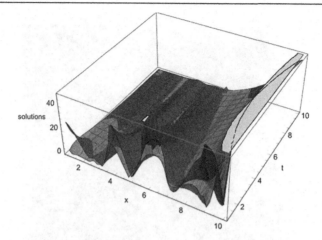

Figure 11.10: Numerical simulation for the flow in the matrix rock and fractures obtained with parameters $\theta = 0.45$, $d = 1.45$.

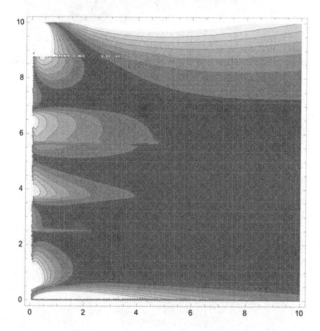

Figure 11.11: Numerical simulation for the flow in the fractures with parameters $\theta = 0.45$, $d = 1.45$.

ing continues; of course, this is creating dissimilarity in hydraulic head involving the aquifer and the aquitard. Accordingly, the subsurface water in the aquitard will begin moving perpendicularly downhill to associate with the water within the aquifer. The aquifer is therefore

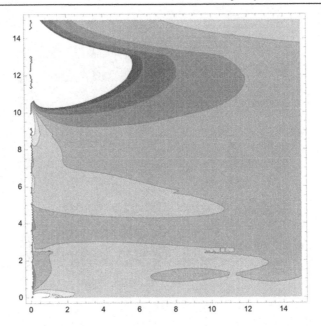

Figure 11.12: **Numerical simulation for the flow in the matrix rock with parameters** $\theta = 0.45$, $d = 1.45$.

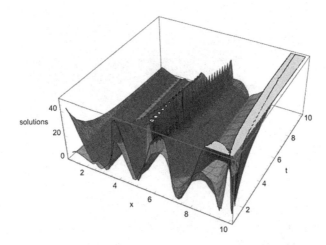

Figure 11.13: **Numerical simulation for the flow in the matrix rock and fractures obtained with parameters** $\theta = 0.35$, $d = 1.35$.

moderately invigorated by decreasing percolation from the aquitard. As pumping carries on, the proportion of the total ejection resulting from this percolation increases. Within a given phase of pumping, symmetry will be observed between the discharge rate of the pumping and

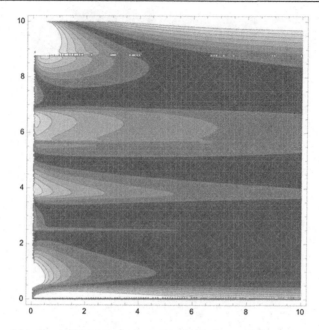

Figure 11.14: Numerical simulation for the flow in the fractures with parameters $\theta = 0.35$, $d = 1.35$.

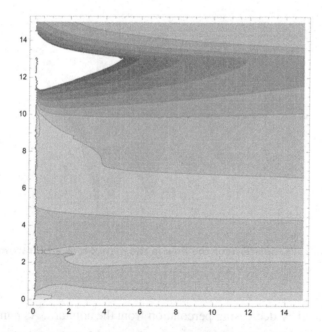

Figure 11.15: Numerical simulation for the flow in the matrix rock with parameters $\theta = 0.35$, $d = 1.35$.

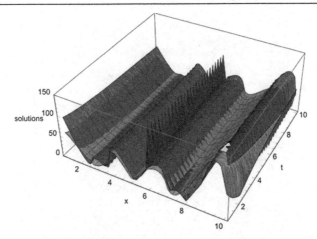

Figure 11.16: Numerical simulation for the flow in the matrix rock and fractures obtained with parameters $\theta = 0.25$, $d = 1.25$.

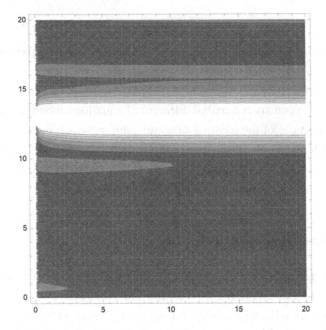

Figure 11.17: Numerical simulation for the flow in the fractures with parameters $\theta = 0.25$, $d = 1.25$.

the recharge rate by perpendicular flow from beginning to the end of the aquitard. This balanced condition will be maintained providing the water bench in the aquitard is kept constant. As said before, to understand and predict this dynamical system, an eminent hydrologist M.S.

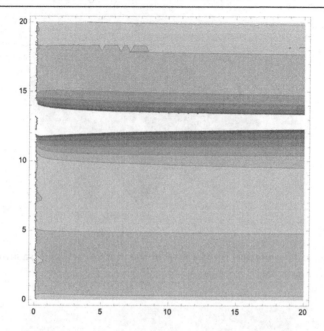

Figure 11.18: Numerical simulation for the flow in the matrix rock with parameters $\theta = 0.25$,
$d = 1.25$.

Hantush was the first to construct a partial differential equation recounting such phenomenon.
On the other hand, because of the twisting of some aquifers which occur as the nature evolves,
the Hantush equation is not able to accurately replicate for the effect of the changes in the
mathematical formulation. The rationale of this section is consequently dedicated to the mathematical model with non-local operators with variable orders.

11.5.1 Definition and Problem Modification

Some already presented definition of fractional differential operators with variable orders is
given here to help readers who are not used to this concept. Let $f : \mathbb{R} \to \mathbb{R}$, $x \to f(x)$ denote a continuous function but necessarily differentiable; Let $\alpha(x)$ be a continuous function in
(0, 1]. Then its variational order differential is defined as

$$D_0^{\alpha(x)}(f(x)) = \frac{1}{\Gamma(1 - \alpha(x))} \int_0^x (x - t)^{-\alpha(t)} \frac{df(t)}{dt} dt. \tag{11.38}$$

The above derivative is called the Caputo variational order differential operator; additionally,
the derivative of the constant is zero.

11.5.2 Problem Formulation

Groundwater representations portray the subsurface movement and transportation processes by means of mathematical equations supported by positive simplifying suppositions. These statements characteristically engage the bearing of flow, geometry of the aquifer, and the heterogeneity or anisotropy of sediments or bedrock within the aquifer. These are geological formations through which the groundwater flow is changing in time and space. The simplest oversimplification of groundwater flow equation, which while we're on the subject is also in accord with factual physics of the observable fact, is to assume that water level is not in a steady but transient state. In 1935, Theis [183] was the first to extend a modus operandi for unsteady-state stream that brought in the time factor and the storativity. He distinguished that at a time when a well trenchant a wide-ranging confined aquifer is pumped at a constant rate, the pressure of the discharge extends externally with time. The rate of decline of head, multiplied by the storativity and summed over the area of influence, equals the discharge. The unsteady-state (or Theis) equation, which was derived from the analogy between the flow of groundwater and the conduction of heat, is perhaps the most widely used partial differential equation in groundwater investigations:

$$U D_t H(s, t) = T D_{ss} H(s, t) + \frac{1}{s} D_s H(s, t). \tag{11.39}$$

The above equation is classified under parabolic equations. However, very few geological formations are completely impermeable to fluids. Leakage of the water could thus occur, should a confined aquifer be over- or under-lain by another aquifer. The behavior of such an aquifer, often referred to as a leaky or semi-confined aquifer, needs thus not be the same as that of a confined aquifer. Although the nature of a semi-confined aquifer differs from that of a true aquifer, it is still possible to use the basic principles of confined flow to arrive at the governing equation for such an aquifer. This is in particular true in those situations where the confining layer between the two aquifers is not too thick and the flow is mainly in the vertical direction. According to Hantush and Jacob [172,173], the drawdown due to pumping a leaky aquifer can be described by the following equation:

$$U D_t u(s, t) = T D_{ss} u(s, t) + \frac{1}{s} D_s u(s, t) + \frac{u(s, t)}{\lambda^2}, \tag{11.40}$$

where u is the drawdown or change in the level of water, U is the specific storativity of the aquifer, and T is the transmissivity:

$$\lambda^2 = \frac{B}{C'},$$

with C and C' the hydraulic conductivities of the main and confining layers, respectively; B and d the thicknesses of the main and confining layers, respectively; and Θ the discharge rate of the pumping. This partial differential equation describing the movement of water through the geological formation during the pumping is subjected to the following initial and boundary conditions:

$$u(s, 0) = H_0, \quad \lim_{s \to \infty} H(s, t) = H_0, \quad \Theta = \frac{2\pi^{n/2}}{\Gamma(n/2)} s_b^{n-1} Cd^{3-n} \partial_u(s_b, t).$$

However, when we consider the diffusion process in the porous medium, if the medium structure or external field changes with time, in this situation, the ordinary integer order and constant-order fractional diffusion equation model cannot be used to properly characterize such a phenomenon [180,184]. This is the case of the groundwater flow in the deformable aquifer, the medium through which the flow changes with time and space [165,166]. Note that Hantush equation cannot handle this case. One of the purposes of this work is therefore devoted to the discussion underpinning the description of water flowing through a deformable leaky aquifer, on one hand. In order to include explicitly the variability of the medium through which the flow takes place, the standard version of the partial derivative with respect to time is replaced here with variable-order (VO) fractional to obtain

$$U D_t^{\alpha(x,t)} u(s, t) = T D_{ss} u(s, t) + \frac{1}{s} D_s u(s, t) + \frac{u(s, t)}{\lambda^2}, \quad 0 < \alpha(x, t) \leq 1. \qquad (11.41)$$

11.5.3 Numerical Solution of the Modified Model by Means of Crank–Nicolson Scheme for the Modified Equation [169]

Environmental phenomena, such as groundwater flow described by variation order derivative, are highly complex phenomena, which do not lend themselves readily to analysis of analytical models. The discussion presented in this section will therefore be devoted to the derivation of numerical solution to the modified Hantush equation (11.41). Solving difficult equations with numerical scheme has been a passionate exercise for many scholars [167–171,175,177–179, 182,185,186]. However, there exist numerous schemes in the literature [167,168,171,177, 179,185,186]. Some of these numerical techniques are very accurate while approximating solutions of difficult equations. These numerical methods yield approximate solutions to the governing equation through the discretization of space and time [165]. Within the discredited problem domain, the variable internal properties, boundaries, and stresses of the system are approximated. Deterministic, distributed-parameter, numerical models can relax the rigid idealized conditions of analytical models or lumped-parameter models, and they can therefore be more realistic and flexible for simulating fields conditions [165]. The work in [165] has extended the groundwater flow model to the concept of time-fractional variable order derivative; the modified equation was solved employing the well-established Crank–Nicolson numerical scheme. The finite difference schemes for constant-order time- or space-fractional diffusion equations have been widely studied in [167,170,175,178,182,186]. Recently, Sun et al. [181]

studied the solution of the advection dispersion equation with time-fractional variable order derivative. The study of the implicit difference approximation scheme for constant-order time-fractional diffusion equations was presented in [168]. Recently, the weighted average finite difference method was introduced [185]. The matrix approach for fractional diffusion equations was proposed [179], and a flexible numerical scheme for the discretization of the space–time fractional diffusion equation can be found in [171]. In recent times, the numerical scheme for variable-order space-fractional advection–dispersion equation was considered [186]. The investigation of the explicit scheme for variable-order nonlinear space-fractional diffusion equation was done (see [177]). Before performing the numerical methods, we assume that (11.40) has a unique and sufficiently smooth solution. To establish the numerical schemes for the above equation, we let $x_l = lh$, $0 \leq l \leq M$, $Mh = L$, $t_k = k\tau$, $0 \leq k \leq N$, $N\tau = T$, where h is the step and τ is the time size, and M and N are grid points. We introduce the Crank–Nicolson scheme as follows. Firstly, the discretization of first- and second-order space derivative is stated as

$$\frac{\partial u}{\partial s} = \frac{1}{2}\left[\left(\frac{u(s_{l+1}, t_{k+1}) - u(s_{l-1}, t_{k+1})}{2(h)}\right) + \left(\frac{u(s_{l+1}, t_k) - u(s_{l-1}, t_k)}{2(h)}\right)\right] + O(h), \quad (11.42)$$

$$\begin{aligned}\frac{\partial^2 u}{\partial s^2} &= \frac{1}{2}\left[\left(\frac{u(s_{l+1}, t_{k+1}) - 2u(s_l, t_{k+1}) + u(s_{l-1}, t_{k+1})}{(h)^2}\right)\right. \\ &\quad \left. + \left(\frac{u(s_{l+1}, t_k) - 2u(s_l, t_k) + u(s_{l-1}, t_k)}{(h)^2}\right)\right] + O(h),\end{aligned} \quad (11.43)$$

$$u = \frac{1}{2}(u(s_l, t_{k+1}) + u(s_l, t_k)). \quad (11.44)$$

The Crank–Nicolson scheme for the VO time-fractional diffusion model can be stated as follows:

$$\begin{aligned}\frac{\partial^{\alpha_l^{k+1}} u(s_l, t_{k+1})}{\partial t^{\alpha_l^{k+1}}} &= \frac{\tau^{-\alpha_l^{k+1}}}{\Gamma(2 - \alpha_l^{k+1})}(u(s_l, t_{k+1}) - u(s_l, t_k)) \\ &\quad + \sum_{j=1}^{k}\left[u(s_l, t_{k+1-j}) - u(s_l, t_{k-j})\right] \times \left[(j+1)^{1-\alpha_l^{k+1}} - (j)^{1-\alpha_l^{k+1}}\right]\right).\end{aligned} \quad (11.45)$$

Now replacing (11.42), (11.43), (11.44), and (11.45) in (11.41), we obtain the following:

$$\begin{aligned}\left[\frac{U\tau^{-\alpha_1^{k+1}}}{\Gamma(2 - \alpha_1^{k+1})}\left(u(s_l, t_{k+1}) - u(s_l, t_k) + \sum_{j=1}^{k}\left[u(s_l, t_{k+1-j}) - u(s_l, t_{k-j})\right]\right.\right. \\ \left.\left. \times \left[(j+1)^{1-\alpha_l^{k+1}} - (j)^{1-\alpha_l^{k+1}}\right]\right)\right]\end{aligned}$$

$$= T \left[\frac{1}{2} \left[\left(\frac{u(s_{l+1}, t_{k+1}) - 2u(s_l, t_{k+1}) + u(s_{l-1}, t_{k+1})}{(h)^2} \right) \right. \right.$$

$$+ \left. \left(\frac{u(s_{l+1}, t_k) - 2u(s_l, t_k) + u(s_{l-1}, t_k)}{(h)^2} \right) \right] \right]$$

$$+ \frac{1}{s_l} \left[\frac{1}{2} \left[\left(\frac{u(s_{l+1}, t_{k+1}) - u(s_{l-1}, t_{k+1})}{2(h)} \right) + \left(\frac{u(s_{l+1}, t_k) - u(s_{l-1}, t_k)}{2(h)} \right) \right] \right]$$

$$+ \frac{1}{2\lambda^2} (u(s_l, t_{k+1}) + u(s_l, t_k)). \tag{11.46}$$

For simplicity, let us put

$$b_j^{l,k+1} = (j+1)^{1-\alpha_l^{k+1}} - (j)^{1-\alpha_l^{k+1}}, \quad S_l^{k+1} = \frac{\Gamma(2 - \alpha_l^{k+1})}{Uh^2} T, \quad F_l^{k+1} = \frac{\Gamma(2 - \alpha_l^{k+1})}{Uh};$$

$$\tag{11.47}$$

$$\lambda_j^{l,k+1} = b_{j-1}^{l,k+1} - b_j^{l,k+1}, \quad \Gamma_l^{k+1} = \frac{\Gamma(2 - \alpha_l^{k+1})}{2\lambda^2}.$$

Eq. (11.46) becomes

$$u_l^{k+1} \left(1 + 2S_l^{k+1} - \Gamma_l^{k+1} \right) = u_{l+1}^{k+1} \left(S_l^{k+1} + \frac{F^{k+1}}{s_l} \right) + u_{l-1}^{k+1} \left(S_l^{k+1} - \frac{F^{k+1}}{s_l} \right)$$

$$+ u_{l+1}^k \left(S_l^{k+1} - \frac{F^{k+1}}{s_l} \right) + u_l^{k+1} \left(1 + 2S_l^{k+1} \right) + \sum_{j=1}^{k} \left(u_l^{k+1-j} - u_l^{k-j} \right) \lambda_j^{l,k+1} F_l^{k+1}$$

$$+ u_l^k \left(1 - \Gamma_l^{k+1} + S_l^{k+1} \right). \tag{11.48}$$

11.5.4 Stability Analysis of the Crank–Nicolson Scheme by Means of Fourier Method

In this section, we will analyze the stability conditions of the Crank–Nicolson scheme for the Hantush equation for a deformable aquifer. Let $\Psi_l^k = u_l^k - U_l^k$, where U_l^k is the approximate solution at the point (x_l, t_k) for $k = 1, 2, \ldots, N$ and $l = 1, 2, \ldots, M - 1$. In addition, $\Psi^k = [\Psi_1^k, \Psi_2^k, \ldots, \Psi_{M-1}^k]^T$ and the function $\Psi^k(x)$ is chosen to be

$$\Psi^k(x) = \begin{cases} \Psi_l^k & \text{if } x_l - \frac{h}{2} < x \leq x_l + \frac{h}{2}, \ l=1,2,\ldots,M-1 \\ 0 & \text{if } L - \frac{h}{2} < x \leq L \end{cases} \tag{11.49}$$

Then, the function $\Psi^k(x)$ can be expressed in Fourier series as follows:

$$\Psi^k(x) \ = \ \sum_{m=-\infty}^{m=\infty} \delta_m(m) \exp\left[\frac{2i\pi mk}{L}\right],$$

$$\delta_k(x) \ = \ \frac{1}{L} \int_0^L \Psi^k(x) \exp\left[\frac{2i\pi mx}{L}\right] dx. \tag{11.50}$$

It was established in [168] that

$$\left\|\Psi^2\right\|_2^2 = \sum_{m=-\infty}^{m=\infty} \|\delta_k(m)\|^2.$$

Observe that for all $k, l \geq 1, 0 \leq 1 - \alpha_l^{k+1} < 1$, and in addition, according to the problem in point, the velocity seepage V, the dispersion coefficient D, the retardation factor R, and the radioactive decay constant λ are positive constants. Then the following properties of the coefficients S_l^{k+1}, F_l^{k+1}, $\lambda_j^{l,k+1} b_l^{k+1}$ and Γ_l^{k+1} can be established.

1. F_l^{k+1}, S_l^{k+1}, and Γ_l^{k+1} are positive for all $l = 1, 2, \ldots, M - 1$,
2. $0 < \lambda_j^{l,k} \leq \lambda_{j-1}^{l,k} \leq 1$ for all $l = 1, 2, \ldots, M - 1$,
3. $0 \leq b_j^{l,k} \leq 1, \sum_{j=0}^{k-1} b_{j+1}^{l,k+1} = 1 - \lambda_k^{l,k+1}$ for all $l = 1, 2, \ldots, M - 1$.

It is a common tradition in groundwater investigations to choose a point on the centerline of the pumped boreholes as a reference for the observations, and therefore, neither the drawdown nor its derivatives will vanish at the origin, as required [165]. In such a situation, the distribution of the piezometric head in the aquifer is a decreasing function of the distance from the boreholes, the expression $(1/s_l) \rightarrow 0$ [165]. Under this situation, the error committed while approximating the solution of the generalized advection dispersion equation with Crank–Nicolson scheme can be presented as follows:

$$\Psi_l^{k+1}\left(1 + 2S_l^{k+1} - \Gamma_l^{k+1}\right) \ = \ \Psi_{l+1}^{k+1}\left(S_l^{k+1}\right) + \Psi_{l-1}^{k+1}\left(S_l^{k+1}\right) + \Psi_{l+1}^k\left(S_l^{k+1}\right) \tag{11.51}$$
$$+ \Psi_l^k\left(1 + 2S_l^{k+1} - \Gamma_l^{k+1}\right)$$
$$+ \sum_{j=1}^k \left(\Psi_l^{k+1-j} - \Psi_l^{k-j}\right) \lambda_j^{l,k+1} F_l^{k+1}.$$

If we assume that Ψ_l^k in (11.51) can be put in the delta exponential form as follows [181]:

$$\Psi_l^k = \delta_k \exp[i\psi lk], \tag{11.52}$$

where ψ is a real spatial wave number, now replacing the above equation (11.52) in (11.51) we obtain

$$\left[1 + 4S_l^1 \sin^2\left(\frac{\psi h}{2}\right) - 2\sin^2\left(\frac{\psi h}{2}\right)\Gamma_l^1\right]\delta_1 =$$

$$\left[1 - 4S_l^1 \sin^2\left(\frac{\psi h}{2}\right) - 2\sin^2\left(\frac{\psi h}{2}\right)\Gamma_l^1\right]\delta_0 \text{ for } k = 0,$$

$$\left[1 + 4S_l^{1+k} \sin^2\left(\frac{\psi h}{2}\right) - 2\sin^2\left(\frac{\psi h}{2}\right)\Gamma_l^{1+k}\right]\delta_{k+1} = \tag{11.53}$$

$$\left[1 - 4S_l^{1+k} \sin^2\left(\frac{\psi h}{2}\right) - 2\sin^2\left(\frac{\psi h}{2}\right)\Gamma_l^{1+k}\right]\delta_k$$

$$+ \sum_{j=0}^{k-1} \lambda_{j+1}^{l,k+1}\delta_{k-j} + \lambda_k^{l,k+1}\delta_0, \text{ for } k = 1, 2, \ldots, N-1.$$

Eq. (11.53) can be written in the following form:

$$\delta_1 = \frac{[1 - 4S_l^1 \sin^2(\psi h/2) - 2\sin^2(\psi h/2)\Gamma_l^1]\delta_0}{[1 + 4S_l^1 \sin^2(\psi h/2) - 2\sin^2(\psi h/2)\Gamma_l^1]},$$

$$\delta_{k+1} = \left(\left[1 - 4S_l^{1+k}\sin^2\left(\frac{\psi h}{2}\right) - e_l^{l,k+1} - 2\sin^2\left(\frac{\psi h}{2}\right)\Gamma_l^{1+k}\right]\delta_k\right.$$

$$+ \sum_{j=0}^{k-1} \lambda_{j+1}^{l,k+1}\delta_{k-j} + \lambda_k^{l,k+1}\delta_0\right)$$

$$\times \left[1 + 4S_l^{1+k}\sin^2\left(\frac{\psi h}{2}\right) - 2\sin^2\left(\frac{\psi h}{2}\right)\Gamma_l^{1+k}\right]^{-1}. \tag{11.54}$$

Our next concern here is to show that for all $k = 1, 2, \ldots, N-1$ the solution of (11.53) satisfies the following condition:

$$|\delta_k| < |\delta_0|. \tag{11.55}$$

To achieve this, we make use of the recurrence technique on the natural number k. For $k = 1$, and bear in mind that d_l^{k+1}, b_l^{k+1} are positive for all $l = 1, 2, \ldots, M-1$, then we obtain

$$\frac{\delta_1}{\delta_0} = \left|\frac{[1 - 4S_l^1 \sin^2(\psi h/2) - 2\sin^2(\psi h/2)\Gamma_l^1]}{[1 + 4S_l^1 \sin^2(\psi h/2) - 2\sin^2(\psi h/2)\Gamma_l^1]}\right| < 1. \tag{11.56}$$

Assuming that the property is verified for $m = 2, 3, \ldots, k$, then

$$|\delta_{k+1}| = \left|\left(\left[1 - 4S_l^{1+k}\sin^2\left(\frac{\psi h}{2}\right) - e_l^{l,k+1} - 2\sin^2\left(\frac{\psi h}{2}\right)\Gamma_l^{1+k}\right]\delta_k +\right.\right.$$

$$\sum_{j=0}^{k-1}\lambda_{j+1}^{l,k+1}\delta_{k-j} + \lambda_k^{l,k+1}\delta_0\right) \times \left[1 + 4S_l^{1+k}\sin^2\left(\frac{\psi h}{2}\right) - 2\sin^2\left(\frac{\psi h}{2}\right)\Gamma_l^{1+k}\right]^{-1}\right|. \quad (11.57)$$

Making use of the triangular inequality, we obtain

$$|\delta_{k+1}| \leq \left(\left|1 - 4S_l^{1+k}\sin^2\left(\frac{\psi h}{2}\right) - e_l^{l,k+1} - 2\sin^2\left(\frac{\psi h}{2}\right)\Gamma_l^{1+k}\delta_k\right||\delta_k| \quad (11.58)\right.$$

$$+ \left|\sum_{j=0}^{k-1} p_{j+1}^{l,k+1}\delta_{k-j}\right| + |e_k^{l,k+1}\delta_0|\right)$$

$$\times \left|1 + 4S_l^{1+k}\sin^2\left(\frac{\psi h}{2}\right) - 2\sin^2\left(\frac{\psi h}{2}\right)\Gamma_l^{1+k}\right|^{-1}.$$

By using the recurrence hypothesis, we obtain

$$|\delta_{k+1}| \leq \left(\left(\left|1 - 4S_l^{1+k}\sin^2\left(\frac{\psi h}{2}\right) - 2\sin^2\left(\frac{\psi h}{2}\right)\Gamma_l^{1+k}\right| + \left|\sum_{j=0}^{k-1}\lambda_{j+1}^{l,k+1}\right|\right)\right.$$

$$\times \left|1 + 4S_l^{1+k}\sin^2\left(\frac{\psi h}{2}\right) - 2\sin^2\left(\frac{\psi h}{2}\right)\Gamma_l^{1+k}\right|^{-1}\right)|\delta_0|,$$

$$|\delta_{k+1}| \leq \left(\left|1 + 4S_l^{1+k}\sin^2\left(\frac{\psi h}{2}\right) - 2\sin^2\left(\frac{\psi h}{2}\right)\Gamma_l^{1+k}\right|\right. \quad (11.59)$$

$$\times \left|1 + 4S_l^{1+k}\sin^2\left(\frac{\psi h}{2}\right) - 2\sin^2\left(\frac{\psi h}{2}\right)\Gamma_l^{1+k}\right|^{-1}\right)|\delta_0|,$$

$$|\delta_{k+1}| \leq |\delta_0|,$$

which completes the proof that starts at (11.55) and ends at (11.59).

11.5.5 Convergence Analysis of the Crank–Nicolson Scheme

If we assume that $u(s_l, t_k)$ for $l = 1, 2, \ldots, M, k = 1, 2, \ldots, N - 1$ is the exact solution of our problem at the point (s_l, t_k), by letting $Y_l^k = u(s_l, t_k) - U_l^k$ and $Y^k = (0, Y_1^k, H_2^k, \ldots, H_{M-1}^k)$ substituting this in (11.51), we obtain

$$Y_l^1(1 + 2S_l^1 - \Gamma_l^1) - Y_{l+1}^1(S_l^1) - Y_{l-1}^1(S_l^1) = R_l^1 \quad \text{for } k = 0, \tag{11.60}$$

$$Y_l^{1+k}(1 + 2S_l^{1+k} - \Gamma_l^{1+k}) - Y_{l+1}^{1+k}(S_l^{1+k}) - Y_{l-1}^{1+k}(S_l^{1+k})$$

$$= R_l^{1+k} + \sum_{j=0}^{k-1} Y_l^{k-j} \lambda_{j+1}^{l,k+1} \quad \text{for } k \geq 1.$$

Here,

$$R_l^{1+k} = u(s_l, t_{k+1}) - \sum_{j=0}^{k-1} u(s_l, t_{k-j})\lambda_{j+1}^{l,k+1} + b_1^{l,k+1} u(s_l, t_0) +$$

$$\left(\Gamma_l^{1+k} - S_l^{1+k}\right)[u(s_{l+1}, t_{k+1}) - 2u(s_l, t_{k+1}) + u(s_{l-1}, t_{k+1})]. \tag{11.61}$$

From (11.43) and (11.44), we have

$$\frac{\partial^2 u(s_l, t_{k+1})}{\partial s^2} + h^2\vartheta_1 = \frac{1}{2}\left(\frac{u(s_{l+1}, t_{k+1}) - 2u(s_l, t_{k+1}) + u(s_{l-1}, t_{k+1})}{h^2}\right.$$

$$\left. + \frac{u(s_{l+1}, t_k) - 2u(s_l, t_k) + u(s_{l-1}, t_k)}{h^2}\right), \tag{11.62}$$

$$\frac{\partial^{\alpha_l^{k+1}} u(s_l, t_{k+1})}{\partial t^{\alpha_l^{k+1}}} + \tau\vartheta_2 = \frac{\tau^{-\alpha_l^{k+1}}}{\Gamma(2 - \alpha_l^{k+1})}\left(u(s_l, t_{k+1}) - u(s_l, t_k)\right.$$

$$\left. + \sum_{j=1}^{k}[u(s_l, t_{k+1-j}) - u(s_l, t_k - j)]\lambda_j^{l,k}\right).$$

From the above, we have that

$$R_l^{k+1} \leq C\left(\tau^{1+\alpha_l^{k+1}} + h^2\tau^{\alpha_l^k}\right),$$

where C_1, C_2, and C are constants. Taking into account Caputo-type fractional derivative, the detailed error analysis on the above schemes can refer to the work done in [181] and further work done in [176,181].

Lemma 11.5.1. *The following inequality* $\|Y^{k+1}\|_\infty \leq C(\tau^{1+\alpha_l^{k+1}} + h^2\tau^{\alpha_l^{k+1}})(Y_j^{l,k+1})^{-1}$ *is true for* $(k = 0, 1, 2, \ldots, N-1)$ *where* $\|Y^k\|_\infty = \max_{1 \leq l \leq M-1}(Y^k)$, C *is a constant. In addition,*

$$\varphi^\varepsilon(a, b) = \begin{cases} \min\limits_{1 \leq l \leq M-1} a_l^{k+1} & \text{if } \tau < 1 \\ \max\limits_{1 \leq l \leq M-1} a_l^{k+1} & \text{if } \tau > 1 \end{cases}. \tag{11.63}$$

This can be achieved by means of the recurrence technique on the natural number k. When $k = 0$, we have the following:

$$|Y_l^1| \le (S_l^1)|Y_{l+1}^1| + (S_l^1)|Y_{l-1}^1| = |G_l^1| \le \vartheta \left(\tau^{1+\alpha_i^{k+1}} + h^2 \tau^{\alpha_i^k} \right) (\lambda_j^{l,k+1})^{-1}. \qquad (11.64)$$

Now suppose that $\|Y^{i+1}\|_\infty \le C \left(\tau^{1+\alpha_i^{i+1}} + h^2 \tau^{\alpha_i^i} \right) \left(\lambda_j^{l,k+1} \right)^{-1}$, $i = 1, 2, \ldots, N-2$. Then,

$$
\begin{aligned}
\left| Y_l^{1+k} \right| &\le \left| Y_l^{1+k} \left(1 + 2S_l^{1+k} - \Gamma_l^{1+k} \right) - Y_{l+1}^{k+1} \left(S_l^{1+k} \right) - Y_{l-1}^{1+k} \left(S_l^{1+l} \right) \right| \\
&\le \left(S_{l+1}^{1+k} \right) \left| Y_l^{k+1} \right| + \left(S_l^{1+k} \right) \left| Y_{l-1}^{k+1} \right| + \left(1 + 2S_l^{1+k} - \Gamma_l^{1+k} \right) \left| Y_l^{k+1} \right| \\
&= \left| R_l^{k+1} + \sum_{i=1}^{k} (Y_l^{k-i}) \lambda_j^{l,k+1} \right| \le \left| R_l^{k+1} \right| + \sum_{i=1}^{k} \left| Y_l^{k-i} \right| \lambda_j^{l,k+1} \\
&\le C \left(\tau^{1+\alpha_i^{k+1}} + h^2 \tau^{\alpha_i^k} \right) + \sum_{i=1}^{k} \left\| Y_l^{k-i} \right\|_\infty \lambda_j^{l,k+1} \qquad (11.65) \\
&\le C \left(\tau^{1+\alpha_i^{k+1}} + h^2 \tau^{\alpha_i^k} \right) \left(\lambda_j^{l,k+1} + \lambda_0^{l,k+1} - \lambda_j^{l,k+1} \right) \times \left(\lambda_j^{l,k+1} \right)^{-1} \\
&\le \vartheta \left(\tau^{1+\alpha_i^{k+1}} + h^2 \tau^{\alpha_i^k} \right) \left(\lambda_0^{l,k+1} \right) \left(\lambda_j^{l,k+1} \right)^{-1} \\
&\le \vartheta \left(\tau^{1+\alpha_i^{k+1}} + h^2 \tau^{\alpha_i^k} \right) \left(\lambda_j^{l,k+1} \right)^{-1},
\end{aligned}
$$

which completes the proof of Lemma 11.5.1.

Theorem 11.5.2. *The Crank–Nicolson scheme is convergent, and there exists a positive constant ϑ such that*

$$\left| U_l^k - u(x_l, t_k) \right| \le C(\tau + h^2), \; l = 1, 2, \ldots, M-1, \; k = 1, 2, \ldots, N.$$

An interesting and detailed research can be found on the solvability of the Crank–Nicolson scheme in the work [165,166,169]. Therefore, the details of the proof of Theorem 11.5.2 will not be presented here.

CHAPTER 12

Modeling Groundwater Pollution With Variable Order Derivatives

12.1 Properties of Soils and Validity of Variable Order Derivatives

Although all aquifers do not necessarily contain freshwater, to a certain extent, groundwater exists at almost every location within the shallow subsurface of the Earth. The crust of the Earth can be divided into two parts. These are namely the saturated zone or phreatic zone, of which examples include aquifers and aquitard; and the unsaturated zone, which is also known as the Vadose zone, where one would find voids of air which contain some water but still allow space for more water. The term saturated infers that the pressure head of the water exceeds that of atmospheric pressure. This means the gauge pressure exceeds zero. On the other hand, the term water table defines a surface where the pressure head equates that of atmospheric pressure. This means the gauge pressure is equal to zero. Above the water table one would find unsaturated conditions where the water, which does not completely fill the voids of the aquifer, is under suction. This is where a negative pressure head is found, keeping in mind that there can never be a negative absolute pressure, although gauge pressure can be negative. Furthermore, the unsaturated zone's water content is held in place by means of surface adhesive forces. This water content rises beyond the water table at the zero-gauge-pressure isobar through capillary action. This saturates a small portion above the phreatic surface known as the capillary fringe. This is known as tension saturation which is dissimilar to saturation on a basis of water. Water content in the capillary fringe is reduced as the distance from the phreatic surface increases. Moreover, the head of the capillary relies on size of the soil pore. The capillary head in large pores, such as in sandy soils, would be lower than what it would be for smaller pores found in clayey soils. The general capillary rise for clayey soils is below 1.80 m (6 feet) but it may range from 0.3 to 10 m (1 to 30 feet). This capillary rise in tubes having a small diameter entails the same physical process. Within a large-diameter open pipe, such as a well penetrating an aquifer, the level to which water rises is known as the water table. Furthermore, groundwater can be found in underground rivers of which an example includes a cave within which there is free flow of water underground. This can be found in karst topography where there are areas of eroded limestone. These areas are only a small portion of the Earth's entire area. Usually the pore spaces within subsurface rock are saturated with water, making an analogy to a kitchen sponge. These subsurfaces (aquifers) can undergo abstraction for agricultural, industrial, or municipal use. If rock material of low porosity is

Fractional Operators with Constant and Variable Order
with Application to Geo-Hydrology
DOI: 10.1016/B978-0-12-809670-3.00012-6

significantly fractured, it may yield a good aquifer associated with fissure flow. This occurs provided there is sufficient hydraulic conductivity for water movement. This means porosity is a fundamental property, however, on its own, it cannot determine the rock unit's capability of acting as an aquifer. Groundwater in unconsolidated aquifers is generated from pore spaces found between gravel, sand, and silt particles. When an aquifer is confined by means of low permeable layers, the decreased water pressure in sand and gravel results in water draining slowly from the adjacent confining layers. When these confining layers are made up of compressible silt or clay, water lost to the aquifer decreases the water pressure in the confining layer, leading to compression from the overlying rock unit's weight. In considerable cases, the compression may be observed as a means of subsidence on the groundwater. Unfortunately, a lot of the subsidence from groundwater abstraction is permanent as there is very little elastic rebound. Subsequently, subsidence is not only permanent but there is rather a case of a compressed aquifer which has a permanent decreased capacity for water to be held. The study of flow and transport processes in the subsurface involves the resolution of the governing hydraulic equations using appropriate geometry, initial and boundary conditions of the system, and an acceptable description of the geology. The distribution of the geologic characteristics properties is the source of the challenging nonhomogeneous conditions often faced by professionals and scientists, in different fields related to the earth subsurface. In groundwater sciences, the distribution of the hydraulic properties of the rocks, and its effect on the flow and associated solute transport properties, is often of concern. The focus here will be neither on the variety and classes of heterogeneity, the reasons of the geologic characteristics' variability, nor the different processes involved in flow and solute transport dispersion, diffusion, sorption, and precipitation, etc. in the subsurface. It will rather be on available approaches and methods used to quantify heterogeneity, as well as model such processes. Many approaches have been developed to deal with the occurrence of spatial heterogeneity in groundwater studies. Geological and geophysical methods used in groundwater exploration are based on the detection of underground global physical heterogeneities. These methods are helpful in locating anomalies, which may be associated with water conducting layers and preferential path, to decide about well locations or discover springs. The access to the subsurface for direct measurements of aquifers' properties through aquifer pumping test is limited to boreholes (percussion and core) and tunnels, which are not representative of the whole aquifer. This constraint makes it impossible to completely handle the variations in the organization of the flow and associated solute transport in the subsurface. Instead of completely accounting for such spatial variability, attempts have been made to use the limited direct measurements to come to models that could represent and predict the effect of underlying structures on the behavior of the aquifers. Hydraulic tests in boreholes have played a huge role in quantifying the subsurface flow, and they remain the most useful way to investigate the properties of the subsurface. During an aquifer hydraulic test, one or more known stresses of pumping or injection are applied in a production borehole to the system being studied, and the responses of the system are measured in suitable placed observation boreholes. The general mathematical model describing the flow of groundwater is a partial differential hydraulic equation in time and three-dimensional space governing the flow of groundwater in the saturated zone. Although most mathematical models can be solved analytically for most simple case, or numerically for those considered to be complex resources; it is worthy to mention that the analytical element method elaborate semi-analytical method, and the boundary integral method (combination of analytical and numerical methods), among others, can be used to solve such a governing differential equations. In the following review, we will however focus on the existing analytical and numerical means. Analytical models represent mathematically (equation) exact solutions to the hydraulic equation for one/two-dimensional flow problems by simplifying assumptions. Although they cannot handle spatial and temporal variability, analytical models are very useful tools, as they can be solved by hand or by simple computer programs (Flow Calculation, etc.). They provide rough approximations for many applications and they usually do not involve calibration to observed data. They can also suit most simple and low-complexity modeling studies. Heterogeneities in groundwater flow structure were first handled by estimating "the equivalent homogeneous properties or equivalent uniform-properties stratum," as suggested by Morris Muskat in 1949. This was done by averaging local values (around wells) calculated with developed analytical solutions for either steady state as by Thiem (1906), aquifers pumping as by Theis (1935), or injection tests. Inspired by work done by precursors, for instance Cardwell and Parsons (1945), Landau and Lifschitz (1960), on the averaging of random variables, Matheron (1967) showed the usefulness of the geometric mean for the determination of an equivalent homogeneous permeability from steady-state tests for two-dimensional parallel flow. Meier et al. (1999) demonstrated empirically that the geometric mean is the long-term average that a well test produces, in transient state tests. Most of the present existing analytical solutions for the interpretation of field tests data are based on this concept. They are used to infer the transmissivity (assuming confining layers) of the aquifer. The calculated transmissivity is the product of the saturated thickness (D) of the aquifer and the equivalent hydraulic conductivity ($T = KD$) over the thickness. If these solutions may represent the heterogeneity (by average) over the thickness (D), they fail to define the vertical distribution of conductivities over the thickness (D) and become problematic where saturated thickness varies like in unconfined layers by Neuman (1973) and Jacob (1944). These models do approximately represent the flow dynamic only for homogeneous case. Whereas there are abundant manufacturing applications of constant-order fractional derivatives with real orders, for instance in control, diffusion or viscoelasticity, a comprehensive bibliographical evaluation falls undoubtedly outside the scope of this section as new concepts of complex orders are far less used; they have however been used to develop the so-called third generation Crone robust controllers. Variable-order derivatives are a more recent area: while the definition given in Riemann–Liouville sense was studied in the 1990s by some researchers, the seminal paper in what the comparison of different possible definitions, addressing mainly variable negative orders but the consideration for positive orders only was recent. Applications are given to viscoelasticity to the processing of geographical data to signature verification and to diffusion. Applications to adaptive control are likely feasible. Possible numerical implementations of variable-order fractional derivatives pass over the study of the memory that the operator has of previous values of the differentiation order.

12.2 Modeling With Caputo Variable Order

The concept of variable order with differentiation has been proven to be very powerful mathematical tool to model anonymous diffusion processes due to their ability to include into mathematical model the effect of heterogeneity of the geological formation by means of which the pollution is moving. For a very complex system it was shown that using the concept of fractional differentiation with constant fractional order could not be applied only when the heterogeneity is uniform. Therefore, if the heterogeneity of the soil is not uniform one will have to employ a very complex mathematical operator to replicate the observed facts into mathematical formulation. Thus in this section, we will be concerned with generalization of advection dispersion model using the concept of variable order differential operators. The operator used in this section is in Caputo sense and also we will consider the one-dimensional model consisting of a considerably long heterogeneous isotropic porous medium with a steady-state uniform flow including a seepage velocity V. A complex compound is introduced commencing on one end of the model for a stage of time t_0 implying that the input concentration varies as an exponential function of time. The significance of that chemical absorption at any moment during the process t and at a distance x beginning with the injection boundary, permitting the decomposing and adsorption, may be acquired from the mathematical solution of the following set of equations [142] (more details for this model can be found in [142]). Consider

$$D\frac{\partial^2 C}{\partial x^2} - v\frac{\partial C}{\partial x} - \lambda RC = R\frac{\partial C}{\partial t} + f(x,t), \qquad (12.1)$$

subject to the initial and boundary conditions:

$$C(x,0) = 0, \ C(0,t) = c_0 \exp(-\alpha t), \ C_x(\infty,t) = 0. \qquad (12.2)$$

For this equation the parameter D is the dispersion coefficient, V is the speed with which the pollution is migrating within the geological formation, R accounts for the retardation due to the properties of the matrix soil, λ accounts for the radioactive decay constant, c_0 is considered the initial input of the concentration, α is a positive constant, and finally $f(x, t)$ is any source and sink in the system. On the other hand, for the model of groundwater pollution movement, $f(x, t)$ is neglected as we assume that there is no sink function within the system under investigation. Consequently, in this section also the sink function will be neglected. Having in mind that the heterogeneity of geological formation could be included into mathematical formulation using the concept of variable order differentiation, in this section, the classical differentiation of the advection model will be replaced with the variable order in Caputo sense. And the modified equation is given as follows:

$$f(x, t) + D\frac{\partial^2 C}{\partial x^2} - v\frac{\partial C}{\partial x} - \lambda RC = RD_0^{\alpha(t,x)}C, \qquad (12.3)$$

subject to the initial and boundary conditions:

$$C(x, 0) = 0, \ C(0, t) = c_0 \exp(-\gamma t), \ C_x(\infty, t) = 0. \qquad (12.4)$$

In this equation also $\alpha(t, x)$ is a continuous function in $(0, 1]$ that can be measured from the field observation facts. With this new model or modified model the analytical solution was only introduced in [187]; also the existence of the exact solution is presented here.

12.3 Numerical Solution

The numerical approximation of a variable order operator goes along with predictable solutions of the principal equation all the way through the discretization of space and time components. Within the boundaries of the discretization domain, the geological formations properties including: the variable internal properties, boundaries, and stresses of the system are approximated. Deterministic, distributed parameter, and numerical models can relax the rigid idealized conditions of analytical models or lumped-parameter models, and they can therefore be more realistic and flexible for simulating field conditions [187]. The finite difference schemes for constant-order time or space fractional diffusion equations have been widely studied [187]. Before performing the numerical methods, we assume that (12.4) has a unique and sufficiently smooth solution. To establish the numerical schemes for the above equation, we let $x_l = lh, 0 \le l \le M, Mh = L, t_k = k\tau, 0 \le k \le N$, and $N\tau = T$, and we let h be the step and τ be the time size and M and N be grid points.

12.3.1 Crank–Nicolson Scheme

We bring in the Crank–Nicolson idea as follows [187,162]. At the outset, the discretization of first- and second-order space derivative is stated as

$$\frac{\partial C}{\partial x} = \frac{1}{2}\left(\frac{C\left(x_{l+1}, t_{k+1}\right) - C\left(x_{l-1}, t_{k+1}\right)}{2\left(h\right)}\right) \tag{12.5}$$

$$+ \left(\frac{C\left(x_{l+1}, t_k\right) - C\left(x_{l-1}, t_k\right)}{2\left(h\right)}\right) + O(h),$$

$$\frac{\partial^2 C}{\partial x^2} = \frac{1}{2}\left(\frac{C\left(x_{l+1}, t_{k+1}\right) - 2C\left(x_l, t_{k+1}\right) + C\left(x_{l-1}, t_{k+1}\right)}{\left(h\right)^2}\right) \tag{12.6}$$

$$+ \left(\frac{C\left(x_{l+1}, t_k\right) - 2C\left(x_l, t_k\right) + C\left(x_{l-1}, t_k\right)}{\left(h\right)^2}\right) + O(h^2),$$

$$C = \frac{1}{2}\left(C\left(x_l, t_{k+1}\right) + C\left(x_l, t_k\right)\right). \tag{12.7}$$

The Crank–Nicolson scheme for the VO time fractional diffusion model can be stated as follows:

$$\frac{\partial^{\alpha_l^{k+1}} C\left(x_l, t_{k+1}\right)}{\partial t^{\alpha_l^{k+1}}} = \tag{12.8}$$

$$\frac{\tau^{-\alpha_l^{k+1}}}{\Gamma\left(2 - \alpha_l^{k+1}\right)}\left(\begin{array}{c} C\left(x_l, t_{k+1}\right) - C\left(x_l, t_k\right) \\ + \sum_{j=1}^{k}\left[C\left(x_l, t_{k+1-j}\right) - C\left(x_l, t_{k-j}\right)\right] \\ \times \left[(j+1)^{1-\alpha_l^{k+1}} - (j)^{1-\alpha_l^{k+1}}\right] \end{array}\right) + O\left(\tau\right).$$

Now replacing Eq. (12.5), Eq. (12.6), Eq. (12.7) and Eq. (12.8) in Eq. (12.3) we obtain the following:

$$b_l^{k+1}\left[\left(C_{l+1}^{k+1} - 2C_l^{k+1} + C_{l-1}^{k+1}\right) + \left(C_{l+1}^k - 2C_l^k + C_{l-1}^k\right)\right] \tag{12.9}$$

$$+ c_l^{k+1}\left[\left(C_{l+1}^{k+1} - C_{l-1}^{k+1}\right) + \left(C_{l+1}^k - C_{l-1}^k\right)\right]$$

$$+ d_l^{k+1}\left(C_l^{k+1} - C_l^k\right)$$

$$= C_l^{k+1} - C_l^k + \sum_{j=1}^{k}\left[C_l^{k+1-i} - C_l^{k-i}\right]e_j^{l,k+1}$$

$$+ \tau^{\alpha_l^{k+1}}\Gamma(2 - \alpha_l^{k+1})f_l^{k+1}.$$

Here

$$b_l^{k+1} = \frac{D\tau^{\alpha_l^{k+1}}\Gamma\left(2 - \alpha_l^{k+1}\right)}{2R\left(h\right)^2} \tag{12.10}$$

$$c_l^{k+1} = -\frac{v\tau^{\alpha_l^{k+1}}\Gamma\left(2-\alpha_l^{k+1}\right)}{2hR}$$

$$d_l^{k+1} = \frac{\lambda\tau^{\alpha_l^{k+1}}\Gamma\left(2-\alpha_l^{k+1}\right)}{2}$$

$$e_j^{l,k+1} = (j+1)^{1-\alpha_l^{k+1}} - (j)^{1-\alpha_l^{k+1}}$$

$$p_j^{l,k+1} = e_{j-1}^{l,k+1} - e_j^{l,k+1}$$

for $j = 1, 2, ..., N$ and $j = 1, 2, ..., M-1$. It is important to point out that the sum term on the right-hand side of Eq. (12.9) automatically vanishes when $k = 0$. Then, Eq. (12.9) can be reformulated as:

$$b_l^1\left[\left(C_{l+1}^1 - 2C_l^1 + C_{l-1}^1\right) + \left(C_{l+1}^0 - 2C_l^0 + C_{l-1}^0\right)\right] \qquad (12.11)$$
$$+ c_l^1\left[\left(C_{l+1}^1 - C_{l-1}^1\right) + \left(C_{l+1}^0 - C_{l-1}^0\right)\right]$$
$$+ d_l^1\left(C_l^1 - C_l^0\right)$$
$$= \tau^{\alpha_l^1}\Gamma(2-\alpha_l^1)f_l^1,$$

for $k = 0$,

$$b_l^{k+1}\left[\left(C_{l+1}^{k+1} - 2C_l^{k+1} + C_{l-1}^{k+1}\right) + \left(C_{l+1}^k - 2C_l^k + C_{l-1}^k\right)\right] \qquad (12.12)$$
$$+ c_l^{k+1}\left[\left(C_{l+1}^{k+1} - C_{l-1}^{k+1}\right) + \left(C_{l+1}^k - C_{l-1}^k\right)\right]$$
$$+ d_l^{k+1}\left(C_l^{k+1} - C_l^k\right)$$
$$= C_l^{k+1} - C_l^k + \sum_{j=1}^{k}\left[C_l^{k+1-i} - C_l^{k-i}\right]e_j^{l,k+1}$$
$$+ \tau^{\alpha_l^{k+1}}\Gamma(2-\alpha_l^{k+1})f_l^{k+1}, \text{ for } k \geq 1.$$

The above discretization can be rewritten also in matrix form.

12.3.2 Stability Analysis of the Crank–Nicolson Scheme

In this section, we will analyze the stability conditions of the Crank–Nicolson scheme for the generalized advection dispersion equation. Let $\zeta_l^k = c_l^k - C_l^k$; here C_l^k is the approximate solution at the point (x_l, t_k), $(k = 1, 2, ..., N, l = 1, 2, ..., M-1)$ and in addition

$\zeta_l^k = [\zeta_1^k, \zeta_2^k, ..., \zeta_{M-l}^k]^T$ and the function $\zeta^k(x)$ is chosen to be

$$\zeta^k(x) = \begin{cases} \zeta_l^k \text{ if } x_l - \frac{h}{2} < x \le x_l + \frac{h}{2}, \ l = 1, 2, ..., M-1, \\ 0 \text{ if } L - \frac{h}{2} < x \le L. \end{cases} \tag{12.13}$$

Then the function $\zeta^k(x)$ can be expressed in Fourier series as follows:

$$\zeta^k(x) = \sum_{m=-\infty}^{m=\infty} \delta_m(m) \exp\left[\frac{2i\pi mk}{L}\right], \tag{12.14}$$

$$\delta_k(x) = \frac{1}{L} \int_0^L \rho^k(x) \exp\left[\frac{2i\pi mk}{L}\right] dx.$$

It was established by [187] that

$$\left\|\zeta^2\right\|_2^2 = \sum_{m=-\infty}^{m=\infty} \|\delta_k(m)\|^2. \tag{12.15}$$

Observe that, for all $k, l \ge 1, 0 \le 1 - \alpha_l^{k+1} < 1$, in addition, according to the problem in point, the velocity seepage v, the dispersion coefficient D, the retardation factor R, and the radioactive decay constant λ are positive constants. Then the following properties of the coefficients $c_l^{k+1}, d_l^{k+1}, p_j^{l,k+1}$ and b_l^{k+1} can be established:

(1) d_l^{k+1}, b_l^{k+1} are positive for all $l = 1, 2, ..., M - 1$; (2) c_l^{k+1} are negative for all $l = 1, 2, ..., M - 1$; (3) $0 < e_j^{l,k} \le e_{j-1}^{l,k} \le 1$ for all $l = 1, 2, ..., M - 1$; (4) $0 \le p_j^{l,k} \le 1$, $\sum_{j=0}^{k-1} p_{j+1}^{l,k+1} = 1 - e_k^{l,k+1}$ for all $l = 1, 2, ..., M - 1$. The error analysis due to the numerical approximation while performing the solution of the more extended advection dispersion model by means of the Crank–Nicolson scheme can be presented as follows:

$$b_l^{k+1}\left(\left(\zeta_{l+1}^{k+1} - 2\zeta_l^{k+1} + \zeta_{l-1}^{k+1}\right) + \left(\zeta_{l+1}^k - 2\zeta_l^k + \zeta_{l-1}^k\right)\right) \tag{12.16}$$

$$+c_l^{k+1}\left[\left(\zeta_{l+1}^{k+1} - \zeta_{l-1}^{k+1}\right) + \left(\zeta_{l+1}^k - \zeta_{l-1}^k\right)\right] \tag{12.17}$$

$$+d_l^{k+1}\left(\zeta_l^{k+1} - \zeta_l^k\right) = \tag{12.18}$$

$$\zeta_l^{k+1} - \zeta_l^k + \sum_{j=1}^k \left[\zeta_l^{k+1-i} - \zeta_l^{k-i}\right]e_j^{l,k+1} \tag{12.19}$$

$$+\tau^{\alpha_l^{k+1}}\Gamma(2 - \alpha_l^{k+1})f_l^{k+1}. \tag{12.20}$$

If we assume that ζ_l^k in (12.15) can be put in the delta exponential form as follows:

$$\zeta_l^k = \delta_k \exp\left[i\varphi l k\right], \tag{12.21}$$

where φ is a real spatial wave number, replacing (12.21) in (12.15) we obtain for $k = 0$:

$$\left[1 + 4b_l^1 \sin^2\left(\frac{\varphi h}{2}\right) - 2d_l^1 \sin^2\left(\frac{\varphi h}{2}\right)\right]\delta_1 \tag{12.22}$$

$$= \left[1 - 4b_l^1 \sin^2\left(\frac{\varphi h}{2}\right) - 2d_l^1 \sin^2\left(\frac{\varphi h}{2}\right)\right]\delta_0,$$

and for $k = 1, 2, ..., N - 1$:

$$\left[1 + 4b_l^{1+k} \sin^2\left(\frac{\varphi h}{2}\right) - 2d_l^{1+k} \sin^2\left(\frac{\varphi h}{2}\right)\right]\delta_{k+1} \tag{12.23}$$

$$\left[1 - 4b_l^{k+1} \sin^2\left(\frac{\varphi h}{2}\right) - 2d_l^{k+l} \sin^2\left(\frac{\varphi h}{2}\right) - e_1^{l,k+1}\right]\delta_k$$

$$+ \sum_{j=0}^{k-1} p_{j+1}^{l,k+1}\delta_{k-j} + e_k^{l,k+1}\delta_0.$$

Eq. (12.18) can be written in the following form:

$$\delta_1 = \frac{\left[1 - 4b_l^1 \sin^2\left(\frac{\varphi h}{2}\right) - 2d_l^1 \sin^2\left(\frac{\varphi h}{2}\right)\right]\delta_0}{\left[1 + 4b_l^1 \sin^2\left(\frac{\varphi h}{2}\right) - 2d_l^1 \sin^2\left(\frac{\varphi h}{2}\right)\right]}, \tag{12.24}$$

$$\delta_{k+1} = \frac{\left[1 - 4b_l^{k+1} \sin^2\left(\frac{\varphi h}{2}\right) - 2d_l^{k+l} \sin^2\left(\frac{\varphi h}{2}\right) - e_1^{l,k+1}\right]\delta_k + \sum_{j=0}^{k-1} p_{j+1}^{l,k+1}\delta_{k-j} + e_k^{l,k+1}\delta_0}{\left[1 + 4b_l^{1+k} \sin^2\left(\frac{\varphi h}{2}\right) - 2d_l^{1+k} \sin^2\left(\frac{\varphi h}{2}\right)\right]}.$$

Our next concern here is to show that for all $k = 1, 2, ..., N - 1$ the solution of (12.24) satisfies the following condition:

$$|\delta_k| < |\delta_0|. \tag{12.25}$$

Proof. To achieve this, we make use of the recurrence technique on the natural number k. For $k = 1$ and remembering that d_l^{k+1}, b_l^{k+1} are positive for all $l = 1, 2, ..., M - 1$, then we obtain

$$\frac{|\delta_1|}{|\delta_0|} = \frac{\left[1 - 4b_l^1 \sin^2\left(\frac{\varphi h}{2}\right) - 2d_l^1 \sin^2\left(\frac{\varphi h}{2}\right)\right]}{\left[1 + 4b_l^1 \sin^2\left(\frac{\varphi h}{2}\right) - 2d_l^1 \sin^2\left(\frac{\varphi h}{2}\right)\right]} < 1. \tag{12.26}$$

Assuming that for $m = 2, 3, ..., k$ the property is verified, then

$$|\delta_{k+1}| = \left| \frac{\left[1 - 4b_l^{k+1} \sin^2\left(\frac{\varphi h}{2}\right) - 2d_l^{k+l} \sin^2\left(\frac{\varphi h}{2}\right) - e_1^{l,k+1}\right]\delta_k + \sum_{j=0}^{k-1} p_{j+1}^{l,k+1}\delta_{k-j} + e_k^{l,k+1}\delta_0}{\left[1 + 4b_l^{1+k} \sin^2\left(\frac{\varphi h}{2}\right) - 2d_l^{1+k} \sin^2\left(\frac{\varphi h}{2}\right)\right]} \right|. \tag{12.27}$$

Making use of the triangular inequality, we obtain

$$|\delta_{k+1}| \leq \frac{\left|1 - 4b_l^{k+1}\sin^2\left(\frac{\varphi h}{2}\right) - 2d_l^{k+l}\sin^2\left(\frac{\varphi h}{2}\right) - e_1^{l,k+1}\right||\delta_k| + \left|\sum_{j=0}^{k-1} p_{j+1}^{l,k+1}\delta_{k-j}\right| + \left|e_k^{l,k+1}\delta_0\right|}{\left|1 + 4b_l^{1+k}\sin^2\left(\frac{\varphi h}{2}\right) - 2d_l^{1+k}\sin^2\left(\frac{\varphi h}{2}\right)\right|}. \tag{12.28}$$

Using the recurrence hypothesis, we have

$$|\delta_{k+1}| \leq \frac{1 - 4b_l^{k+1}\sin^2\left(\frac{\varphi h}{2}\right) - 2d_l^{k+l}\sin^2\left(\frac{\varphi h}{2}\right) + \left|\sum_{j=0}^{k-1} p_{j+1}^{l,k+1}\right|}{\left|1 + 4b_l^{1+k}\sin^2\left(\frac{\varphi h}{2}\right) - 2d_l^{1+k}\sin^2\left(\frac{\varphi h}{2}\right)\right|}|\delta_0|,$$

$$|\delta_{k+1}| \leq \left(\frac{\left|1 - 4b_l^{1+k}\sin^2\left(\frac{\varphi h}{2}\right) - 2d_l^{1+k}\sin^2\left(\frac{\varphi h}{2}\right)\right|}{\left|1 + 4b_l^{1+k}\sin^2\left(\frac{\varphi h}{2}\right) - 2d_l^{1+k}\sin^2\left(\frac{\varphi h}{2}\right)\right|}\right)|\delta_0|, \tag{12.29}$$

$$|\delta_{k+1}| \leq |\delta_0|,$$

and this completes the proof.   □

12.3.3 Convergence Analysis of the Crank–Nicolson Scheme

If we assume that $c(x_l, t_k)$, $(k = 1, 2, ..., N-1, l = 1, 2, ..., M)$ is the exact solution of our problem at the point (x_l, t_k), by letting $\omega_l^k = c(x_l, t_k) - C_l^k$ and $\omega_l^k = (0, \omega_1^k, \omega_2^k, ..., \omega_{M-1}^k)$ and substituting this into Eq. (12.12) we get:

$$b_l^1 \left[\left(\omega_{l+1}^1 - 2\omega_{l-1}^1 \right) \right] + c_l^1 \left[\left(\omega_{l+1}^1 - \omega_{l-1}^1 \right) \right] + d_l^1 \left(\omega_l^1 \right)$$

$$= F_l^1, \text{ for } k = 0.$$

$$b_l^{k+1} \left[\left(\omega_{l+1}^{k+1} - 2\omega_l^{k+1} + \omega_{l-1}^k \right) \right] \tag{12.30}$$

$$+ c_l^{k+1} \left[\left(\omega_{l+1}^{k+1} - \omega_{l-1}^{k+1} \right) \right] + d_l^{k+1} \omega_l^{k+1}$$

$$= F_l^{k+1} + \sum_{i=1}^k \left(\omega_l^{k-i} \right) e_j^{l,k+1}, \text{ for } k \geq 1.$$

Here

$$F_l^{k+1} = d_l^{k+1} \left(-C \left(x_l, t_{k+1} \right) + C \left(x_l, t_k \right) \right) \tag{12.31}$$

$$+ \sum_{i=1}^k C \left(x_l, t_{k-i+1} \right) e_j^{l,k+1}$$

$$+ b_l^{k+1} \left(-C \left(x_{l+1}, t_k \right) - 2C \left(x_l, t_k \right) - C \left(x_{l-1}, t_k \right) \right)$$

$$- c_l^{k+1} \left(C \left(x_{l+1}, t_k \right) - C \left(x_{l-1}, t_k \right) \right).$$

From Eq. (12.5), Eq. (12.6) and Eq. (12.8):

$$\frac{\partial C \left(x_l, t_{k+1} \right)}{\partial x} + V_1 = \frac{1}{2} \left(\frac{C \left(x_{l+1}, t_{k+1} \right) - C \left(x_{l-1}, t_{k+1} \right)}{2 \left(h \right)} \right) \tag{12.32}$$

$$+ \left(\frac{C \left(x_{l+1}, t_k \right) - C \left(x_{l-1}, t_k \right)}{2 \left(h \right)} \right),$$

$$\frac{\partial^2 C \left(x_l, t_{k+1} \right)}{\partial x^2} + h^2 V_2 = \frac{1}{2} \left(\begin{array}{c} C \left(x_{l+1}, t_{k+1} \right) \\ -2C \left(x_l, t_{k+1} \right) + C \left(x_{l-1}, t_{k+1} \right) \\ \hline \left(h \right)^2 \end{array} \right) \tag{12.33}$$

$$+ \left(\frac{C \left(x_{l+1}, t_k \right) - 2C \left(x_l, t_k \right) + C \left(x_{l-1}, t_k \right)}{\left(h \right)^2} \right),$$

$$\frac{\partial^{\alpha_l^{k+1}} C \left(x_l, t_{k+1} \right)}{\partial t^{\alpha_l^{k+1}}} + \tau V_3 \tag{12.34}$$

$$= \frac{\tau^{-\alpha_l^{k+1}}}{\Gamma \left(2 - \alpha_l^{k+1} \right)} \left(\begin{array}{c} C \left(x_l, t_{k+1} \right) - C \left(x_l, t_k \right) \\ + \sum_{j=1}^k \left[C \left(x_l, t_{k+1-j} \right) - C \left(x_l, t_{k-j} \right) \right] \\ \times \left[\left(j + 1 \right)^{1-\alpha_l^{k+1}} - \left(j \right)^{1-\alpha_l^{k+1}} \right] \end{array} \right).$$

From the above we have that

$$F_l^{k+1} \leq V \left(\tau^{1+\alpha_l^{k+1}} + h^2 \tau^{\alpha_l^k} + h\tau^{\alpha_l^k} \right), \tag{12.35}$$

where V_1, V_2, V_3 and V are constants. Taking into account Caputo type fractional derivative, the detailed error analysis of the above schemes can refer to the work by Diethelm et al. [187] and further work by Li and Tao [187].

Lemma: The following inequality

$$\left\| \omega^{k+1} \right\|_{\infty} \leq V \left(\tau^{1+\alpha_l^{k+1}} + h^2 \tau^{\alpha_l^k} + h\tau^{\alpha_l^k} \right) \left(e_j^{l,k+1} \right)^{-1}$$

is true for $(k = 1, 2, ..., N - 1)$ where

$$\left\| \omega^k \right\|_{\infty} = \max_{1 \leq l \leq M-1} \left(\omega^k \right)$$

and V is a constant. In addition,

$$\alpha^{k+1} = \begin{cases} \min_{1 \leq l \leq M-1} \alpha_l^{k+1}, & \text{if } \tau < 1, \\ \max_{1 \leq l \leq M-1} \alpha_l^{k+1}, & \text{if } \tau > 1. \end{cases} \tag{12.36}$$

Proof. This can be achieved by means of the recurrence technique on the natural number k. When $k = 0$, one has the following:

$$\begin{aligned} \left| \omega_l^1 \right| &\leq \left(c_l^1 + b_l^1 \right) \left| \omega_{l+1}^1 \right| \\ &\quad + \left(c_l^1 - 2b_l^1 \right) \left| \omega_{l-1}^1 \right| + d_l^1 \left| \omega_l^1 \right| \\ &= \left| F_l^1 \right| \\ &\leq V \left(\tau^{1+\alpha_l^{k+1}} + h^2 \tau^{\alpha_l^k} + h\tau^{\alpha_l^k} \right) \left(e_j^{l,k+1} \right)^{-1}. \end{aligned} \tag{12.37}$$

Now suppose that

$$\left\| \omega^{i+1} \right\|_{\infty} \leq V \left(\tau^{1+\alpha_l^{k+1}} + h^2 \tau^{\alpha_l^k} + h\tau^{\alpha_l^k} \right) \left(e_j^{l,k+1} \right)^{-1}, i = 1, 2, ..., N - 1. \tag{12.38}$$

Then

$$\begin{aligned} \left| \omega_l^{k+1} \right| &\leq \left| \begin{array}{l} b_l^{k+1} \left[\left(\omega_{l+1}^{k+1} - 2\omega_l^{k+1} + \omega_{l-1}^k \right) \right] \\ + c_l^{k+1} \left[\left(\omega_{l+1}^{k+1} - \omega_{l-1}^{k+1} \right) \right] + d_l^{k+1} \omega_l^{k+1} \end{array} \right| \\ &\leq \left(b_l^{k+1} + c_l^{k+1} \right) \left| \omega_{l+1}^{k+1} \right| + \left(b_l^{k+1} - c_l^{k+1} \right) \left| \omega_{l-1}^{k+1} \right| \end{aligned} \tag{12.39}$$

$$+ \left(d_l^{k+1} - 2b_l^{k+1} \right) \left| \omega_l^{k+1} \right|$$

$$= \left| F_l^{k+1} + \sum_{i=1}^{k} \left(\omega_l^{k-i} \right) e_j^{l,k+1} \right|$$

$$\leq \left| F_l^{k+1} \right| + \sum_{i=1}^{k} \left| \omega_l^{k-i} \right| e_j^{l,k+1}$$

$$\leq V \left(\tau^{1+\alpha_l^{k+1}} + h^2 \tau^{\alpha_l^k} + h\tau^{\alpha_l^k} \right) + \sum_{i=1}^{k} \left| \omega_l^{k-i} \right|_\infty e_j^{l,k+1}$$

$$\leq V \left(\tau^{1+\alpha_l^{k+1}} + h^2 \tau^{\alpha_l^k} + h\tau^{\alpha_l^k} \right)$$
$$\times \left(e_j^{l,k+1} + e_0^{l,k+1} - e_j^{l,k+1} \right) \left(e_j^{l,k+1} \right)^{-1}$$

$$\leq V \left(\tau^{1+\alpha_l^{k+1}} + h^2 \tau^{\alpha_l^k} + h\tau^{\alpha_l^k} \right) \left(e_0^{l,k+1} \right) \left(e_j^{l,k+1} \right)^{-1},$$

$$\left| \omega_l^{k+1} \right| \leq V \left(\tau^{1+\alpha_l^{k+1}} + h^2 \tau^{\alpha_l^k} + h\tau^{\alpha_l^k} \right) \left(e_j^{l,k+1} \right)^{-1}.$$

This completes the proof. □

Theorem: The Crank–Nicolson scheme is convergent, and there exists a positive constant V such that

$$C_l^k - c \left(x_l, t_k \right) \leq V \left(\tau + h + h^2 \right), \tag{12.40}$$
$$k = 1, 2, ..., M - 1, \ l = 1, 2, ..., N.$$

This concludes the proof.

12.4 Groundwater Flow Model With Space Time Riemann–Liouville Fractional Derivatives Approximation

This part investigates the effect uncertainty has on the predictive accuracy of flow through a porous medium. There are two different issues existing within groundwater research. On the one hand, it is the appropriateness of a groundwater model for the geometry through which groundwater flows. On the other hand, it is how theoretically expected values deviate from observed values. Researchers within the geohydrology field are familiar within the concepts of doubt and uncertainty, and this is due to the inability to precisely understand and model phenomena existing within aquifers. Moreover, there is doubt and uncertainty in historical and current theoretical knowledge in groundwater investigations. With that being said, doubt and uncertainty become important to understand. We believe it is imperative, extending beyond the theories used to infer occurrences within a groundwater system. Doubt is evidently a phenomenon that groundwater flow models should analytically incorporated in their mathematical formulations. Furthermore, uncertainty within geohydrology is generated from various sources. Neglecting uncertainty may result in incorrect findings and distorted outputs. Usually there exist different sources of uncertainty in model outputs. For instance, uncertainty can be related to inadequate knowledge of inaccurate model inputs. It can also be structural uncertainty which relates to how the model is mathematically interpreted. Assessing and presenting the effects of uncertainty is regarded as significant in analyzing complex systems. On the simplest level, these analyses can be associated with the study of functions. For explicitly including the potential effect of uncertainty into mathematical model, this work introduces the uncertainties in groundwater models as a function of both time and space, we consider

$$u = u(\underline{x}, t) \tag{12.41}$$

to be uncertain function of a given dynamical system. The above function will be used in the next section to construct a new concept of derivative.

12.5 Modified Groundwater Flow Model by Means of Riemann–Liouville Approximate Fractional Differentiation

For clarity, here we consider the modification of the classical model for groundwater flow in the case of density independent flow within a uniform and homogeneous groundwater system. To do this modification, we make use of Riemann–Liouville fractional derivatives which were introduced and applied by several other authors. These derivatives are defined as

$$
\begin{aligned}
D^{1+\varepsilon_t} f &= D^{\mu_t}_{+,t} f \\
&= \left(\frac{d}{dt}\right)^n \int_0^t \left[\frac{f(\tau)}{\Gamma\left(n - \mu_\tau(\tau)(t-\tau)^{\mu_\tau - n + 1}\right)} \right] d\tau,
\end{aligned}
\tag{12.42}
$$

$$
\begin{aligned}
D^{1+\varepsilon_x} f &= D^{\mu_x}_{+,x} f \\
&= \left(\frac{d}{dx}\right)^n \int_0^x \left[\frac{f(\tau)}{\Gamma\left(n - \mu_\tau(\tau)(x-\tau)^{\mu_\tau - n + 1}\right)} \right] d\tau.
\end{aligned}
$$

Here, Γ is the Euler gamma function; $n = \lceil \mu \rceil + 1$, where $\lceil \mu \rceil$ is the integer part of μ for $\mu \geq 0$, that is $n - 1 \leq \mu < n$ and $n = 0$ for $\mu < n$. Following Eq. (12.42) we have that $\mu_t = 1 + \varepsilon_t$ and $\mu_x = 1 + \varepsilon_x$. The integral operator defined above for fractional exponents μ_t and μ_x depending on coordinates and time can be expressed in terms of ordinary derivative and integral or $|\varepsilon| \leq 1$. For this matter, generalized Riemann–Liouville fractional derivatives satisfy the approximate relations. Consider

$$D^{1+u_t} f \;\cong\; (1+u_t)\frac{\partial f}{\partial t} + \frac{\partial u_t}{\partial t} f, \tag{12.43}$$

$$D^{1+u_x} f \;\cong\; (1+u_x)\frac{\partial f}{\partial x} + \frac{\partial u_x}{\partial x} f.$$

The above relations ensure the possibility of describing the flow system. This includes the effect uncertainties have on the behavior of a physical system, through the use of partial differential and integral equations. Now we examine some properties of the above derivative operator.

(i) Addition. If u_x, $f(x)$, and $g(x)$ are differentiable in the opened interval **I**, then

$$D^{1+u_x}\left[f(x)+g(x)\right] \;\cong\; D^{1+u_x}\left[f(x)\right]+D^{1+u_x}\left[g(x)\right] \tag{12.44}$$

$$D^{1+u_x}\left[f(x)+g(x)\right] \;\cong\;$$

$$(1+u_x)\frac{\partial\left[f(x)+g(x)\right]}{\partial x} + \frac{\partial u_x}{\partial x}\left[f(x)+g(x)\right]$$

$$(1+u_x)\frac{\partial\left[f(x)\right]}{\partial x} + \frac{\partial u_x}{\partial x}\left[f(x)\right]$$

$$+(1+u_x)\frac{\partial\left[g(x)\right]}{\partial x} + \frac{\partial u_x}{\partial x}\left[g(x)\right]$$

$$\cong\; D^{1+u_x}\left[f(x)\right]+D^{1+u_x}\left[g(x)\right].$$

(ii) Division. If u_x and $\frac{1}{f(x)}$ are differentiable on the open interval **I**, then

$$D^{1+u_x}\left[\frac{1}{f(x)}\right] \cong \frac{\left[-(1+u_x)f'(x)+u'_x f(x)\right]}{f^2(x)} \;=\; \tag{12.45}$$

$$\frac{f'(x)}{f^2(x)} - \frac{u_x f'(x)}{f^2(x)} + \frac{u'_x f(x)}{f^2(x)}. \tag{12.46}$$

(iii) Multiplication. If u_x, $f(x)$, and $g(x)$ are differentiable on the open interval **I**, then

$$D^{1+u_x}\left[f(x).g(x)\right] \;\cong\; g(x).f'(x)+f(x)g'(x) \tag{12.47}$$

$$+\left(gf'+fg'\right)(x)u_x + u'_x\left(f(x).g(x)\right).$$

(iv) Power. If u_x and $f(x)$ are differentiable on the open interval **I,** then

$$D^{1+u_x}\left[(f(x))^n\right] \cong nf'f^{n-1} + u_x nf'f^{n-1} + u'_x f^n, \quad n \geq 1. \tag{12.48}$$

(v) If u_x and $f(x)$ are twice differentiable on the open interval **I,** then

$$D^{1+u_x}\left[D^{1+u_x}\left[f(x)\right]\right] \tag{12.49}$$

$$\cong (1+u_x)\left[(1+u_x)\frac{\partial^2 f}{\partial x^2} + 3\frac{\partial f}{\partial x}\frac{\partial u_x}{\partial x} + \frac{\partial^2 u_x}{\partial x^2}f\right] + \frac{\partial u_x}{\partial x}f.$$

It is imperative to see that if $u_x = 0$, we obtained the properties of normal derivatives. Recent research studies propose flow which is affected by the geometry of bedding parallel fractures. In doing an attempt to bypass this issue, a model in which the geometry of the groundwater system is defined as a fractal, was introduced by Barker. Similarly, a concept of non-integer fractional derivative was introduced by other authors for investigating a radially symmetric form of (12.41). This was done by replacement of the classical first-order derivative of the piezometric head, by means of a complementary fractional derivative. Their findings revealed that the fractal and fractional have a close relationship. Therefore, in order to incorporate the fractal dimension into the mathematical expression of the modified groundwater flow equation, the constant fractal dimension α is next introduced. In the case of density-independent flow in a uniform homogeneous groundwater system, the classical model for groundwater flow is then reformulated in the following way:

$$D^{1+u_t(r,t)}\Phi(r,t) = \frac{K}{S_0}D^{1+u_t(r,t)}\left[rD^{1+u_t(r,t)}\Phi(r,t)\right] + f(r,t), \tag{12.50}$$

where K is the hydraulic conductivity of the aquifer, S_0 the specific storativity of the aquifer, f the strength of any sources or sink, here it will be neglected, and finally $\Phi(r,t)$ is the piezometric head. In order to ensure the physical and mathematical requirements are met, we enforce the uncertainties function to be a positive function in a way that

$$0 < u_x < 1. \tag{12.51}$$

Eq. (12.50) ensures flow through the geological formation, and the effect uncertainties have on the behavior of the physical system can be described using partial differential and integral equations. Despite this, no analytical solution for this equation exists. In fact, it is difficult to determine the analytical solution. Subsequently, the following approximation to simplify (12.50) is required:

$$D^{1+u_t(r,t)}\left[rD^{1+u_t(r,t)}\Phi(r,t)\right] \tag{12.52}$$

$$\cong (1+u_r)\begin{bmatrix} r(1+u_r)\frac{\partial^2\Phi(r,t)}{\partial r^2}+r\frac{\partial^2 u_r}{\partial r^2}\Phi(r,t) \\ +\frac{\partial u_r}{\partial r}\Phi(r,t)+(1+u_r)\frac{\partial\Phi(r,t)}{\partial r} \\ +3r\frac{\partial u_r}{\partial r}\frac{\partial\Phi(r,t)}{\partial r} \end{bmatrix}$$

$$+r\left(\frac{\partial u_r}{\partial r}\right)^2\Phi(r,t).$$

Making use of (12.50), (12.51), and (12.52), we obtain the following equation:

$$(1+u_t)\frac{\partial\Phi(r,t)}{\partial t}+\frac{\partial u_t\Phi(r,t)}{\partial t} \tag{12.53}$$

$$\cong \frac{K}{S_0}(1+u_r)\begin{bmatrix} r(1+u_r)\frac{\partial^2\Phi(r,t)}{\partial r^2}+r\frac{\partial^2 u_r}{\partial r^2}\Phi(r,t) \\ +\frac{\partial u_r}{\partial r}\Phi(r,t)+(1+u_r)\frac{\partial\Phi(r,t)}{\partial r} \\ +3r\frac{\partial u_r}{\partial r}\frac{\partial\Phi(r,t)}{\partial r} \end{bmatrix}.$$

Since uncertainties additions to unit are small, the right- and left-hand sides of (12.53) can be divided by $(1+u_t)$ to obtain the following approximate equation:

$$\frac{\partial\Phi(r,t)}{\partial t} = \frac{K}{S_0}\begin{Bmatrix} \frac{1}{r}\frac{\partial\Phi(r,t)}{\partial r}+\frac{\partial^2\Phi(r,t)}{\partial r^2} \\ +(2u_r-u_t)\frac{\partial^2\Phi(r,t)}{\partial r^2} \\ +(1+u_r-u_t)\frac{\partial^2 u_r}{\partial r^2}\Phi(r,t) \\ +\frac{u_r-u_t}{r}\frac{\partial u_r}{\partial r}\Phi(r,t) \\ +\frac{2u_r-u_t}{r}\frac{\partial\Phi(r,t)}{\partial r} \\ +3(1+u_r-u_t)\frac{\partial u_r}{\partial r}\frac{\partial\Phi(r,t)}{\partial r} \\ +(1+u_r-u_t)\left(\frac{\partial u_r}{\partial r}\right)^2\Phi(r,t) \end{Bmatrix} \tag{12.54}$$

$$-u_t\frac{\partial u_t}{\partial t}\Phi(r,t),$$

where u_r^2 term is omitted due to its being very small. Moreover, since we are dealing with approximation, there is no need to consider it in this case. For simplicity (12.54) can be reformulated as

$$\frac{\partial\Phi(r,t)}{\partial t} = \frac{K}{S_0}\left[\frac{1}{r}\frac{\partial\Phi(r,t)}{\partial r}+\frac{\partial^2\Phi(r,t)}{\partial r^2}\right] \tag{12.55}$$

$$+F(u_r,u_t,\Phi(r,t)).$$

Here the additional term can be roughly approximated to

$$
\begin{aligned}
F\left(u_r, u_t, \Phi\left(r, t\right)\right) \;=\;& \left(2u_r - u_t\right) \frac{\partial^2 \Phi\left(r, t\right)}{\partial r^2} \\
& + 3\left(1 + u_r - u_t\right) \frac{\partial u_r}{\partial r} \frac{\partial \Phi\left(r, t\right)}{\partial r} \\
& - u_t \left(\frac{\partial^2 u_r}{\partial r^2} + \frac{1}{r}\frac{\partial u_r}{\partial r}\right) \Phi\left(r, t\right) \\
& + u_r \left(\frac{\partial u_t}{\partial r} + \frac{\partial u_t}{\partial t}\right) \Phi\left(r, t\right)
\end{aligned}
\tag{12.56}
$$

and $\Phi\left(r, t\right)$ satisfies the equation of classical model for groundwater flow in the case of density independent flow occurring in a uniform and homogeneous groundwater system. It is imperative to note that the modified equations, (12.54) and (12.54), differ from the normal form of the groundwater flow equations. These differences are seen in three properties: The new operator accounts for variation in piezometric head as well as the uncertainties function given below as

$$
\begin{aligned}
\omega\left(u_r, u_t, \Phi\left(r, t\right)\right) \;=\;& \left(2u_r - u_t\right) \frac{\partial^2 \Phi\left(r, t\right)}{\partial r^2} \\
& + 3\left(1 + u_r - u_t\right) \frac{\partial u_r}{\partial r} \frac{\partial \Phi\left(r, t\right)}{\partial r}.
\end{aligned}
\tag{12.57}
$$

Secondly, the "force"

$$
F = \frac{\partial^2 u_r}{\partial r^2} + \frac{1}{r}\frac{\partial u_r}{\partial r}
\tag{12.58}
$$

appears due to the coordinate dependence of uncertainty function. Lastly, there is a derivative-free term depending solely on the uncertainties time function

$$
B = \frac{\partial u_t}{\partial r} + \frac{\partial u_t}{\partial t}
\tag{12.59}
$$

and it is proportional to the piezometric head Φ which depends on the coefficient sign, the retardation, or improvement of flow through the porous medium. It is noteworthy that the terms in (12.54) involving fractional additions, F and B, to both the time and space dimensions are small. Furthermore, this equation can be solved approximately by means of changing the function Φ by Φ_0, and satisfying the standard groundwater flow equation which is the left-side of (12.54) and (12.54), in terms of u. Now, assuming this type of change occurs in the mathematical expression (12.55), Eqs. (12.54) and (12.54) become

$$\frac{\partial \Phi(r,t)}{\partial t} = \frac{K}{S_0}\left[\frac{1}{r}\frac{\partial \Phi(r,t)}{\partial r} + \frac{\partial^2 \Phi(r,t)}{\partial r^2}\right] \tag{12.60}$$

$$+ \begin{bmatrix} (2u_r - u_t)\frac{\partial^2 \Phi(r,t)}{\partial r^2} \\ +3(1+u_r-u_t)\frac{\partial u_r}{\partial r}\frac{\partial \Phi(r,t)}{\partial r} \\ -u_t\left(\frac{\partial^2 u_r}{\partial r^2} + \frac{1}{r}\frac{\partial u_r}{\partial r}\right)\Phi(r,t) \\ +u_r\left(\frac{\partial u_t}{\partial r} + \frac{\partial u_t}{\partial t}\right)\Phi(r,t) \end{bmatrix}.$$

Prior to solving the above equation, the additional function in the modified equation should be related to a physical problem occurring within a groundwater system. A few determin-istic models account for properties within a porous medium as lumped parameters. This in essence infers it is treated as a black-box model. This model however does not represent hydraulic properties which are heterogeneous. On the note of heterogeneity, heterogeneity alongside variability in hydraulic properties exists within all geological systems. Moreover, it is now seen as having a fundamental role in affecting groundwater flow as well as solute transport. As a result, preference is given to application of distributed parameter models, as these represent more realistic distributions of hydraulic properties. There is significant vari-ation in most geological formations, and these variations are in both horizontal and vertical directions. This infers that geological formations are not very often homogeneous. Hetero-geneity does not only occur at the big scale in a given aquifer, but also as individual layers that are compressed between. Moreover, their grain size could show variation in horizon-tal directions; they could be comprised of lenses having other grain sizes; or they could also be discontinued through faulting or scour-and-fill structures. Furthermore, the heterogene-ity of geohydrological properties of porous sedimentary groundwater systems at different scales are controlled by the distribution of sedimentary facies. The paths of groundwater flow across these groundwater systems are determined by the arrangement of individual facies, as well as their porosity and permeability. Subsequently, the ability to predict geo-hydrological heterogeneity as a result of facies changes assists in improving solutions of flow and diffusion issues arising in these types of groundwater systems. When investiga-tions are done on real groundwater systems, it becomes impossible to model groundwater flow at a scale where the effects of fine-scaled sedimentary heterogeneity should be ac-counted for. In fact, for this to occur, there is a necessity of precise knowledge on the sed-imentary systems; and this however cannot be achieved through sparse data from a few boreholes. Moreover, this would be unreasonable for the necessary computing power. As a result, fine scaled heterogeneity is generally "up-scaled" and the real heterogeneous system is replaced at a large scale with an equal and habitually anisotropic system having param-eters allowing reproduction of the average flow of the real heterogeneous system. In this research the function $F\left(u_r, u_t, \Phi_o\left(r, t\right)\right)$ will be considered to account for the effect of heterogeneity and variability of the geological formation within which the groundwater flow occurs.

12.6 Derivation of Solutions of the Modified Groundwater Flow

Numerical methods give off approximate solutions to the governing equation through space and time discretization. Moreover, within the discretized problem domain, there is approximation of the variable internal properties, boundaries, as well as stresses. The rigid idealized conditions of analytical models or lumped-parameter models can be relaxed through deterministic, distributed-parameter, and numerical models. As a result, they may be more realistic and flexible for simulation of field conditions. The next focus of this research is to present a solution of Eq. (12.60). This is obtained through making use of two techniques which are, namely, the Green function and the variational iteration method. We firstly start with the variational iteration method.

12.6.1 Variational Iteration Method

Values of the variational iteration method as well as its applications for a variety of categories of differentials equations are documented in several research papers in the literature. We are following work by Theis in 1935, in which he suggested an analytical solution to the normal groundwater flow equation. This solution can be approximated as

$$\Phi_o\left(r, t\right) = \frac{Q}{4\pi T}\left\{e^{-}\left(\frac{r^2 S}{4Tt}\right)\ln\left[1 + \frac{\alpha 4Tt}{r^2 S}\right]\right\} \tag{12.61}$$

where Q is the constant discharge rate, T is the transmissivity of the aquifer, and $\Phi_o\left(r, t\right)$ the piezometric head. It follows that the right side of (12.54) and (12.54) is known. On the basis of the above equation and knowing the function $u(x, t)$, a solution can be derived for (12.54) and (12.54) where the unknown is the function $\Phi_1\left(r, t\right)$. For simplicity, we put $h(r, t) = F\left(u_r, u_t, \Phi_o\left(r, t\right)\right)$, and (12.54) and (12.54) become

$$\frac{\partial \Phi_1\left(r, t\right)}{\partial t} = \frac{K}{S_0}\left[\frac{1}{r}\frac{\partial \Phi_1\left(r, t\right)}{\partial r} + \frac{\partial^2 \Phi_1\left(r, t\right)}{\partial r^2}\right] + h(r, t). \tag{12.62}$$

Solving (12.62) using the variational iteration method, we put (12.62) in the form

$$\frac{K}{S_0}\left[\left(\Phi_1\left(r, t\right)\right)_{2r} + \frac{1}{r}\left(\Phi_1\left(r, t\right)\right)_r\right] - h(r, t) - \left(\Phi_1\left(r, t\right)\right)_r = 0 \tag{12.63}$$

The correction functional can be approximately expressed for this matter as follows:

$$\Phi_{1,n+1}(r,t) = \Phi_{1n}(r,t)$$

$$+ \int_0^t \lambda(\tau) \left\{ \begin{array}{c} \left[(\Phi_{1n}(r,t))_{2r} + \frac{1}{r}(\Phi_1(r,t))_r \right] \\ -h(r,t) - \frac{\partial^m \Phi_{n1}(r,\tau)}{\partial \tau^m} \end{array} \right\} d\tau, \tag{12.64}$$

where λ is a general Lagrange multiplier, which can be recognized optimally by means of variation assumption, and $(\widehat{\Phi_1(r,t)})_{2r}$, $(\widehat{\Phi_1(r,t)})_r$, and $\widehat{h(r,\tau)}$ are considered as constrained variations. Making the above functional stationary

$$\delta\Phi_{1,n+1}(r,t) = \delta\Phi_{1n}(r,t)$$

$$+ \delta \int_0^t \lambda(\tau) \left\{ \frac{\partial^m \Phi_{n1}(r,\tau)}{\partial \tau^m} \right\} d\tau \tag{12.65}$$

capitulates the following Lagrange multipliers, giving up to the following Lagrange multipliers: $\lambda = -1$ for the case where $m = 1$ and $\lambda = t - \tau$ for $m = 2$. For this matter if $m = 1$, we obtained the following iteration formula:

$$\Phi_{1,n+1}(r,t) = \Phi_{1n}(r,t)$$

$$- \int_0^t \left\{ \begin{array}{c} \frac{K}{S_0} \left[(\Phi_{1,n}(r,\tau))_{2r} + \frac{1}{r}(\Phi_{1,n}(r,\tau))_r \right] \\ -h(r,\tau) - (\Phi_{1,n}(r,\tau))_{\tau} \end{array} \right\} d\tau. \tag{12.66}$$

Hence we start with

$$\Phi_{1,0}(r,t) = \Phi_1(r,0) = 0. \tag{12.67}$$

This infers that before water is pumped out of a borehole, the water level in the groundwater system is the same. This is considered to be zero level. It is worth noting that if the zeroth component $\Phi_0(r,t)$ is defined, then there can be a complete determination of the remaining components $n \geq 1$, in such a way that each term is determined through use of the previous terms. Subsequently, entire series solutions is determined. Finally, the solution $\Phi_0(r,t)$ is approximated by the truncated series

$$\Phi_{1N}(r,t) = \sum_{n=0}^{N-1} \Phi_{1n}(r,t), \tag{12.68}$$

$$\lim_{N \to \infty} \Phi_{1N}(r,t) = \Phi(r,t).$$

Next we follow with the second component,

$$\Phi_{1,1}(r, t) = \int_0^t h(r, \tau)d\tau. \tag{12.69}$$

To calculate $\Phi_{1,1}(r, t)$ we first need to explicitly define the function $u(r, t)$. The following function we define here actually has no physical meaning. Nonetheless, it is used as an example. To make things simpler, we assume that $u_t = 1$ and $u_r = 0.5$ and the function $h(r, t)$ becomes

$$\begin{aligned}
h(r,t) &= \exp\left[-\frac{r^2 S}{4Tt}\right]\left(Q\ln\left[1 + \frac{\alpha 4Tt}{r^2 S}\right] - 1\right) \\
&\times \left(-\frac{16Qt^2 T\alpha^2}{\pi r^6 S^2 \left(1 + \frac{\alpha 4Tt}{r^2 S}\right)^2}\right) \\
&+ \frac{6Qt\alpha}{\pi r^2 T \left(1 + \frac{\alpha 4Tt}{r^2 S}\right)} \\
&+ \frac{2Q\alpha}{\left(1 + \frac{\alpha 4Tt}{r^2 S}\right)} + \frac{Qr^2 S^2 \ln\left[1 + \frac{\alpha 4Tt}{r^2 S}\right]}{16\pi t^2 T^3} \\
&- \frac{QS\ln\left[1 + \frac{\alpha 4Tt}{r^2 S}\right]}{8\pi t T^2}.
\end{aligned} \tag{12.70}$$

In this matter two components of the decomposition series were obtained, whereby $\Phi(r, t)$ was evaluated to have the following expansion:

$$\Phi_1(r, t) = \Phi_{10}(r, t) + \Phi_{11}(r, t) + ... \tag{12.71}$$

12.6.2 Solution by Means of Green Function Methods

For solving (12.54) and (12.54), there is a construction of an appropriate Green's function for this case in point. Let (R, τ_1) be the Green's function to be constructed, where $R = |r - r_0|$ and $\tau_1 = |t - t_0|$. G is chosen in a way that it satisfies homogeneous boundary conditions which correspond to the boundary conditions. It is essential to notice that the homogeneous solution of (12.54) and (12.54) is similar to the diffusion equation when one subsequently replaces $\Phi_1(r, t)$ by $\Psi(r, t)$, the Green's function dealt with here is the Green's function for the diffusion equation. Moreover, since the groundwater system is regarded as being infinite, the Green's function for the flow equation for an infinite groundwater system is given by

$$G(R, \tau_1) = \frac{4\pi T}{S_0} \left(\frac{1}{2\sqrt{\pi \tau_1}} \right)^2 \exp\left[-\frac{T^2 R^2}{4 S_0 \tau_1} \right] k(\tau_1). \qquad (12.72)$$

Here the function $k(\tau_1)$ is to be determined by using the boundary condition. The above equation satisfies a fundamental integral property valid for $n = 2$. Consider

$$\int G(R, \tau_1) dS = \frac{4\pi S_0^2}{T^2}, \quad \tau_1 > 0. \qquad (12.73)$$

This equation is the mathematical formulation to groundwater flow. At a time and position, the piezometer is introduced in the well tapping the groundwater system. Moreover, the water pumped from the groundwater system through the well migrates through the porous medium. However, it does so in such a way that the total amount of water in the groundwater system reduces as time passes, provided no recharge occurs. Since (12.55) still holds, it can be seen that

$$G(R, \tau_1) \rightarrow \frac{4\pi S_0^2}{T^2} \delta(R), \quad \tau_1 \rightarrow 0. \qquad (12.74)$$

Additionally, the Green's function used for this intention is a solution to the following equation:

$$\frac{\partial G(R, \tau_1 \mid R_0, \tau_{10})}{\partial t} \qquad (12.75)$$

$$= \frac{K}{S_0} \left[\begin{array}{c} \frac{1}{r} \frac{\partial G(R,\tau_1|R_0,\tau_{10})}{\partial r} + \frac{\partial^2 G(R,\tau_1|R_0,\tau_{10})}{\partial r^2} \\ -4\pi \delta(R) \delta(\tau_1) \end{array} \right].$$

The general solution of (12.54) and (12.54) can then be given as a function of Green's function, as

$$\Phi_1(r, t) = \iint \frac{4\pi T}{S_0} \left(\frac{1}{2\sqrt{\pi \tau_1}} \right)^2 \qquad (12.76)$$

$$\times \exp\left[-\frac{T^2 R^2}{4 S_0 \tau_1} \right] k(\tau_1) h(R, \tau_1) dR d\tau_1.$$

Here

$$h(r,t) = \exp\left[-\frac{r^2 S}{4Tt}\right]\left(Q\ln\left[1+\frac{\alpha 4Tt}{r^2 S}\right]-1\right) \qquad (12.77)$$

$$\times\left(-\frac{16Qt^2 T\alpha^2}{\pi r^6 S^2\left(1+\frac{\alpha 4Tt}{r^2 S}\right)^2}\right)$$

$$+\frac{6Qt\alpha}{\pi r^2 T\left(1+\frac{\alpha 4Tt}{r^2 S}\right)}$$

$$+\frac{2Q\alpha}{\left(1+\frac{\alpha 4Tt}{r^2 S}\right)}+\frac{Qr^2 S^2\ln\left[1+\frac{\alpha 4Tt}{r^2 S}\right]}{16\pi t^2 T^3}$$

$$-\frac{QS\ln\left[1+\frac{\alpha 4Tt}{r^2 S}\right]}{8\pi t T^2}.$$

Due to lack of experimental data for this issue, no graphical representation is given in this research. There is a need for modeling the function of uncertainties introduced in this work and apply it to computational simulations and the analytical solutions of the modified groundwater flow equation. Comparisons can then also be made with experimental data. On the other hand, using the standard solution, aquifers parameters can be determined, and used to determine values of the function $u(x,t)$. This is not done here. This work entailed modifying the standard groundwater flow equation though replacing the standard derivative with Riemann–Liouville fractional derivatives approximation. The modified equations (12.54) and (12.54) deviate from the standard groundwater flow equation in three different properties. Firstly, a new operator accounts for the variation in piezometric head and uncertainties function. Secondly, the "force" is seen as a result of the coordinate dependence of uncertainty function. Lastly, a derivative-free term depends solely on the uncertainties time function exists. Furthermore, the modified groundwater flow equation accounts for flow through a porous medium as well as for the effect of variability on groundwater system. It also accounts for heterogeneity of a groundwater system. To add, the Green's function and variational iteration method were used to solve the modified groundwater flow equation.

12.7 Application of the Modified Riemann–Liouville Variable Order on the Advection Dispersion Equation

The issue dealt with in this work involves the modification of the advection dispersion equation by means of incorporating the potential effect of heterogeneity or variability of a groundwater system into the governing equation. The model considered here is a model of one dimension which is comprised of a porous medium defined as an infinitely long, homogeneous, and isotropic. Additionally, it has a steady-state uniform flow with a seepage velocity v. Furthermore, a certain chemical enters the model from one end for a time period t_0 in a way that the inserted chemical's concentration differs as an exponential function of time. Using the solution of the following set of equations, the value of the chemical concentration at any time t as well as at a distance x, allowing the occurrence of decay and adsorption, may be obtained. The mathematical model expressing the groundwater transport is given as follows:

$$D\frac{\partial^2 C}{\partial x^2} - v\frac{\partial C}{\partial x} - \lambda RC = R\frac{\partial C}{\partial t} + f(x,t) \tag{12.78}$$

subject to the initial and boundary conditions:

$$C(x,0) = 0, \quad C(0,t) = c_0 \exp(-\alpha t) \tag{12.79}$$

and

$$C_x(\infty, t) = 0, \tag{12.80}$$

where D is the dispersion coefficient, v the seepage velocity, R the retardation factor, λ the radioactive decay constant, c_0 the initial concentration, α is a positive constant, and $f(x,t)$ is any source and sink in the system. However, since it is thought that there is no source nor sink within the groundwater system of interest, in the case of groundwater pollution, the function $f(x,t)$ is always neglected [145]. As a result, in our case, the function will be set to zero. The above equation does not account for the effect of heterogeneity, variability, nor uncertainties in these groundwater systems. Subsequently, with the intention of accounting for these three concepts in a geological formation, and in a mathematical equation, the ordinary derivative is replaced by an uncertainty function equation in Eq. (12.78), so the following is obtained:

$$DD^{u_x}\left[D^{u_x}\left[C(x,t)\right]\right] - vD^{u_x}\left[C(x,t)\right] - \lambda RC(x,t) \tag{12.81}$$

$$= RD^{u_t}\left[C(x,t)\right]$$

$$D\left\{ (1+u_x)\left[(1+u_x)\frac{\partial^2 C}{\partial x^2} + 3\frac{\partial C}{\partial x}\frac{\partial u_x}{\partial x} + \frac{\partial^2 u_x}{\partial x^2}C\right] \atop +\frac{\partial u_x}{\partial x}C \right\}$$

$$- v\left[(1+u_x)\frac{\partial c}{\partial x} + \frac{\partial u_x}{\partial x}C\right] - \lambda RC(x,t)$$

$$= R\left[(1+u_t)\frac{\partial c}{\partial t} + \frac{\partial u_t}{\partial t}C\right].$$

The previous mathematical equation is defined as the generalized hydrodynamic advection dispersion equation. It is an equation accounting for the value of a chemical concentration at any time t and at a distance x, from the boundary at which the chemical concentration is injected. This equation allows for the decay and adsorption, as well as the potential effect of heterogeneity, variability, and uncertainty of a geological formation within which the concentration is measured.

12.7.1 Analysis and Possible Solutions

Let us put Eq. (12.79) in an appropriate form which may be easily used within a possible analytical solution. Assuming the heterogeneity, variability, or uncertainty of a groundwater system with respect to time are very small in a way that the addition to unity is small, then, Eq. (12.79) can be divided on both sides by $(1 + u_t)$. In doing so, Eq. (12.79) can be roughly approximated to:

$$D(x,t)\frac{\partial^2 c}{\partial x^2} - v(x,t)\frac{\partial c}{\partial x} - \lambda RC + F(x)C + B(x,t)C = R\frac{\partial c}{\partial t}$$

$$F(x) = D\frac{\partial^2 u_x}{\partial x^2} - v\frac{\partial u_x}{\partial x}$$

$$B(x,t) = \frac{\partial u_x}{\partial x} - R\frac{\partial u_t}{\partial t}$$

$$D(x,t) = (1 + u_x - u_t)D$$

$$v(x,t) = (1 + u_x - u_t)v.$$

12.7.2 Analysis

$F(x)$ can be defined as a force of heterogeneity, variability and uncertainty of the geological formation at respective positions acting on the constant dispersion coefficient as well as seepage velocity. To add, it is a force proportional to the chemical concentration value. $B(x,t)$ defines the proportion allowing the value of the concentration to recall its flow path in the groundwater system as well as the time retardation took place since the chemical concentration was injected into the system. $D(x,t)$ is clearly the dispersion function and $v(x,t)$ defines the seepage velocity function, where both functions are at each point and time in the geological formation. Furthermore, there are four properties causing variation between the modified and standard equations. First of all, the dispersion coefficient is dependent on time as well as coordinates, and this is a result of the effect of heterogeneity, variability, and uncertainty of the geological formation within which the chemical concentration is being dispersed with memory. This ultimately depends on the time and coordinates. The second property is the seepage velocity coefficient which is dependent on time and coordinates, and this is a result of the effect of heterogeneity, variability, and uncertainty of the geological formation within which the chemical concentration is being transported with memory. The third of the four properties, the force of heterogeneity, variability, and uncertainty of the geological formation at each position in the system, acts on a constant dispersion coefficient and seepage velocity; and is proportional to the chemical concentration value. The last property is the functional proportion allowing for the value of the chemical concentration to recall its route through which it migrated through the geological formation. Additionally, it remembers the time as retardation occurs, from the location at which the chemical concentration was injected.

12.7.3 Possible Analytical Solution

In order to solve Eq. (12.80), consideration should first be given to some approximation. Subsequently, a suitable method should be selected for solving the non-linear partial differential equation. As written by V.M. Alexandrov in the foreword of "Asymptotology: Ideas, Methods, and Applications," the most romantic field of recent mathematics entails asymptotic methods [114]. Furthermore, despite the vast growth and application in computer science and numerical simulations, respectively, there will still remain a significant role played by non-numerical issues. Several different perturbation techniques have widespread application for solving non-linear problems. In this case, an asymptotic solution to Eq. (12.80) is found using the variational iteration. In order to solve Eq. (12.80), the approximations to follow are given consideration. Eq. (12.80) is firstly reformulated as:

$$D\frac{\partial^2 C}{\partial x^2} - v\frac{\partial C}{\partial x} - \lambda RC - R\frac{\partial C}{\partial t} = K(u_t, u_x C). \tag{12.82}$$

Therefore, this equation can be solved approximately by changing the function C to C_0, satisfying Eq. (12.80), in terms involving u (or in some of these terms). Moreover, if this type of change occurs in all the terms, the following hydrodynamic advection dispersion equation is obtained:

$$\begin{aligned} D\frac{\partial^2 C}{\partial x^2} - v\frac{\partial C}{\partial x} - \lambda RC - R\frac{\partial C}{\partial t} &= K(u_t, u_x C_0) \\ &= K(x, t). \end{aligned} \tag{12.83}$$

Variational iteration method

The variational iteration method has been favored by many, and thus applied to several different types of nonlinear problems. The primary property of this method lies in its capability and flexibility in accurately and appropriately solving nonlinear equations. Furthermore, it was recently noted that the variational iteration method, as well as other analytical methods, is considered as effective technique for solving several different nonlinear problems, without the general restrictive assumptions. Using the variational iterative method, Eq. (12.82) is solved in the following way:

$$D(C(x, t))_{2x} - v(C(x, t))_x - \lambda RC(x, t) - K(x, t) = 0. \tag{12.84}$$

The correction functional for Eq. (12.84) can be approximately expressed as follows:

$$C_{n+1}(x, t) = C_n(x, t) + \int_0^t \lambda_1(\tau) \left[\begin{array}{c} D\frac{\partial^2 \widehat{C}_n}{\partial x^2} - v\frac{\partial \widehat{C}_n}{\partial x} \\ -\lambda R\widehat{C}_n - R\frac{\partial^m C}{\partial \tau^m} - \widehat{K(x, \tau)} \end{array} \right] d\tau, \tag{12.85}$$

where λ_1 is a general Lagrange multiplier [114] which can be recognized optimally by means of variation assumption; here $\left(\widehat{C(x, \tau)}\right)_{2x}$, $\widehat{K(x, \tau)}$ and $\widehat{C}_n(x)$ are considered as constrained variations. Making the above functional stationary, we obtain:

$$\delta C_{n+1}(x, t) = \delta C_n(x, t) + \delta \int_0^t \lambda_1(\tau) \left[-R \frac{\partial^m C}{\partial \tau^m} \right] d\tau. \tag{12.86}$$

Capitulating the next Lagrange multipliers results in the following Lagrange multipliers: $\lambda_1 = -1$ for the case where $m = 1$ and $\lambda_1 = x - \tau$ for $m = 2$. For this matter $m = 1$, we obtain the following iteration formula:

$$C_{n+1}(x, t) = C_n(x, t) - \int_0^t \left[\begin{array}{c} D \frac{\partial^2 c_n}{\partial x^2} - v \frac{\partial c_n}{\partial x} \\ -\lambda R C_n - R \frac{\partial^m C}{\partial \tau^m} - K(x, t) \end{array} \right] d\tau. \tag{12.87}$$

It is imperative to note that if the zeroth component $C_0(x, t)$ is defined, the remaining components, $n \geq 1$, can be completely determined in a way that each term is determined using the previous terms. Subsequently, the entire series solutions are determined. Finally, the solution $P(r, t)$ is approximated by the truncated series

$$C_N(x, t) = \sum_{n=0}^{N-1} C_n(x, t) \tag{12.88}$$

and

$$\lim_{N \to \infty} C_N(r, t) = C(x, t).$$

■ **Example**

Assume the average variability in space and time of the geological formation of a groundwater system is governed by the following equation:

$$u(x) = 0.5 \cos(\pi x) \text{ and } u(t) - 1. \tag{12.89}$$

Following the discussion previously given in Eq. (12.82), we obtain, the non-homogeneous part of (12.82):

$$\begin{aligned} K(x, t) &= (1 - 0.5 \cos(\pi x)) \frac{\partial^2 C_0}{\partial x^2} \\ &\quad + v (0.5 \cos(\pi x) - 1) \frac{\partial C_0}{\partial x} \end{aligned} \tag{12.90}$$

$$+ \left(D \frac{\partial^2 0.5 \cos(\pi x)}{\partial x^2} - v \frac{\partial 0.5 \cos(\pi x)}{\partial x} \right) C_0$$

$$+ \frac{\partial 0.5 \cos(\pi x)}{\partial x} C_0,$$

here

$$C_0(x, t) = \frac{c_0 \exp(-\alpha t)}{2} \left[\begin{array}{c} \exp \frac{x(q_r - u_r)}{2 D_R} \\ \times Erfc \frac{x - u_r t}{2\sqrt{D_R t}} + \exp \frac{x(q_r + u_r)}{2 D_R} \\ \times Erfc \frac{x + u_r t}{2\sqrt{D_R t}} \end{array} \right] \tag{12.91}$$

$$u_r = \sqrt{q_r^2 + 4 D_r (\lambda - \alpha)}, \quad D_r = \frac{D}{R} \quad \text{and} \quad q_r = \frac{v}{R}$$

$$Erfc(x) = \frac{2}{\pi} \int_x^\infty \exp\left(-x^2\right) dx$$

is known as complementary error function and α is a positive constant. Here the trajectory of the chemical concentration can be traced. However, the time of retardation cannot be determined. Following the discussion previously given we obtained the following recursive formula:

$$C_{n+1}(x, t) = C_n(x, t) - \int_0^t \left[\begin{array}{c} D \frac{\partial^2 C_n}{\partial x^2} - v \frac{\partial C_n}{\partial x} \\ -\lambda R C_n - R \frac{\partial^m C}{\partial \tau^m} - K(x, t) \end{array} \right] d\tau. \tag{12.92}$$

For simplicity, we choose the first component to be zero such that the second component can be determined as

$$C_{11}(x, t) = \int_0^t K(x, t) d\tau. \tag{12.93}$$

In this case, two components of the decomposition series were obtained of which $P(x, t)$ was evaluated to have the following expansion:

$$C_1(x, t) = C_{11}(x, t) + C_{10}(x, t) + \cdots. \tag{12.94}$$

∎

12.8 A Model of the Groundwater Flowing Within a Leaky Aquifer Using the Concept of Local Variable Order Derivative

For the first time, Pythagoras realized that mathematics tools can be used to describe the pattern of real physical problems. Later on, Diophantus of Alexandria realized that these natural patterns can be described by mean of mathematical equations [143]. The notion has been used intensively in the circle of mathematics; however the idea of motion was not yet introduced by that time. In the year 18th, Sir Isaac Newton and Gottfried W. Leibniz independently introduced the concept of motion leading to the concept of derivative. Since then, this concept has been used in almost all the branches of sciences to model real-world problems [144,146, 145]. It is perhaps important to note that the big challenge in this process is to include into mathematical formula all the details surrounding the physical problem under observation. It happens to appear that the Newtonian concept of derivative cannot satisfy all the complexity of the natural occurrences. For instance, how do we explain accurately the movement of water within the leaky aquifer? In an attempt to answer this question, Hantush has proposed an equation based on the model proposed by Theis. Although this model has being used by many hydrogeologists, it is worth noting that the model does not take into account all the details surrounding the movement of water through a leaky geological formation. A first attempt to enhance model was to introduce the concept of derivative with fractional order [144]. This model has improved the description of this physical problem in a certain extent. Nonetheless, to be accurate, when dealing with complex systems, even the concept of fractional order derivative has some limitations, for instance it is not possible to accurately model the trap of water under matrix rocks. We shall mention that a mathematical model will be considered accurate if and only if the numerical representation of the mathematical solution is in good agreement with the observed facts. If not, there are two questions that need to be answered: the first one is to know if the experimental data were accurately measured. The second one will be to know if the mathematical equation is accurately implemented. If the second question appears to be negative, then the model needs to be revised. In the case of leaky aquifer model with non-integer and integer order derivatives has failed to do the job. The aim of our paper is to revise this model by introducing the concept of variable order derivative, which so far appears to be the best concept for complex systems.

12.8.1 Groundwater Flow Equation Using the Local Variable Order Derivative

The initial proposed groundwater equation within the leaky aquifer that was proposed by Hantush is given by:

$$\frac{\partial^2 S(r,t)}{\partial r^2} + \frac{1}{r}\frac{\partial S(r,t)}{\partial r} - \frac{S(r,t)}{B^2} = \frac{S}{T}\frac{\partial S(r,t)}{\partial t}. \tag{12.95}$$

The above equation then was modified by Atangana [144] as follows

$$\frac{\partial^2 S(r,t)}{\partial r^2} + \frac{1}{r}\frac{\partial S(r,t)}{\partial r} - \frac{S(r,t)}{B^2} = \frac{S}{T}\frac{\partial^\alpha S(r,t)}{\partial t^\alpha} \tag{12.96}$$

$$\frac{\partial^\alpha S(r,t)}{\partial t^\alpha} = \frac{1}{\Gamma(1-\alpha)} \int_0^t (t-x)^{1-\alpha} \frac{\partial S(r,x)}{\partial r} dx, 0 < \alpha \leq 1.$$

As we said before, the above model was also unable to describe accurately the complexity of the geological formation. Therefore in order to further include into mathematical formula the complexity of the aquifer through which the flow takes place, we shall propose the following version. Here the function $l(r; t)$ accounts for the complexity associated with the leaky aquifer, $S(r; t)$ is the change of level of water, S is the storativity, T is the transmissivity, and B is the factor that accounts for the leakage. We shall show some useful properties of the local variable order derivative.

$$\frac{\partial^2 S(r,t)}{\partial r^2} + \frac{1}{r}\frac{\partial S(r,t)}{\partial r} - \frac{S(r,t)}{B^2} = \frac{S}{T} {}_0^A D_\alpha^t S(r,t) \tag{12.97}$$

$$_0^A D_\alpha^t S(r,t) = \lim_{\epsilon \to 0} \frac{S(r, t + \epsilon(t + \frac{1}{\Gamma(1-l(r,t))})^{1-l(r,t)}) - S(r,t)}{\epsilon}$$

Here the function $l(r; t)$ accounts for the complexity associated with the leaky aquifer, $S(r; t)$ is the change of level of water, S is the storativity, T is the transmissivity, and B is the factor that accounts for the leakage. We shall show some useful properties of the local variable order derivative.

Theorem 12.8.1. *Assume that $f(x,y)$ is a function with existing $\partial_x^{l(x)}[\partial_y^{o(y)}(f(x,y))]$ and $\partial_y^{o(y)}[\partial_x^{l(x)}(f(x,y))]$ which is continuous over the domain $D \subset \mathbb{R}_2$. Then [145]:*

$$\partial_x^{l(x)}[\partial_y^{o(y)}(f(x,y))] = \partial_y^{o(y)}[\partial_x^{l(x)}(f(x,y))]. \tag{12.98}$$

Theorem 12.8.2. *Assume that a given function, say $f : [a, \infty) \to \mathbb{R}$, is $o(x)$-differentiable at a given point, say $x_0 \geq a$, then f is also continuous at x_0.*

Proof. Assume that f is $o(x)$-differentiable, then

$$_0^A D_x^{o(x)} f(x_0) = \lim_{\epsilon \to 0} \frac{f(x_0 + \epsilon(x_0 + \frac{1}{\Gamma(1-o(x))})^{1-o(x)}) - f(x_0)}{\epsilon}. \tag{12.99}$$

□

Theorem 12.8.3. *Assume that f is $o(x)$-differentiable on an open interval $(a; b)$, then*

1. *If ${}_0^A D_x^{o(x)}(f(x)) < 0$ for all $x \in (a,b)$ then f is decreasing there,*
2. *If ${}_0^A D_x^{o(x)}(f(x)) > 0$ for all $x \in (a,b)$ then f is increasing there,*
3. *If ${}_0^A D_x^{o(x)}(f(x)) = 0$ for all $x \in (a,b)$ then f is f is constant there.*

Definition 12.8.4. *Let f be a given continuous function, then we propose that the anti-variable derivative of f is*

$$\,_a^A I_x^{o(x,t)}(f(x)) = \int_a^x \left(t + \frac{1}{\Gamma(1-o(x,t))}\right)^{o(x,t)-1} f(t)dt. \tag{12.100}$$

The above operator is the inverse operator of the proposed fractional derivative. We shall underpin this statement by the following theorem.

Theorem 12.8.5. *Fundamental theorem of local variable calculus:* $\,_0^A D_x^{o(x)}[\,_O^A I_x^{o(,t)}(f(x))] = f(x)$ *for all $x \ge a$ with f a given continuous and differentiable function.*

Proof. Let f be a continuous function, then by definition, if we let $\,_O^A I_x^{o(,t)}(f(x)) = F(x)$, we have

$$
\begin{aligned}
\,_0^A D_x^{o(x)}[\,_O^A I_x^{o(,t)}(f(x))] &= \lim_{\epsilon \to 0} \frac{F\left(x + \epsilon(x + \frac{1}{\Gamma(1-o(x))})^{1-o(x)}\right) - F(x)}{\epsilon} \\
&= \left(x + \frac{1}{\Gamma(1-o(x))}\right)^{1-o(x)} \frac{dF(x)}{dx} \\
&= \left(x + \frac{1}{\Gamma(1-o(x))}\right)^{1-o(x)} \frac{d}{dx}\int_a^x \left(t + \frac{1}{\Gamma(1-o(t))}\right)^{1-o(t)} f(t)dt \\
&= \left(x + \frac{1}{\Gamma(1-o(x))}\right)^{1-o(x)} \left(x + \frac{1}{\Gamma(1-o(x))}\right)^{1-o(x)} f(x) \\
&= f(x).
\end{aligned}
$$

This completes the proof. □

12.8.2 Construction of a Possible Special Solution

The aim of this section is to construct a possible solution of the novel groundwater flow within a leaky aquifer. To construct a solution to the new equation, we employ the $o(x; t)$-Laplace operator defined in the following.

Definition 12.8.6. *Let g be a function defined in $(0, \infty)$, then we define the $o(x; t)$-Laplace transform of f as*

$$L_{o(x)}(f(x))(s) = \int_0^\infty \left(t + \frac{1}{\Gamma(1-o(x,t))}\right)^{o(x,t)-1} e^{-st} f(t)dt. \tag{12.101}$$

We shall give some properties of the above operator. The above operator satisfies the following properties, $F(s)$ being the Laplace transform of $f(t)$

$$L_{o(x,t)}\left({}_0^A D_x^{o(x,t)}(\frac{df(x)}{dx})\right)(s) = s^2 F(s) - sf(0) - f(0).$$

The proposed operator satisfies the following properties:

1. Linearity

$$L_{o(x)}\left(af(x) + bg(x)\right)(s) = aL_{(o(x))}(f(x))(s) + bL_{o(x)}(g(x)))(s),$$

2. Time delay

$$L_{o(x)}\left({}_0^A D_x^{o(x,t)}\{f(x-a).\delta(x-a)\}\right)(s) = se^{-sa} F(s),$$

3. First derivative

$$L_{o(x)}\left({}_0^A D_x^{o(x,t)}(\frac{df}{dx})\right)(s) = s^2 F(s) - sf(0) - f(0),$$

4. N order derivative

$$L_{o(x)}\left({}_0^A D_x^{o(x,t)}(\frac{d^n f}{dx^n})\right)(s) = s^{n+1} F(s) - \sum_{j=0}^{n} s^j f^{(n-1)}(0),$$

5. Fractional derivative Caputo type

$$L_{o(x)}\left({}_0^A D_x^{o(x)}(\frac{d^\alpha f}{dx^\alpha})\right)(s) = s^{\alpha+1} F(s) - \sum_{j=0}^{n} s^{\alpha-k} f^{(n-1)}(0), n-1 < \alpha \le n,$$

6. Integral

$$L_{o(x)}\left({}_0^A D_x^{o(x)}(\int_0^x f(t)dt)\right)(s) = F(s),$$

7. Convolution

$$L_{o(x)}\left({}_0^A D_x^{o(x)}(f * g(x))\right)(s) = sF(s)G(s),$$

8. Multiplication by distance

$$L_{o(x)}\left({}_0^A D_x^{o(x)}\{xf(x)\}\right)(s) = -sF'(s),$$

9. Complex shift

$$L_{o(x)}\left({}_0^A D_x^{o(x)}\{e^{-ax} f(x)\}\right)(s) = sF(s+a) - f(0),$$

10. Distance scaling

$$L_{o(x)}\left({}_0^A D_x^{o(x)}\{f(ax)\}\right)(s) = \frac{s}{a}F(\frac{s}{a}) - f(0).$$

Proof. Proof of property 1: By definition, we have the following formula

$$L_{o(x)}\left(af(x) + bg(x)\right)(s) = \int_0^\infty \left(t + \frac{1}{\Gamma(1 - o(x, t))}\right)^{o(x,t)-1} e^{-st}(af(t) + bg(t))dt.$$

Using the linearity of the integral, we obtain the following results

$$aL_{o(x)}\left(f(x)\right)(s) + bL_{o(x)}\left(g(x)\right)(s).$$

This completes the proof of property 1.

Proof of property 2: By definition, we have the following formula

$$L_{o(x)}\left({}_0^A D_x^{o(x,t)}\{f(x - a).\delta(x - a)\}\right)(s)$$

$$= \int_0^\infty \left(t + \frac{1}{\Gamma(1 - o(x, t))}\right)^{1-o(x,t)}\left(t + \frac{1}{\Gamma(1 - o(x, t))}\right)^{o(x,t)-1} e^{-st}(f(t - a).\delta(t - a))dt$$

$$= \int_0^\infty e^{-st}\left(f(t - a).\delta(t - a)\right)dt.$$

Using the properties of Laplace transform operator [145], we obtain the requested result

$$L_{o(x)}\left({}_0^A D_x^{o(x,t)}\{f(x - a).\delta(x - a)\}\right)(s) = se^{-sa}F(s).$$

Proof of property 3: By definition, we have the following

$$L_{o(x)}\left({}_0^A D_x^{o(x,t)}\left(\frac{df(t)}{dt}\right)\right)(s)$$

$$= \int_0^\infty \left(t + \frac{1}{\Gamma(1 - o(x, t))}\right)^{1-o(x,t)}\left(t + \frac{1}{\Gamma(1 - o(x, t))}\right)^{o(x,t)-1} e^{-st}(\frac{df(t)}{dt})dt$$

$$= \int_0^\infty e^{-st}\left(\frac{df(t)}{dt}\right)dt.$$

Using the property of Laplace transform for the first derivative, we obtain the requested results [145]

$$L_{o(x)}\left({}^A_0 D^{o(x,t)}_x\left(\frac{df}{dx}\right)\right)(s) = s^2 F(s) - sf(0) - f(0).$$

The proofs of properties 4 and 5 are similar to the one above.

Proof of property 6: by definition we have

$$L_{o(x)}\left({}^A_0 D^{o(x)}_x\left(\int_0^x f(t)dt\right)\right)(s) = L_{o(x)}\left(\left(t + \frac{1}{\Gamma(1 - o(x,t))}\right)^{1-o(x,t)}\left(\int_0^t f(v)dv\right)'\right).$$

Thanks to the fundamental theorem of calculus, the right-hand side can be transformed to

$$L_{o(x)}\left(\left(t + \frac{1}{\Gamma(1 - o(x,t))}\right)^{1-o(x,t)}\left(\int_0^t f(v)dv\right)'\right)$$

$$= L_{o(x)}\left(\left(t + \frac{1}{\Gamma(1 - o(x,t))}\right)^{1-o(x,t)} f(t)\right)$$

$$= \int_0^\infty e^{-st}\left(f(t)\right)dt.$$

This completes the proof of 6.

Proof of property 7:

$$L_{o(x)}\left({}^A_0 D^{o(x)}_x(f * g(x))\right)(s) = L_{o(x)}\left(\left(t + \frac{1}{\Gamma(1 - o(x,t))}\right)^{1-o(x,t)}(f * g(x))'\right)$$

$$= \int_0^\infty \frac{d}{dt} f * g(t) e^{-st} dt$$

$$= sF(s)G(s).$$

This completes the proof of 7. Note that properties 8, 9 and 10 are obvious.   □

Therefore, applying the above operator on both sides of Eq. (12.97), we obtain the following equation

$$L_{o(x,t)}\left\{\frac{\partial^2 S(r,t)}{\partial r^2} + \frac{1}{r}\frac{\partial S(r,t)}{\partial r} - \frac{S(r,t)}{B^2}\right\}(u) = \frac{S}{T}(u^2 S(r,u) - uS(r,0) - S(r,0)). \quad (12.102)$$

The above equation can be rearranged as follows

$$S(r,u) = \frac{T}{Su^2}\left\{L_{o(x,t)}\left\{\frac{\partial^2 S(r,t)}{\partial r^2} + \frac{1}{r}\frac{\partial S(r,t)}{\partial r} - \frac{S(r,t)}{B^2}\right\}(u)\right\} + \left\{\frac{1}{u} + \frac{1}{u^2}\right\}S(r,0).$$

$$(12.103)$$

We next apply the inverse Laplace transform operator on both sides of the above equation to obtain

$$S(r,t) = L^{-1}\left\{\frac{T}{Su^2}\left\{L_{o(x,t)}\left\{\frac{\partial^2 S(r,t)}{\partial r^2} + \frac{1}{r}\frac{\partial S(r,t)}{\partial r} - \frac{S(r,t)}{B^2}\right\}(u)\right\}\right\} + L^{-1}\left\{\left\{\frac{1}{u} + \frac{1}{u^2}\right\}S(r,0)\right\}.$$

(12.104)

For simplicity, we put

$$g(r,t) = (L)^{-1}\left\{\left\{\frac{1}{u} + \frac{1}{u^2}\right\}S(r,0)\right\}.$$

Then, from Eq. (12.104) one can construct a recursive formula that will be used to generate the special solution of Eq. (12.97). The recursive formula associated with Eq. (12.104) is

$$S_{n+1}(r,t) = L^{-1}\left\{\frac{T}{Su^2}\left\{L_{o(x,t)}\left\{\frac{\partial^2 S_n(r,t)}{\partial r^2} + \frac{1}{r}\frac{\partial S_n(r,t)}{\partial r} - \frac{S_n(r,t)}{B^2}\right\}(u)\right\}\right\} \text{ for } n \geq 1.$$

(12.105)

$$S_0(r,t) = g(r,t).$$

Our next step is to prove the stability of the used iteration method.

12.8.3 Uniqueness of the Solution

Let assume by contradiction that there exist two different special solutions, $S_{sp1}(r,t)$ and $S_{sp2}(r;t)$. Let

$$G(s) = {}_0^A D_\alpha^t\big(S(r,t)\big) = \frac{\partial^2 S(r,t)}{\partial r^2} + \frac{1}{r}\frac{\partial S(r,t)}{\partial r} - \frac{S(r,t)}{B^2}.$$

(12.106)

We will use the inner product to show that product

$$\|S_{sp1} - S_{sp2}\| \ll \epsilon.$$

(12.107)

To achieve this, we evaluate $(G(S_{sp1}) - G(S_{sp2}), \omega)$ for $\omega \in H = \{u, v/\int uv < \infty\}$. However,

$$G(S_{sp1}) - G(S_{sp2}) = \frac{\partial^2}{\partial r^2}\{S_{exp2}(r,t) - S_{exp1}(r,t)\}$$

(12.108)

$$+ \frac{1}{r}\frac{\partial}{\partial r}\{S_{exp2}(r,t) - S_{exp1}(r,t)\}$$

(12.109)

$$+ \frac{1}{B^2}\{S_{exp1}(r,t) - S_{exp2}(r,t)\}.$$

(12.110)

Thus

$$\left(G(S_{sp1}) - G(S_{sp2}), \omega\right) = \left(\frac{\partial^2}{\partial r^2}\{S_{exp2}(r, t) - S_{exp1}(r, t)\}, \omega\right) \tag{12.111}$$

$$+ \left(\frac{1}{r}\frac{\partial}{\partial r}\{S_{exp2}(r, t) - S_{exp1}(r, t)\}, \omega\right) \tag{12.112}$$

$$+ \left(\frac{1}{B^2}\{S_{exp1}(r, t) - S_{exp2}(r, t)\}, \omega\right). \tag{12.113}$$

Let us now evaluate the first component

$$\left(\frac{\partial^2}{\partial r^2}\{S_{exp2}(r, t) - S_{exp1}(r, t)\}, \omega\right). \tag{12.114}$$

Because actual levels are indeed bounded, we can find a constant M such that $(S_{sp1}, S_{sp2}) < M^2$. If we apply Cauchy–Schwartz inequality, we arrive at the following

$$\left(\frac{\partial^2}{\partial r^2}\{S_{exp2}(r, t) - S_{exp1}(r, t)\}, \omega\right) \le \left\|\frac{\partial^2}{\partial r^2}\{S_{exp2}(r, t) - S_{exp1}(r, t)\}\right\|\|\omega\|. \tag{12.115}$$

It is possible to find positive constant ω_1, ω_2 such that

$$\left\|\frac{\partial^2}{\partial r^2}\{S_{exp2}(r, t) - S_{exp1}(r, t)\}, \omega\right\| \le \omega_1\omega_2\left\|\{S_{exp2}(r, t) - S_{exp1}(r, t)\}\right\|, \tag{12.116}$$

$$\left(\frac{\partial^2}{\partial r^2}\{S_{exp2}(r, t) - S_{exp1}(r, t)\}, \omega\right) \le \omega_1\omega_2\left\|\{S_{exp2}(r, t) - S_{exp1}(r, t)\}\right\|\|\omega\|. \tag{12.117}$$

To proceed we still make use of Schwartz inequality to evaluate

$$\left(\frac{1}{r}\frac{\partial}{\partial r}\{S_{exp2}(r, t) - S_{exp1}(r, t)\}, \omega\right).$$

The following is obtained

$$\left(\frac{1}{r}\frac{\partial}{\partial r}\{S_{exp2}(r, t) - S_{exp1}(r, t)\}, \omega\right) \le \left\|\frac{1}{r}\frac{\partial}{\partial r}\{S_{exp2}(r, t) - S_{exp1}(r, t)\}\right\|\|\omega\|. \tag{12.118}$$

There exists a positive constant O_1 such that

$$\left(\frac{1}{r}\frac{\partial}{\partial r}\{S_{exp2}(r, t) - S_{exp1}(r, t)\}, \omega\right) \le \frac{O_1}{r_1}\left\|S_{exp2}(r, t) - S_{exp1}(r, t)\right\|\|\omega\|,$$

$$\left(\frac{\partial^2}{\partial r^2}\{S_{exp2}(r,t) - S_{exp1}(r,t)\}, \omega\right) + \left(\frac{1}{r}\frac{\partial}{\partial r}\{S_{exp2}(r,t) - S_{exp1}(r,t)\}, \omega\right)$$

$$+ \left(\frac{1}{B^2}\{S_{exp1}(r,t) - S_{exp2}(r,t)\}, \omega\right) \le \left(\frac{O_1}{r_1} + \omega_1\omega_2 + \frac{1}{B^2}\right)\left\|\{S_{exp2}(r,t) - S_{exp1}(r,t)\right\|\|\omega\|.$$

Since the exact solution $S(r,t)$ converges to S_{sp1}, S_{sp2} we can find two large numbers N, M such that

$$\|S - S_{sp1}\| \ll \frac{\epsilon}{2\left(\frac{O_1}{r_1} + \omega_1\omega_2 + \frac{1}{B^2}\right)\|\omega\|} \quad \text{for } N,$$

and

$$\|S - S_{sp2}\| \ll \frac{\epsilon}{2\left(\frac{O_1}{r_1} + \omega_1\omega_2 + \frac{1}{B^2}\right)\|\omega\|} \quad \text{for } M.$$

Considering now $m = \max(N, M)$, and noticing that

$$\|S_{sp1} - S_{sp2}\| \le \|S - S_{sp2}\| + \|S - S_{sp1}\|,$$

we have then

$$\|S_{sp1} - S_{sp2}\| \ll \frac{\epsilon}{\left(\frac{O_1}{r_1} + \omega_1\omega_2 + \frac{1}{B^2}\right)\|\omega\|}. \tag{12.119}$$

Substituting the above we then obtain

$$\left(G(S_{sp1}) - G(S_{sp2}), \omega\right) \ll \epsilon. \tag{12.120}$$

Since ϵ can be made arbitrarily small, we ultimately have that

$$\|(S_{sp1} - S_{sp2},)\| = 0 \Longrightarrow S_{sp1} = S_{sp2}.$$

12.8.4 Stability Analysis of the Method

Using the inner product with the operator G defined earlier in Eq. (12.108), we establish the stability of the method. The aim is to prove that we can find a positive number k such that

$$\left(G(S) - G(S_1), S - S_1\right) \le L\|S - S_1\|^2.$$

Proof. From the definition in Section 12.8.4 we have:

$$\big(G(S) - G(S_1), S - S_1\big) = \Big(\frac{\partial^2}{\partial r^2}\{S(r,t) - S_1(r,t)\}, S - S_1\Big)$$

$$+ \Big(\frac{1}{r}\frac{\partial}{\partial r}\{S(r,t) - S_1(r,t)\}, S - S_1\Big)$$

$$+ \Big(\frac{1}{B^2}\{S(r,t) - S_1(r,t)\}, S - S_1\Big).$$

Looking at the first term of that addition we have

$$\Big(\frac{\partial^2}{\partial r^2}\{S(r,t) - S_1(r,t)\}, S(r,t) - S_1(r,t)\Big) \le \Big\| \frac{\partial^2}{\partial r^2}\{S(r,t) - S_1(r,t)\} \Big\| \|S(r,t) - S_1(r,t)\|.$$

$$(12.121)$$

There exist positive constants f_1, f_2 such that

$$\Big\| \frac{\partial^2}{\partial r^2}\{S(r,t) - S_1(r,t)\} \Big\| \le f_1 f_2 \|S(r,t) - S_1(r,t)\|,$$

$$\Big(\frac{\partial^2}{\partial r^2}\{S(r,t) - S_1(r,t)\}, S - S_1\Big) \le f_1 f_2 \|S(r,t) - S_1(r,t)\|^2.$$

Looking at the second term of that addition we have

$$\Big(\frac{1}{r}\frac{\partial}{\partial r}\{S(r,t) - S_1(r,t)\}, S(r,t) - S_1(r,t)\Big) \le \Big\| \frac{1}{r}\frac{\partial}{\partial r}\{S(r,t) - S_1(r,t)\} \Big\| \|S(r,t) - S_1(r,t)\|.$$

$$(12.122)$$

Similarly there exists a positive constant g_1 such that

$$\Big(\frac{1}{r}\frac{\partial}{\partial r}\{S(r,t) - S_1(r,t)\}, S(r,t) - S_1(r,t)\Big) \le \frac{g_1}{r_1} \|S(r,t) - S_1(r,t)\|^2,$$

$$\Big(\frac{\partial^2}{\partial r^2}\{S(r,t) - S_1(r,t)\}, S(r,t) - S_1(r,t)\Big) + \Big(\frac{1}{r}\frac{\partial}{\partial r}\{S(r,t) - S_1(r,t)\}, S(r,t) - S_1(r,t)\Big)$$

$$+ \Big(\frac{1}{B^2}\{S(r,t) - S_1(r,t)\}, S(r,t) - S_1(r,t)\Big) \le \Big(\frac{g_1}{r_1} + f_1 f_2 + \frac{1}{B^2}\Big)\|S(r,t) - S_1(r,t)\|^2.$$

Thus

$$\big(G(S) - G(S_1), S - S_1\big) \le \Big(\frac{g_1}{r_1} + f_1 f_2 + \frac{1}{B^2}\Big)\|S(r,t) - S_1(r,t)\|^2.$$

By letting $L = \frac{g_1}{r_1} + f_1 f_2 + \frac{1}{B^2}$, we have

$$\big(G(S) - G(S_1), S - S_1\big) \le L\|S(r,t) - S_1(r,t)\|^2.$$

We now only have to prove

$$\big(G(S) - G(S_1), S - S_1\big) \leq H \|S(r,t) - S_1(r,t)\|^2 \|S - S_1\| \|W\|,$$

$$\big(G(S) - G(S_1), W\big) = \left(\frac{\partial^2}{\partial r^2}\{S(r,t) - S_1(r,t)\}, W\right) + \left(\frac{1}{r}\frac{\partial}{\partial r}\{S(r,t) - S_1(r,t)\}, W\right)$$
$$+ \left(\frac{1}{B^2}\{S(r,t) - S_1(r,t)\}, W\right), \quad (12.123)$$

$$\left(\frac{\partial^2}{\partial r^2}\{S(r,t) - S_1(r,t)\}, W\right) \leq \left\|\frac{\partial^2}{\partial r^2}\{S(r,t) - S_1(r,t)\}\right\| \|W\|.$$

We can find positive constants m_1, m_2 such that

$$\left\|\frac{\partial^2}{\partial r^2}\{S_{exp2}(r,t) - S_{exp1}(r,t)\}\right\| \leq m_1 m_2 \|S(r,t) - S_1(r,t)\|,$$

$$\left(\frac{\partial^2}{\partial r^2}\{S_{exp2}(r,t) - S_{exp1}(r,t)\}, W\right) \leq m_1 m_2 \|S(r,t) - S_1(r,t)\| \|W\|.$$

Similarly we have

$$\left(\frac{1}{r}\frac{\partial}{\partial r}\{S(r,t) - S_1(r,t)\}, W\right) \leq \left\|\frac{1}{r}\frac{\partial}{\partial r}\{S(r,t) - S_1(r,t)\}\right\| \|W\|. \quad (12.124)$$

We can also find N_1 such that

$$\left(\frac{1}{r}\frac{\partial}{\partial r}\{S(r,t) - S_1(r,t)\}, W\right) \leq \frac{N_1}{r_1} \|S(r,t) - S_1(r,t)\| \|W\|.$$

Assembling all the equations together we get

$$\left(\frac{\partial^2}{\partial r^2}\{S(r,t) - S_1(r,t)\}, W\right) + \left(\frac{1}{r}\frac{\partial}{\partial r}\{S(r,t) - S_1(r,t)\}, W\right) + \left(\frac{1}{B^2}\{S(r,t) - S_1(r,t)\}, W\right)$$
$$\leq \left(\frac{N_1}{r_1} + m_1 m_2 + \frac{1}{B^2}\right) \|S(r,t) - S_1(r,t)\| \|W\|.$$

We complete the proof by taking

$$H = \frac{N_1}{r_1} + m_1 m_2 + \frac{1}{B^2}$$

and writing

$$(G(S) - G(S_1), S - S_1) \leq H \|S - S_1\| \|W\|. \quad (12.125)$$

$\square$

CHAPTER 13

Groundwater Recharge Model With Fractional Differentiation

As indicated in literature, without groundwater recharge, there is no groundwater resource. We shall inform readers that are not aware of this process that groundwater recharge also called deep drainage or deep percolation is a dynamical process of a hydrology during which water moves downward from surface to subsurface. We must note that the recharge is the main method through which water from surface enters an aquifer. It is known that the process takes place in a specific place called vadose zone, which is found below plant roots and also it is usually expressed in terms of a flux to the water table surface. It is worth noting that the deep drainage process which recharges groundwater occurs naturally by rain and melted snow for instance. Lately, it was found out that the deep drainage process can be implemented by humans, this includes paving, development and also/or logging. Sustainable management of the resource is therefore a call and requirement for groundwater recharge estimation on, for instance, a catchment, regional and/or global scale. In the literature nowadays one can find several groundwater recharge methods, also in this work, we will mention the following: unsaturated zone methods, saturated water balance methods, tracer methods, and groundwater modeling methods [139,133]. However, as all the mathematical models used to describe real-world problems, the literature regarding these methods exists and also indicates that these methods all have limitations and also have associated degree of uncertainty, in particular in dependence on parameters which vary significantly in space causing recharge estimates to only be site-specific. However, it is important to also note that the groundwater recharge should still be estimated on a catchment to regional scale, and so researchers often move in the direction of using some techniques, for instance, the remote sensing and geographic information system techniques. Although some good results are obtained via these techniques, nonetheless, for the practical and accuracy purposes, one can still be in a position to argue that using digital spatial data may be associated with a high degree of uncertainty [139,133]. Therefore, there is a need for a new approach to groundwater recharge estimation. As mentioned, groundwater recharge models cannot be reliable on a regional scale because of the significance of spatial variability in the factors influencing recharge [134,132]. This gives raise to the concept of heterogeneity which in the context of a groundwater system is the variability of the geological formation. They are known as aquifer properties as result of an aquifer system's structure and composition. Furthermore, heterogeneity can be compli-

Fractional Operators with Constant and Variable Order with Application to Geo-Hydrology
DOI: 10.1016/B978-0-12-809670-3.00013-8

cated by the concept of viscoelasticity, which is a property of both viscosity and elasticity that enhances heterogeneity [138,137,136,135]. This property has been of interest to water-related fields such as engineering and when wanting to understand molecular transport in biological and environmental processes [134,132]. It is reasonable to say that by incorporating viscoelasticity in groundwater recharge models, there will be reduced uncertainty and increased reliability in groundwater recharge estimates. Several reasons justify this statement. Firstly, viscosity is associated with an effect on hydraulic properties, that is to say when viscosity increases, hydraulic conductivity K decreases [140–142]. Moreover, when geology materials' response to change in effective stress is time-dependent, the effect of viscoelasticity can be seen in specific storage estimates [134,132]. Finally, viscoelastic effects cause drawdown curves to vary slightly in the middle stages, where it gives an analogy to double porosity and an unconfined aquifer having a delayed yield. Essentially, the aforementioned signifies how aquifers properties are influenced by the effect of viscoelasticity; and since these properties are required in recharge models, this calls for a recharge model taking into account this property. This issue can be addressed in the context of mathematics by means of fractional differentiation including: firstly, the Caputo derivative is a fractional operator very often used in an issue within the real world, because it keeps memory of past values as the curve changes over time with the fractional order integration. In addition, it accounts for the effect of elasticity. Secondly, the Caputo–Fabrizio derivative has the ability to take into account heterogeneity on different scales. Finally, the Atangana–Baleanu derivative which is based on the Mittag–Leffler function is another derivative which appears to address even real-world problems with even greater complexity, as it accounts for both heterogeneity and viscoelasticity. In light of the phenomena the aforementioned derivatives respectively account for, one could apply these derivatives to a real-world problem such as groundwater recharge, which requires more accurate estimation.

13.1 Motivation for the New Development

Existing groundwater recharge estimation methods appear to mainly generate site-specific groundwater recharge estimates. In doing so, they fail to give reliable recharge estimates on a regional to global scale. This is due to failure of taking into account the concepts of heterogeneity and viscoelasticity. Accordingly, this study will direct at developing a new approach to groundwater recharge estimation by means of taking into account these concepts. The methodological approach entailed obtaining an exact solution to a selected groundwater recharge equation by applying the Laplace and inverse Laplace transforms. Upon applying an uncertainty analysis and statistical analysis of the parameters within the solution, it was found that storativity and drainage resistance both require accurate estimation when estimating recharge from the selected equation. Following this, the Caputo derivative, Caputo–Fabrizio derivative, and the Atangana–Baleanu derivative were applied and an exact solution was obtained for each derivative. Upon applying a numerical simulation for each of these solutions, the results depict the behavior of a particular real-world problem. The recharge within a heterogeneous and viscoelastic geological formation is well described with the concept of differentiation with the generalized Mittag–Leffler law or the Atangana–Baleanu fractional derivative. The recharge by means of elastic geological formation can be modeled by means of the Caputo and Caputo–Fabrizio fractional differentiation.

13.2 Analytical Solution Using the Green Function Approach

This section is presenting an analysis of groundwater recharge equation, the following groundwater recharge equation taken from the Earth Model considered, the results presented here are those obtained from the work by Atangana and Jessica [220]:

$$S\frac{dH}{dt} = R\frac{H}{DR} \tag{13.1}$$

with initial condition $H(0)$ which can be a constant or a fraction of space. To solve the above equation, we apply the Laplace transform on both sides of the equation to obtain [220]:

$$S\left[P\overline{H}(v) - H(0)\right] = \frac{R}{P} - \frac{\overline{h}(v)}{DR}. \tag{13.2}$$

Rearranging, we obtain:

$$\overline{H}(v) = \frac{R}{v\left(vS + \frac{1}{DR}\right)} + \frac{Sh(0)}{vS + \frac{1}{DR}} = \frac{R}{v\left(vS + \frac{1}{DR}\right)} + \frac{h(0)}{v + \frac{1}{DR}}. \tag{13.3}$$

The exact solution of the recharge equation is derived using the Laplace transform operation. To accommodate for the reader who is not familiar with this operation, we present first the definition of the Laplace transform.

$$L\left(f(t)\right)(v) = \int_0^\infty e^{-vt} f(t)dt \tag{13.4}$$

We also present some properties of the Laplace transformation

$$L\left(af(t) + bh(t)\right)(v) = aL\left(f(t)\right)(v) + bL\left(h(t)\right)(v), \tag{13.5}$$

$$L\left(\frac{dh}{dt}\right)(s) = L\left(f(t)\right)(v) - f(0). \tag{13.6}$$

Now applying the inverse Laplace transformation, we obtain:

$$\left[DR - \exp\left[-\frac{1}{DRS}t \right] DR \right] R + H(0) \exp\left[-\frac{1}{DR}t \right]. \tag{13.7}$$

Therefore, the exact solution of the groundwater recharge equation using the Laplace transformation is:

$$H(t) = R.DR + \exp\left[-\frac{1}{DRS}t \right](H(0) - R.DR). \tag{13.8}$$

For an understanding of the extent of the effect of uncertainty of both S and DR on R in the considered equation it is necessary to understand the parameters S and DR. S is a dimensionless property known as the volume of water an aquifer system will store or release from storage per unit surface area per unit change in hydraulic head [220]. As indicated, the mathematical equations used to obtain this parameter differ for unconfined and confined aquifer systems, whereby for unconfined aquifer systems it is specific yield (Sy) and for confined aquifer systems it is storativity (S). This is due to the physical mechanism controlling releases and storage of water. Nonetheless, in relation to groundwater recharge, S/Sy controls the amount of water eventually researching the water table. DR is a lumped, site-specific parameter, and is given by the considered equation [220]:

$$DR = \frac{L^2}{\beta T} \tag{13.9}$$

where DR is drainage resistance (days), L is length of the flow path (m), $\beta = 2$ for radial flow or $\beta = 4$ for parallel flow, and T is transmissivity ($m2/d$). T is given as the rate at which water is transmitted through a unit width of an aquifer system. Therefore it is given by

$$T = KB \tag{13.10}$$

where B is aquifer thickness (m). The study done by Atangana and Jessica [220] indicates that, in light of the mentioned above, it is fundamental to make correct assumptions regarding the aquifer type and the geological nature of the system of interest, when estimating groundwater recharge. This is due to T, K, and S differing for different geology and aquifer types. Using the exact solution given in the recharge equation and applying a selected range of values for both S and DR, the corresponding hydraulic head change over time is given. Some numerical simulations were presented in their studies but will not be mentioned here, however the interested reader can found them in [220]. The former depicts hydraulic head in their work [220], a change over time for S ranging from 0.0021 to 0.35, with a constant DR of 100 days and recharge of 40%. On the other hand, the latter depicts hydraulic head change over time for DR ranging from 10 to 500 days, with a constant S of 0.02 and recharge of 40%, see numerical simulation therein [220].

Water level fluctuation signifies groundwater recharge or groundwater discharge. It was also indicated that in the scenarios presented above, water level fluctuations are depicted as a basis of variation in S and DR with a constant recharge of 40%. The results depict that minor changes in both S and DR have a significant effect on the change in hydraulic head. To expand, with an increasing S and constant DR, the hydraulic head undergoes a more gradual increase over time. In contrast, with increasing DR and constant S, the hydraulic head has a more rapid and peaked rise. This is noteworthy when DR changes from a value of 200 days to 500 days [220]. Furthermore, when recharge occurs, slow water migration into the subsurface causes a rapid rise into the capillary fringe, allowing the soil pores to become saturated. Moreover, once fully saturated, the water table reaches ground surface and there is an occurrence of surface runoff. This entire process occurs provided there is no exploitation of the aquifer system. This concept can be related to the change in hydraulic head over time reaching a constant, regardless of the change in S or change in DR. To expand, at about 120 days to 130 days, regardless of the value of S, the hydraulic head reaches a point where no change occurs. This is assumed due to the aquifer being recharged by the hypothetical 40% while no water loss abstraction and/or evapotranspiration occurs [220]. In other words, it could mean that the aquifer has reached its capacity to store water. To add, the value in days where all DR values yield a constant hydraulic head is significantly lower (about 30 days) for the selected DR ranges. In light of the aforementioned, it becomes imperative to understand the error associated with these parameters because it is these errors which indicate the effect of a value of either S or DR, or both, on the hydraulic head which is further related to groundwater recharge. The following section presents a statistical analysis of the results presented above [220].

13.3 Analysis of Uncertainties Within the Scope of Statistics

In this section, we devote the work to the study underpinning the uncertainties on parameters used in the mathematical equation as indicated in [220]. The aim of this study is to avoid all the errors obtained from field measurements [220]. It is important to note that data collected from field observations may differ from one time to another, therefore to get rid of uncertainties, we make use of statistical formula to evaluate uncertainties associated to the collected data. *Harmonic mean:* The mean is a typical value found within a data set over time series, and can often be seen as the operating point of a physical system generating the series of data [220]. Furthermore, the harmonic mean is a kind of mean used when the numbers are defined

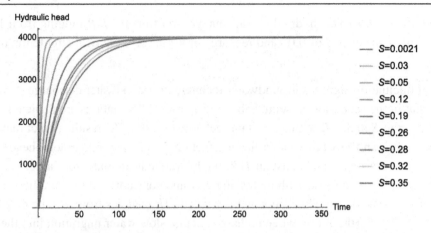

Figure 13.1: Selected range of storativity S and their associated hydraulic head.

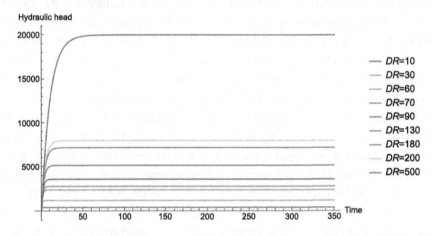

Figure 13.2: Selected range of drainage coefficient DR and their associated hydraulic head.

in relation to some unit; when a sample contains extreme values; and when more stability is needed regarding outliers. It is given as

$$H_x = \frac{n}{\sum\limits_{i=1}^{n} \frac{1}{x_i}} \tag{13.11}$$

where H_x is harmonic mean, n is a number of data points, x_i is the ith sample point. Figs. 13.1 and 13.2 show the harmonic mean of the hydraulic head as a function of time for the coefficient drainage Dr and the storativity.

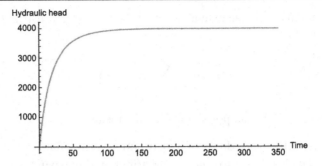

Figure 13.3: Harmonic mean for storativity distribution.

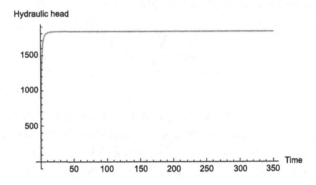

Figure 13.4: Harmonic mean for drainage coefficient.

Standard deviation SD is a measure of dispersion round the mean value, for a generated time series of data. In other words, it gives insight to uncertainty and error based on how concentrated or scattered the samples are from the mean value. SD is given by the square root of the variance as follows:

$$S = \sqrt{\frac{\sum_{i=1}^{n}(X_i - \overline{X})^2}{n - 1}} \tag{13.12}$$

where s is the standard deviation, X_i is a data point, and $\overline{X}$ is the mean X (see Figs. 13.3 and 13.4). *Skewness* is higher-order statistical attribute of a time series. It is essentially used for a measure of symmetry of the probability density function of the amplitude of a time series and the assessment of the departure from normality of the data. To expand, when the number of large and small amplitude values is equal within a time series of data, there is a value of zero skewness. Subsequently, a skewed distribution positive or negative has a mean and median that are not identical. Furthermore, skewness can be quantified using skewness mathematical formula to define the extent to which the time series distribution differs from a normal distribution [220]:

$$G = \frac{1}{ns^3} \sum_{i=1}^{n} (X_i - \overline{X})^3 \tag{13.13}$$

where g is the skewness, Y_i a data point, and $\overline{Y}$ is the mean.

Kurtosis is a measure of the peakedness of a distribution, or in other words how 'heavy-tailed' or 'light-tailed' the data is relative to a normal distribution. To expand, when a data set has a high kurtosis, it is associated with heavy tails, or outliers. Alternatively, when the measure of kurtosis is low, it is associated with a lack of outliers.

$$g = \frac{1}{ns^4} \sum_{i=1}^{n} (Y_i - \overline{Y})^4 \tag{13.14}$$

where g is the kurtosis.

The harmonic mean indicates that the typical hydraulic head values occurring within the generated time series data as a result of change in S and change DR are in the ranges of 0–4000 m and 0–1650 m, respectively. These are given in Figs. 3 and 4, respectively; this can be found in the work done by Atangana and Jessica (see [220]). The SD appears to decrease for the S distribution and increase for the DR distribution over time as depicted in Figs. 5 and 10, respectively; this can be found in the work done by Atangana and Jessica (see [220]). This essentially means that DR has a greater deviation from the mean, and therefore a greater extent of uncertainty and error. Moreover, the SD for DR becomes significantly high after approximately 30 days, which means that the vulnerability to error is greater after 30 days. Furthermore, the graphs generated for skewness (Figs. 7 and 8; this can be found in the work done by Atangana and Jessica [220]) both indicate the data are not normal for both S and DR distributions. Moreover, when skewness is positive, it yields larger error, and so a greater uncertainty in the data. This is significant for the DR distribution, and more so after about 30 days. This supports what was previously said regarding DR having a greater extent of uncertainty and error. As mentioned, a higher kurtosis is associated with a presence of outliers. As a result, the distribution generated for S (Fig. 9; this can be found in the work done by Atangana and Jessica [220]) indicates that a shorter time period in hydraulic head changes would have less uncertainty than a longer period of time. This is because as time increases for this parameter, the value of kurtosis increases (presence of outliers increases). This means that as time increases, the error associated with S increases as well. On the other hand, the value of kurtosis is high already at a much earlier time for the distribution given for DR (Fig. 10; this can be found in the work done by Atangana and Jessica [220]). Ultimately, this means that both S and DR yield data that is not normal, but DR is the parameter associated with greater error and uncertainty. This statistical analysis infers that although both S and DR have associated error and so also uncertainty, it is DR that yields greater amount of error. This makes sense when looking at the initial plots (Figs. 1 and 2; this can be found in the work done by Atangana and Jessica [220]) because DR variation depicted an even greater change in hydraulic head over time, in comparison to S variation. With that being said, it is imperative to correctly distinguish between aquifer systems when using literature as a means for a value of S, if no fieldwork can be done. Even in the case where field work or laboratory work can be done to estimate S, critical consideration should be given to the reliability of the estimates as they have a tendency to be disputed. On the other hand, the aforementioned also suggests that incorrect DR estimates would yield inadequate changes in hydraulic head; and this would ultimately yield unreliable recharge estimates. More importantly, this means that sufficient consideration should be given to estimates of transmissivity, flow type, and flow path length, as all this information is needed to estimate DR. These considerations should be done with much thought for a groundwater system associated with significant heterogeneity, because this phenomenon causes one aquifer system's hydraulic properties to differ significantly from an adjacent aquifer system, and/or even within the same aquifer system. These considerations should be made because an inaccurate S or DR estimate would yield significant error in the eventual recharge estimate (this can be found in the work done by Atangana and Jessica [220]).

13.4 Groundwater Recharge Model With Power Law

In this study, the problem is addressed developing a new groundwater recharge estimation model which accounts for the effect of heterogeneity within a geological formation (this can be found in the work done by Atangana and Jessica [220]):

$$S_0^C D_t^\alpha h(t) = R - \frac{h(t)}{DR} \tag{13.15}$$

where

$$S_0^C D_t^\alpha h(t) = \frac{1}{\Gamma(1-\alpha)} \int_0^t (t-\tau)^{-\alpha} \frac{dh}{d\tau}(t)d\tau. \tag{13.16}$$

The above equation has a memory effect as well as the capability of describing the change in hydraulic head globally. In addition, the fractional order introduced here accounts for heterogeneity of a geological formation. The following solution to Eq. (13.16) is obtained by applying the Laplace transform on both sides:

$$L\left(S_0^C D_t^\alpha H(t)\right)(v) = L(R)(v) - L\left(\frac{H(t)}{DR}\right) \tag{13.17}$$

$$S\left[P^\alpha \tilde{H}(v) - v^{\alpha-1}H(0)\right] = \frac{R}{P} - \frac{\tilde{H}(P)}{DR}$$

$$\tilde{H}(v)\left[SP^\alpha + \frac{1}{DR}\right] = Sv^{\alpha-1}h(0) + \frac{R}{P}$$

$$\tilde{H}(v) = \frac{Sv^{\alpha-1}h(0) + \frac{R}{P}}{SP^\alpha + \frac{1}{DR}}$$

$$\tilde{H}(v) = \frac{Sv^{\alpha-1}h(0)}{Sv^\alpha + \frac{1}{DR}} + \frac{R}{v\left[Sv^\alpha + \frac{1}{DR}\right]}.$$

Now applying the inverse Laplace transformation, we obtain:

$$\tilde{H}(t) = L^{-1}\left(\frac{Sv^{\alpha-1}h(0)}{Sv^\alpha + \frac{1}{DR}}\right) + L^{-1}\left(\frac{R}{v\left[Sv^\alpha + \frac{1}{DR}\right]}\right) \tag{13.18}$$

$$= L^{-1}\left(\frac{v^{\alpha-1}h(0)}{v^\alpha + \frac{1}{DR\alpha}}\right) + L^{-1}\left(\frac{R}{P\left[Sv^\alpha + \frac{1}{DR}\right]}\right)$$

$$= H(0)E\left(-\frac{1}{DR\alpha}t^\alpha\right) + L^{-1}\left(\frac{R}{P\left[Sv^\alpha + \frac{1}{DR}\right]}\right).$$

Let

$$L^{-1}\left(\frac{R}{v}\right) = R, \; L^{-1}\left(\frac{1}{v^\alpha + \frac{1}{DRS}}\right) = E_\alpha\left(-\frac{1}{DRS}t\right). \tag{13.19}$$

Using the convolution theorem the following is obtained:

$$L^{-1}\left(\frac{R}{v\left[Sv^\alpha + \frac{1}{DR}\right]}\right) = \frac{R}{S}\int_0^t E_\alpha\left[-\frac{1}{DRS}(t-\tau)\right]d\tau. \tag{13.20}$$

Therefore the exact solution of the new groundwater equation is the following:

$$H(t) = H(0)E_\alpha\left[-\frac{1}{DR\alpha}t^\alpha\right] + \frac{R}{S}\int_0^t E_\alpha\left[-\frac{1}{DRS}(t-\tau)\right]d\tau. \tag{13.21}$$

Let $y = t - \tau$,

$$H(t) = H(0)E_\alpha\left[-\frac{t^\alpha}{DR\alpha}\right] + \frac{R}{S}\int_0^t E_\alpha\left[-\frac{1}{DRS}y\right]d\tau. \tag{13.22}$$

In this part, the numerical approximation of the used differential operation for the differentiation, namely, the Caputo derivative with fractional order, is firstly given. Let n be a natural number greater than 1, then:

$$\substack{C \\ 0} D_t^\alpha f(t_n) = \frac{1}{\Gamma(1-\alpha)} \int_0^{t_n} (t_n - \tau)^{-\alpha} \frac{d}{d\tau} f(\tau) d\tau. \tag{13.23}$$

Applying the Crank–Nicolson Solution, we obtain:

$$
\begin{aligned}
\substack{C \\ 0} D_t^\alpha f(t_n) &= \frac{1}{\Gamma(1-\alpha)} \int_0^{t_n} (t_n - \tau)^{-\alpha} \frac{f(\Delta\tau + \tau) - f(\tau)}{\Delta\tau} d\tau \tag{13.24} \\[2mm]
&= \frac{1}{\Gamma(1-\alpha)} \sum_{k=0}^{n-1} \int_{t_k}^{t_{k+1}} \frac{f(t_{k+1}) - f(t_k)}{\Delta t}(t_n - \tau)^{-\alpha} d\tau \\[2mm]
&= \frac{1}{\Gamma(1-\alpha)} \sum_{k=0}^{n-1} \frac{f(t_{k+1}) - f(t_k)}{\Delta t} \int_0^{t_n} (t_n - \tau)^{-\alpha} d\tau \\[2mm]
&= \frac{1}{\Gamma(1-\alpha)} \sum_{k=0}^{n-1} \frac{f(t_{k+1}) - f(t_k)}{\Delta t} \int_{t_n - t_k}^{t_n - t_{k+1}} -Y^{-\alpha} dy \\[2mm]
&= \frac{1}{\Gamma(1-\alpha)} \sum_{k=0}^{n-1} \frac{f(t_{k+1}) - f(t_k)}{\Delta t} - \frac{Y^{-\alpha+1}}{1-\alpha} \int_{t_n - t_k}^{t_n - t_{k+1}} \\[2mm]
&= \frac{1}{\Gamma(1-\alpha)} \sum_{k=0}^{n-1} \frac{f(t_{k+1}) - f(t_k)}{\Delta t} \left\{ (t_n - t_k)^{1-\alpha} - (t_n - t_{k+1})^{1-\alpha} \right\} \\[2mm]
&= \frac{1}{\Gamma(1-\alpha)} \sum_{k=0}^{n-1} \frac{f(t_{k+1}) - f(t_k)}{\Delta t} (\Delta t)^{1-\alpha} \left\{ (n-k)^{1-\alpha} - (n-k-1)^{1-\alpha} \right\}.
\end{aligned}
$$

Replacing (13.24) in (13.15), we obtain the following:

$$S\left[\frac{1}{\Gamma(1-\alpha)}\right] \sum_{k=0}^{n} \frac{H(t_{k+1}) - H(t_k)}{\Delta t} (\Delta t)^{1-\alpha} \left\{ (n-k)^{1-\alpha} - (n-k-1)^{1-\alpha} \right\} \tag{13.25}$$

$$= R - \frac{H(t_{n+1}) + H(t_n)}{2}.$$

Let

$$a = \frac{S}{\Gamma(1-\alpha)} (\Delta t)^{-\alpha}, \quad \Phi_\alpha^{n,k} = (n-k)^{1-\alpha} - (n-k-1)^{1-\alpha} \tag{13.26}$$

$$\sum_{k=0}^{n} \{ H(t_{k+1}) - H(t_k) \} a \Phi_\alpha^{n,k}$$

$$= R - \frac{H(t_{n+1}) + H(t_n)}{2}$$

$$\{H(t_{n+1}) - H(t_n)\} a\Phi_\alpha^{n,n} + \sum_{k=0}^{n-1} \{H(t_{k+1}) - H(t_k)\} a\Phi_\alpha^{n-1,k} \qquad (13.27)$$

$$= \quad R - \frac{H(t_{n+1}) + H(t_n)}{2}. \qquad (13.28)$$

Rearrange and the following is obtained:

$$\left(a\Phi_\alpha^{n,n} + \frac{1}{2}\right) H(t_{n+1}) \quad = \quad \left\{a\Phi_\alpha^{n,n} + \frac{1}{2}\right\} H(t_n) + \sum_{k=0}^{n-1} (H(t_{k+1}) - H(t_k)) a\Phi_\alpha^{n-1,k} + R.$$

The above recursive formula can be used to generate the numerical simulation of the model for different values of the fractional order.

13.5 Groundwater Recharge Model With Caputo–Fabrizio Fractional Differentiation

In this section, the analysis of the groundwater recharge model is presented using the law of exponential decay. This model could be used for those geological formations for which the recharge process follows the law of exponential decay. In this section, the modified recharge equation is given as (this can be found in the work done by Atangana and Jessica [220]):

$$S_0^{CF} D_t^\alpha H(t) = R - \frac{H(t)}{DR} \qquad (13.29)$$

where $_0^{CF} D_t^\alpha$ is the well-known Caputo–Fabrizio operator which is defined as:

$$_0^{CF} D_t^\alpha H(t) = \frac{M(\alpha)}{1-\alpha} \int_0^t \frac{d}{d\tau} H(\tau) \exp\left[-\frac{\alpha}{1-\alpha}(t-\tau)\right] d\tau. \qquad (13.30)$$

Based on the literature review, this operator was introduced due to the necessity of employing the behavior of classical viscoelastic material; and in the case of this study, the viscoelastic material represents the geological formation through which recharge occurs. The exact solution of the modified model will be obtained using the Laplace transform. Thus by applying the Laplace transform on both sides, we obtain the following:

$$SL\left(_0^{CF} D_t^\alpha H(t)\right) = L\left(R - \frac{H(t)}{DR}\right). \qquad (13.31)$$

Nevertheless, the Laplace transform of the Caputo–Fabrizio derivative is given as:

$$L\left(_0^{CF} D_t^\alpha H(t)\right) = \frac{v\tilde{H}(v) - H(0)}{v + \alpha(1-v)}. \qquad (13.32)$$

Replacing this in (13.31), we obtain the following:

$$S\frac{v\widetilde{H}(v) - H(0)}{v + \alpha(1 - v)} = \frac{R}{v} - \frac{\widetilde{H}(v)}{DR} \tag{13.33}$$

$$\widetilde{H}(v)\left[\frac{Sv}{v + \alpha(1 - v)} - \frac{1}{DR}\right] = \frac{SH(0)}{v + \alpha(1 - v)} + \frac{R}{v}$$

$$\widetilde{H}(v) = \frac{\frac{SH(0)}{v+\alpha(1-v)} + \frac{R}{v}}{\frac{Sv}{v+\alpha(1-v)} - \frac{1}{DR}}$$

$$= \frac{\frac{SH(0) + vR + \alpha(1-v)R}{v(v+\alpha(1-v))}}{\frac{Sv - v - \alpha(1-v)}{DR(v+\alpha(1-v))}}$$

$$\widetilde{H}(v) = \frac{[vSH(0) + vR + \alpha(1 - v)R] DR}{v [Sv - v - \alpha(1 - v)]}$$

$$= \frac{v[SH(0) + R - \alpha R] DR + \alpha RDR}{v [v (s - 1 + \alpha) - \alpha]}$$

$$= \frac{[SH(0) + R - \alpha R] DR}{v - \frac{\alpha}{s - 1 + \alpha}} \frac{1}{S - 1 + \alpha}$$

$$+ \frac{\alpha RDR}{v [v (s - 1 + \alpha) - \alpha]}.$$

The exact solution is obtained by applying the inverse Laplace on both sides and also using the convolution theorem:

$$H(t) = \frac{[SH(0) + R - \alpha R] DR}{s - 1 + \alpha} \exp\left[\frac{\alpha}{s - 1 + \alpha}t\right] + \frac{\alpha RDR}{s - 1 + \alpha} \int_0^t \exp\left[\frac{\alpha (t - \tau)}{s - 1 + \alpha}\right] d\tau. \tag{13.34}$$

It is important to note that when $\alpha = 1$, the solution of the classical model is obtained. To accommodate for the researchers working in numerical analysis, the finite approximation of the Caputo–Fabrizio derivative is presented: If $n \geq 1$, then

$$_{0}^{CF} D_t^\alpha f(t_n) = \frac{M(\alpha)}{1 - \alpha} \int_0^{t_n} \frac{d}{d\tau} f(\tau) \exp\left[-\frac{\alpha}{1 - \alpha}(t_n - \tau)\right] d\tau \tag{13.35}$$

$$= \frac{M(\alpha)}{(1 - \alpha)} \sum_{k=0}^{n} \int_{t_k}^{t_{k+1}} \frac{f(t_{k+1}) + f(t_k)}{\Delta t} \exp\left[-\frac{\alpha}{1 - \alpha}(t_n - \tau)\right] d\tau \tag{13.36}$$

$$= \frac{M(\alpha)}{(1 - \alpha)} \sum_{k=0}^{n} \frac{f(t_{k+1}) + f(t_k)}{\Delta t} \exp\left[-\frac{\alpha}{1 - \alpha}(t_n - \tau)\right] d\tau \tag{13.37}$$

$$= \frac{M(\alpha)}{(1-\alpha)} \sum_{k=0}^{n} \frac{f(t_{k+1}) + f(t_k)}{\Delta t} \int_{t_n-t_{k+1}}^{t_n-t_k} \exp\left[-\frac{\alpha}{1-\alpha} y\right] dy \tag{13.38}$$

$$= \frac{M(\alpha)}{(1-\alpha)} \sum_{k=0}^{n} \frac{f(t_{k+1}) + f(t_k)}{\Delta t} \frac{1-\alpha}{\alpha} \exp\left[-\frac{\alpha}{1-\alpha} y\right]_{t_n-t_{k+1}}^{t_n-t_k} \tag{13.39}$$

$$= \frac{M(\alpha)}{(1-\alpha)} \sum_{k=0}^{n} \frac{f(t_{k+1}) + f(t_k)}{\Delta t} \left\{ \exp\left[-\frac{\alpha}{1-\alpha}(t_n - t_k)\right] - \exp\left[-\frac{\alpha}{1-\alpha}(t_n - t_{k+1})\right] \right\} \tag{13.40}$$

$$= \frac{M(\alpha)}{(1-\alpha)} \sum_{k=0}^{n} \frac{f(t_{k+1}) + f(t_k)}{\Delta t} \Phi_{n,k}^{\alpha} \tag{13.41}$$

where [19]:

$$\Phi_{n,k}^{\alpha} = \exp\left[-\frac{\alpha}{1-\alpha}(t_n - t_k)\right] - \exp\left[-\frac{\alpha}{1-\alpha}(t_n - t_{k+1})\right]. \tag{13.42}$$

Replacing the first approximation in the original equation, we obtain the following:

$$S\left[\frac{M(\alpha)}{\alpha} \sum_{k=0}^{n} \frac{h(t_{k+1}) + h(t_k)}{\Delta t} \Phi_{n,k}^{\alpha}\right] = R - \frac{h(t_{n+1}) + h(t_n)}{2}. \tag{13.43}$$

To obtain the recursive formula, we reformulate the above equation as follows: Let $a_1 = \frac{M(\alpha)}{\alpha\Delta t}$, then

$$H(t_{n+1})\left[a_1\Phi_{n,n}^{\alpha} + \frac{1}{2}\right] = H(t_n)\left[a_1\Phi_{n,k}^{\alpha} - \frac{1}{2}\right] + a_1 \sum_{k=0}^{n-1} \{h(t_{k+1}) + h(t_k)\} \Phi_{n-1,k}^{\alpha} + R. \tag{13.44}$$

The above is the numerical solution of the recharge equation with the exponential decay law.

13.6 Modeling Groundwater Recharge With Atangana–Baleanu Fractional Differentiation

It is important to note that the geological formations by means of which recharge takes place are sometimes very complex and can be a combination of both heterogeneity and viscoelasticity (this can be found in the work done by Atangana and Jessica [220]). In this situation neither the power law nor the exponential decay law can be used to portray the change in hydraulic head for a given percentage of recharge, thus a more suitable operator for differentiation must be used. Recently and for the purpose of extending the limitation to the power law and exponential decay law, a new operation of differentiation was introduced and used in many fields of science and engineering to model a few real world problems with great success. The new operation is given as (this can be found in the work done by Atangana and Jessica [220]):

$$
{}^{ABC}_{0}D^{\alpha}_{t}f(t) = \frac{AB(\alpha)}{1-\alpha} \int_0^t \frac{d}{d\tau} f(\tau) E_\alpha \left[-\frac{\alpha}{1-\alpha}(t-\tau)^\alpha \right] d\tau.
\tag{13.45}
$$

Thus replacing the time classical derivative by the new operation of differentiation, we obtain the following modified model:

$$
S_0^{ABC} D^{\alpha}_t H(t) = R - \frac{H(t)}{DR}.
\tag{13.46}
$$

This model takes into account the recharge of groundwater in a medium with different layers, heterogeneity, and viscoelasticity. Next, the exact solution is derived, applying once again the Laplace transform:

$$
S \left[\frac{AB(\alpha)}{1-\alpha} \frac{v^\alpha \tilde{H}(v) - v^{\alpha-1}H(0)}{v^\alpha + \frac{\alpha}{1-\alpha}} \right] = \frac{R}{v} - \frac{\tilde{H}(v)}{DR}
\tag{13.47}
$$

$$
\tilde{H}(v) \left[\frac{SAB(\alpha)}{1-\alpha} \frac{s^\alpha}{s^\alpha + \frac{\alpha}{1-\alpha}} + \frac{1}{DR} \right] = \frac{SAB(\alpha)}{1-\alpha} \frac{s^{\alpha-1}h(0)}{s^\alpha + \frac{\alpha}{1-\alpha}} + \frac{R}{s}
$$

$$
\begin{aligned}
\tilde{H}(P) &= \frac{\Phi(\alpha)\frac{v^{\alpha-1}}{v^\alpha+\Phi(\alpha)} + \frac{R}{v}}{\Phi(\alpha)\frac{v^\alpha}{v^\alpha+\Phi(\alpha)} + \frac{1}{DR}} \\[2mm]
&= \frac{\dfrac{\Phi(\alpha)v^{\alpha-1} + v^\alpha R + \Phi(\alpha)}{v\left[v^\alpha + \Phi(\alpha)\right]}}{\dfrac{DR\left(v^\alpha \Phi(\alpha)\right)}{DR\left(v^\alpha \Phi(\alpha)\right) + v^\alpha + \Phi(\alpha)}} \\[2mm]
&= \frac{\left[\Phi(\alpha)v^{\alpha-1} + v^\alpha R + \Phi(\alpha)\right]DR}{v\left[DR\left(v^\alpha \Phi(\alpha)\right) + v^\alpha + \Phi(\alpha)\right]} \\[2mm]
&= \frac{v^{\alpha-1}DR\left(R + \Phi(\alpha)\right)}{v^\alpha + \frac{\Phi(\alpha)}{\Phi(\alpha)DR+1}} \cdot \frac{1}{\Phi(\alpha)DR+1} \\[2mm]
&\quad + \frac{\Phi(\alpha)R}{v\left[v^\alpha + \frac{\Phi(\alpha)}{\Phi(\alpha)DR+1}\right]} \cdot \frac{1}{\Phi(\alpha)DR+1}.
\end{aligned}
$$

Thus applying the inverse Laplace transform and the convolution theorem for Laplace transform, we obtain:

$$H(t) = \frac{DR(R + \Phi(\alpha))}{\Phi(\alpha)DR + 1} E_\alpha \left[-\frac{\Phi(\alpha)}{\Phi(\alpha)DR + 1} t^\alpha \right] \tag{13.48}$$

$$+ \frac{\Phi(\alpha)R}{\Phi(\alpha)DR + 1} \int_0^t E_\alpha \left[-\frac{\Phi(\alpha)}{\Phi(\alpha)DR + 1} (t - \tau) \right] d\tau.$$

Here

$$\Phi_1(\alpha) = \frac{AB(\alpha)h(0)}{1 - \alpha}, \quad \Phi(\alpha) = \frac{\alpha}{1 - \alpha}. \tag{13.49}$$

We now present the numerical solution of the model with the generalized Mittag–Leffler Law. To do this, we first present the first approximation of the Atangana–Baleanu derivative with fractional order. Let n be a positive natural number, we sub-devise the time period as follows

$$t_0 \leq t_1 \leq t_2 \leq \ldots \leq t_{n-1} \leq t_n, \quad \Delta t = t_k - t_{k-1}. \tag{13.50}$$

Thus,

$$^{ABC}_0 D_t^\alpha h(t_n) = \frac{AB(\alpha)}{1 - \alpha} \int_0^{t_n} \frac{d}{d\tau} h(\tau) E_\alpha \left[-\frac{\alpha}{1 - \alpha}(t_n - \tau)^\alpha \right] d\tau \tag{13.51}$$

$$= \frac{AB(\alpha)}{1 - \alpha} \sum_{k=0}^n \int_{t_k}^{t_{k+1}} \frac{h(t_{k+1}) - h(t_k)}{\Delta t} E_\alpha \left[-\frac{\alpha}{1 - \alpha}(t_n - \tau)^\alpha \right] d\tau$$

$$\frac{AB(\alpha)}{1 - \alpha} \sum_{k=0}^n \frac{h(t_{k+1}) - h(t_k)}{\Delta t} \delta_{n,k}^\alpha$$

where

$$\delta_{n,k}^\alpha = \int_{t_k}^{t_{k+1}} E_\alpha \left[-\frac{\alpha}{1 - \alpha}(t_n - \tau)^\alpha \right] d\tau \tag{13.52}$$

$$= (t_n - t_{k+1}) E_{\alpha,2} \left[\frac{\alpha}{1 - \alpha}(t_n - t_{k+1}) \right] - (t_n - t_k) E_{\alpha,2} \left[\frac{\alpha}{1 - \alpha}(t_n - t_k) \right].$$

Therefore replacing this in the main equation, we obtain:

$$S \left[\frac{AB(\alpha)}{1 - \alpha} \sum_{k=0}^n \frac{h(t_{k+1}) - h(t_k)}{\Delta t} \delta_{n,k}^\alpha \right] = R - \frac{h(t_{n+1}) - h(t_n)}{DR}. \tag{13.53}$$

We used the numerical solution presented in previous section to show the graphical representation of the groundwater recharge model for different values of the fractional differentiation. The recursive formula used in these simulations is presented by Eqs. (13.52), (13.42) and (13.29). The parameters used in the simulations are $DR = 200$, $R = 40\%$ and $S = 0.002$.

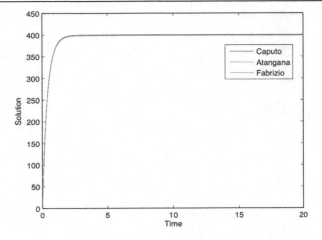

Figure 13.5: Numerical simulation of hydraulic head obtained from the model with Caputo, Caputo–Fabrizio and Atangana–Baleanu fractional derivatives for $\alpha = 1$.

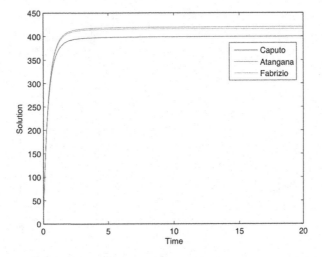

Figure 13.6: Numerical simulation of hydraulic head obtained from the model with Caputo, Caputo–Fabrizio and Atangana–Baleanu fractional derivatives for $\alpha = 0.95$.

The numerical simulations are depicted in Figs. 13.5, 13.6, 13.7, 13.8, 13.9. In numerical simulation for each of these solutions, the results depict the behavior of a particular real world problem. The recharge within a heterogeneous and viscoelastic geological formation is well described with the concept of differentiation with the generalized Mittag–Leffler law or the Atangana–Baleanu fractional derivative. The recharge by mean of elastic geological formation can be modeled by the mean of Caputo and Caputo–Fabrizio (this can be found in the work done by Atangana and Jessica [220]).

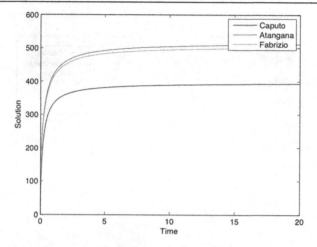

Figure 13.7: Numerical simulation of hydraulic head obtained from the model with Caputo, Caputo–Fabrizio and Atangana–Baleanu fractional derivatives for $\alpha = 0.85$.

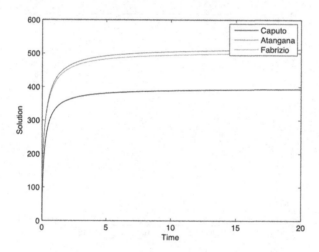

Figure 13.8: Numerical simulation of hydraulic head obtained from the model with Caputo, Caputo–Fabrizio and Atangana–Baleanu fractional derivatives for $\alpha = 0.6$.

The uncertainty analysis revealed that minor changes in both storativity (S) and drainage resistance (DR) yield a significant change in hydraulic head over time. This means that if no accurate measurements are taken during fieldwork, the groundwater recharge estimate, which is dependent on accuracy of hydraulic head, will not be reliable. Furthermore, the statistical analysis revealed that both S and DR within the selected recharge equation yield uncertainty in their distribution of hydraulic head change over time. The uncertainty increased significantly for both parameters after a period of time. This was concluded using the harmonic

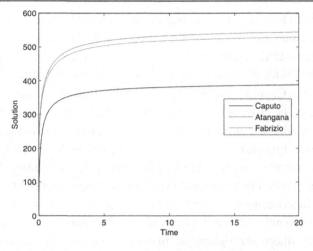

Figure 13.9: Numerical simulation of hydraulic head obtained from the model with Caputo, Caputo–Fabrizio and Atangana–Baleanu fractional derivatives for $\alpha = 0.4$.

mean, standard deviation, kurtosis, and skewness. Although both parameters yield uncertainty, drainage resistance appeared to have a greater extent of uncertainty and error among the two. This means that when using this equation for recharge, considerable thought should be put into the accuracy of these parameters while measuring them from the field observation. Three models were developed to suit each kind of geological formations. The model based on the generalized Mittag–Leffler law (Atangana–Baleanu fractional derivative) is more suitable for all classes of geological formation including the homogeneous, heterogeneous, and viscoelastic soils. The model based on the power law (Caputo fractional derivative) is suitable for elastic, homogeneous soils, and finally the model based on the exponential decay law (Caputo–Fabrizio) is suitable for heterogeneous soils.

13.7 The Model With Eton Approach

Fractional differentiation has an adequate use for investigating real-world scenarios related to geological formations associated with elasticity, heterogeneity, viscoelasticity, and the memory effect. Since groundwater systems exist in these geological formations, modeling groundwater recharge as a real world scenario is a challenging task to carry because existing recharge estimation methods are governed by linear equations which make use of constant field parameters. This is inadequate because in reality these parameters are a function of both space and time. This study therefore concentrates on modifying the recharge equation governing the EARTH model, by application of the Eton approach. Accordingly, this paper presents a modified equation which is non-linear, and accounts for parameters in a way that it is a function of both space and time. To be more specific, herein, recharge and drainage resistance, which are parameters in the equation, became a function of both space and time. Additionally, the study entailed solving the non-linear equation using an iterative method as well as numerical solutions by means of the Crank–Nicolson scheme. The numerical solutions were used alongside the Riemann–Liouville, Caputo–Fabrizio, and Atangana–Baleanu derivatives, so that account was taken for elasticity, heterogeneity, viscoelasticity, and the memory effect. In essence, this paper presents a more adequate model for recharge estimation. Fractional differentiation has been of interest to modeling real world scenarios associated with the "power law non-locality, power law long-term memory, or fractal properties" [163]. It is a fundamental tool for describing the phenomenon of memory [156], and therefore becomes essential to investigating the behavior of porous, heterogeneous, and viscoelastic materials [159–161]. This means that it is a field of research adequately accounting for real-world scenarios which are a function of both space and time. Furthermore, throughout the history of fractional differentiation, improvement has been made so that these real-world scenarios are modeled in the best way possible. To expand, the Riemann–Liouville derivative model scenarios are associated with elasticity and the memory effect. Although it has been said in many published research papers within the field of fractional calculus that using the Laplace transform on the Riemann–Liouville fractional derivative the obtained initial condition are unusual [147,158]. This argument is not supported mathematically because the Riemann–Liouville fractional derivative is obtained from the classical differentiation and also the Cauchy formulation of the multiple integrals. By applying the Laplace transform on the Riemann–Liouville fractional derivative, we obtain a formula with initial condition raised to the power also, which is acceptable and suitable because we are working within the framework of fractional differentiation. It is important to notice that the Caputo fractional derivative is just an adaptation derivative with no fractional integral and was introduced artificially not by following a clear mathematical derivation. Nevertheless, both fractional operators can be used for different purposes. Both these aforementioned derivatives are however associated with a singular function, meaning it does not fully describe the effect of memory [147–149,151,152]. Consequently, the Caputo–Fabrizio derivative was developed. It addresses heterogeneity on different scales and avoids singularity, meaning it enhances the memory effect [162,151,152]. To add, this derivative is based on the convolution of the first order derivative and the exponential function, as well as is associated with viscoelastic material, thermal media, and electromagnetic systems [151]. Application of Caputo–Fabrizio derivative was given in epidemiology [151] and improving the advection dispersion equation [152]. Later, the Atangana–Baleanu derivative was developed. It is based on the Mittag–Leffler function which is associated with non-singular and non-local functions, allowing the behavior of flow in lithological material viscoelastic effects to be addressed [147–149, 151]. Application of this derivative is given for a simple non-linear system [153] as well as modeling groundwater migration through an unconfined aquifer [149]. Furthermore, another study presents how this derivative was combined with a concept of variable order derivative to formulate a new and enhanced approach to account for the memory effect in real-world scenarios. Application was done with the Poisson equation [148], whereby the numerical analysis of this study revealed that the approach is good for modeling real world scenarios with memory. Furthermore, the groundwater recharge equation governing the EARTH model has been used in several previous studies [157,159–161,163,164]. This equation is however non-linear, makes use of constant parameters, and does not account for heterogeneity, viscoelasticity, nor the memory effect. With that being said, it is reasonable to infer that the particular equation is not suitable to investigating groundwater recharge in non-homogeneous and non-uniform geological formations, where recharge and recharge controlling factors become functions of both space and time [154–157,159–161,163,164]. Ultimately, this justifies a need for modifying the particular groundwater recharge equation, and so therefore the aim of this study is to modify the groundwater recharge equation governing the EARTH model using fraction order differentiation.

13.7.1 Application of the Atangana Derivative With Memory

In mathematics, a dynamic scheme is a tuple (D, h, B) with D as a manifold that could be either a locally Banach space or Euclidean space, B the domain for time which is a set non-negative real, and h is an evolution rule $t \to f^t$ the range being of course a diffeomorphism of a manifold to itself. **Definition of the derivative.** Let D be a dynamic system with domain B (time or space), let u be a positively defined function called the uncertainty function of D, within the domain B, then if $h \in D$, then the Atangana derivative with memory of a function h denoted by $U^u(f)$ is defined as:

$$U^{u(t)}(h(t)) = (1 - u(t))h(t) + u(t)h'(t). \tag{13.54}$$

If $u = 1$ the first derivative is recovered (the local derivative); if $u = 0$, the initial function is recovered. This follows the primary law of derivative.

13.7.2 Properties of the Atangana Derivative With Memory

Addition:

$$U^{u(t)}(ah(t) + bf(t)) = aU^{u(t)}(h(t)) + bU^{u(t)}(f(t)) \tag{13.55}$$

Multiplication:

$$U^{u(t)}(fg(t)) = f(t)U^{u(t)}(g(t)) + u(t)g(t)f'(t) \tag{13.56}$$

Using the definition of Atangana derivative with memory, the following is obtained:

$$U^{u(t)}(fg(t)) = (1 - u(t))f(t)g(t) + u(t)(f(t)g(t))' \tag{13.57}$$

$$U^{u(t)}(fg(t)) = (1-u(t))f(t)g(t) + u(t)[f(t)g'(t) + g(t)f'(t)]$$
$$U^{u(t)}(fg(t)) = f(t)U^{u(t)}(g(t)) + u(t)g(t)f'(t)$$

Division:

$$U^{u(t)}(\frac{f(t)}{g(t)}) = \left(\frac{g(t)U^{u(t)}(f(t)) - u(t)f(t)g'(t)}{g^2(t)}\right). \tag{13.58}$$

Lipschitz condition: Let $f(t)$ and $g(t)$ be two functions, then,

$$\left\|U^{u(t)}(g(t)) - U^{u(t)}(f(t))\right\| \tag{13.59}$$
$$\leq \left\|(1-u(t))g(t) + u(t)g'(t) - (1-u(t))f(t) - u(t)f'(t)\right\|$$
$$\leq |1-u(t)|\,\|g(t) - f(t)\| + |u(t)|\,\|g'(t) - f'(t)\|$$
$$\leq a\,\|g(t) - f(t)\| + b\,\|g'(t) - f'(t)\|$$
$$\leq a\,\|g(t) - f(t)\| + b\alpha\,\|g(t) - f(t)\|$$
$$\leq H\,\|g(t) - f(t)\|.$$

The above shows that the Atangana derivative with memory possesses the Lipschitz condition.

13.7.3 Application to the Selected Groundwater Recharge Equation

This section focuses on modifying the selected groundwater recharge equation governing the EARTH model, by incorporating the effect of heterogeneity, variability, and uncertainty in the equation. Considering a scenario where the groundwater system is of homogenous, uniform aquifer properties and rainfall recharge, the addition of water to the water table may be estimated using the selected equation. This equation is given by (13.60). As indicated, the use of this equation has been given in many studies.

$$S\frac{dh(t)}{dt} = R - \frac{h(t)}{DR} \tag{13.60}$$

where S is specific yield, $\frac{dh}{dt}$ is change in water level over a period of time, R is recharge, h is groundwater level, and DR is drainage resistance which is a site-specific parameter. The above equation does not consider the effect of heterogeneity, variability, and uncertainties in a groundwater system. To expand, the equation essentially describes a model using constant storativity and drainage resistance with change in hydraulic head to obtain a certain percentage of recharge. However, constant storativity and drainage resistance are only appropriate provided the geological system within recharge that takes place is homogeneous and uniform. Therefore, for inclusion of these properties into the recharge equation, the ordinary derivative in (13.57) is replaced by the uncertain function equation in (13.51). In doing so, the following is obtained:

$$(1 - u(t))h(t) + u(t)h'(t)S = R - \frac{h(t)}{DR} \tag{13.61}$$

$$u(t)h'(t) = \frac{R}{S} - \frac{h(t)}{DRS} - (1 - u(t))h(t) \tag{13.62}$$

$$u(t)h'(t) = \frac{R}{S} - \frac{h(t)}{DRS} - h(t) + u(t)h(t) \tag{13.63}$$

$$u(t)h'(t) = \frac{R}{S} - h(t)\left(\frac{1}{DRS} + 1 - u(t)\right) \tag{13.64}$$

$$h'(t) = \frac{R}{Su(t)} - \frac{\frac{1}{DRS} + 1 - u(t)}{u(t)}h(t) \tag{13.65}$$

$$h'(t) = R_1(t) - \frac{\frac{1}{s} + DR - DRu(t)}{DRu(t)}h(t) \tag{13.66}$$

$$h'(t) = R_1(t) - \frac{h(t)}{\frac{DRu(t)}{\frac{1}{s} + DR - DRu(t)}} \tag{13.67}$$

$$h'(t) = R_1(t) - \frac{h(t)}{DR_1(t)}. \tag{13.68}$$

The groundwater recharge equation has non-constant parameters. To expand, both the contribution of recharge and the drainage resistance are now functions of time. Physically, the function $R_1(t)$ is the recharge percentage at a specific time "t," and $DR_1(t)$ is the drainage resistance at the time "t." **Remark:** If the uncertain function $u(t) = 1$, we recover the classical equation with constant parameters. Essentially, the modified equation is non-linear, and this means it is more appropriate for modeling real world groundwater recharge problems. Furthermore, since the new equation has non-constant parameters, then the equation cannot be solved using the Laplace transform. Therefore, in this case we use the classical method for ordinary differential equations, to obtain:

$$h(t) = \frac{\int_0^t \exp\left[\int P(t)\right].g(t)dt + c}{\exp\left[\int P(t)\right]} \tag{13.69}$$

where $g(t) = R_1(t)$ and $P(t) = \frac{1}{\frac{DRu(t)}{\frac{1}{s} + DR - DRu(t)}}$. So the exact solution is given by:

$$h(t) = \frac{\int \exp\left[\int P(t)\right].R_1(t)dt + c}{\exp\left[\int \frac{\frac{1}{SDR} + 1 - u(t)}{DRu(t)}dt\right]}. \tag{13.70}$$

One can select the uncertain function to be the Error function. The uncertain function must be selected in a way that at $t = 0$, $u(0) = 0$ and at $t = T$, $u(T) = 1$. Nevertheless, if the uncertain function is complicated in a way that $DR_1(t)$ is not integrable, one can use an iterative or numerical method.

13.7.4 Iterative Method for the New Model

In this section we solve the non-linear groundwater recharge model using an iterative method. To achieve this, we apply the integral on (13.67) to obtain:

$$h(t) = \int_0^t \left(R_1(\tau) - \frac{h(\tau)}{DR_1(\tau)} \right) d\tau. \tag{13.71}$$

We consider the following iterative formula:

$$h_{n+1}(t) = \int_0^t \left(R_1(t) - \frac{h_n(t)}{DR_1(t)} \right) d\tau + h(0). \tag{13.72}$$

The above iterative formula can be used to generate the solution to the new model, provided the initial condition is known. Now we present the stability of the used method: Let Γ be a function defined as:

$$\Gamma(h(t)) = \int_0^t \left(R_1(\tau) - \frac{h(\tau)}{DR_1(\tau)} \right) d\tau. \tag{13.73}$$

Let h_1 and h_2 be two different functions, then:

$$
\begin{aligned}
\|\Gamma h_1 - \Gamma h_2\| &= \left\| \int_0^t R_1(\tau) - \frac{h_1(\tau) - h_2(\tau)}{DR_1(\tau)} d\tau \right\| \\
&\leq \int_0^t \frac{\|h_1(\tau) - h_2(\tau)\|}{\|DR_1(\tau)\|} d\tau \\
&\leq \frac{1}{M} \|h_1(t) - h_2(t)\| T \\
&\leq \frac{T}{M} \|h_1(t) - h_2(t)\|.
\end{aligned}
\tag{13.74}
$$

So Γ possesses the Lipschitz conditions. Now we define the following iterative formula by means of Γ:

$$\Gamma(h_n) = h_{n+1} \text{ then } n \in N$$

$$\|h_{n+1} - h_n\| = \|\Gamma(h_n) - \Gamma(h_{n-1})\| \leq \frac{T}{M} \|h_n - h_{n-1}\| \tag{13.75}$$

$$\leq \frac{T}{M} \|\Gamma(h_{n-1}) - \Gamma(h_{n-2})\| \left(\frac{T}{M}\right)^2 \|h_{n-1} - h_{n-2}\|$$

$$\leq \left(\frac{T}{M}\right)^n \|h_1 - h_0\|.$$

We select $M > T$ so that $\lim\limits_{n \to \infty} \left(\frac{T}{M}\right)^n \to 0$. Under the above conditions, we conclude that $(h_n)_{n \in N}$ is Cauchy sequence in Banach space, therefore it converges. Thus, one can find a function $\tilde{h}(t)$ in a way that $\lim\limits_{n \to n} h_n(t) = \tilde{h}(t)$. In this case

$$\lim_{n \to 0} \Gamma(h_n) = \lim_{n \to \infty} h_{n+1} = \tilde{h}. \tag{13.76}$$

Thus $\tilde{h}$ is a fixed point of Γ.

13.7.5 Numerical Solution Using Crank–Nicolson

In this section we present the numerical solution using the Crank–Nicolson scheme.

$$\frac{h^{j+1} - h^j}{\Delta t} = R_1^j - \frac{1}{DR_1^j} \left(\frac{h^{j+1} + h^j}{2}\right). \tag{13.77}$$

Rearranging, we obtain:

$$\left(\frac{1}{\Delta t} + \frac{1}{2DR_1^j}\right) h^{j+1} = \left(\frac{1}{\Delta t} - \frac{1}{2DR_1^j}\right) h^j + R_1^j. \tag{13.78}$$

The above numerical solution can be used to generate the simulation for different functions $u(t)$. The new non-linear model can further be generalized by replacing the local differential operator with a non-linear differential operator such as $_0^{ABR}D_t^\alpha$, $_0^{CFR}D_t^\alpha$, or $_0^{RL}D_t^\alpha$. By replacing $\frac{d}{dt}$ by $_0^{RL}D_t^\alpha$, we obtain:

$$_0^{RL}D_t^\alpha h(t) = R_1(t) - \frac{h(t)}{DR_1(t)}. \tag{13.79}$$

The above model has the ability to account for elasticity of a geological formation as well as the memory effect. By replacing $\frac{d}{dt}$ by $_0^{CFR}D_t^\alpha$, we obtain:

$$_0^{CFR}D_t^\alpha h(t) = R_1(t) - \frac{h(t)}{DR_1(t)}. \tag{13.80}$$

The above model has the ability to account for heterogeneity as well as the memory effect. By replacing $\frac{d}{dt}$ by $_0^{ABR}D_t^\alpha$, we obtain:

$$_0^{ABR}D_t^\alpha h(t) = R_1(t) - \frac{h(t)}{DR_1(t)}. \tag{13.81}$$

The above model has the ability to account for heterogeneity, memory, and viscoelasticity. These three models will be solved numerically, and to do this, we present the numerical approximation of the three non-local operators. We firstly present $_0^{RL}D_t^\alpha$. By definition:

$$_0^{RL}D_t^\alpha h(t) = \frac{1}{\Gamma(1-\alpha)}\frac{d}{dt}\int_0^t h(t)(x-\tau)^{-\alpha}d\tau \tag{13.82}$$

$$= \frac{d}{dt}F(T).$$

According to Crank–Nicolson, this can be given for $n \geq 1$ as:

$$_0^{RL}D_t^\alpha h(t) = \left[\frac{F(t_{n-1}) - F(t_n)}{\Delta t}\right] \tag{13.83}$$

where:

$$F(t_{n-1}) = \frac{1}{\Gamma(1-\alpha)}\int_0^{t_{n+1}} h(\tau)(t_{n+1}-\tau)^{-\alpha}d\tau \tag{13.84}$$

$$= \frac{1}{\Gamma(1-\alpha)}\sum_{k=0}^n \int_{t_k}^{t_{k+1}} h(t_k)(t_{n+1}-\tau)^{-\alpha}d\tau$$

$$= \frac{1}{\Gamma(1-\alpha)}\sum_{k=0}^n h(t_k)\int \tag{13.85}$$

Let $y = t_{n+1} - \tau, dy = d\tau, y = t_{n+1} - t_{k+1}, y = t_{n+1} - t_k$.

$$F(t_{n+1}) = \frac{1}{\Gamma(1-\alpha)}\sum_{k=0}^n h(t_n)\int_{t_{n+1}-t_{k+1}}^{t_{n+1}-t_k} Y^{-\alpha}dY \tag{13.86}$$

$$= \frac{1}{\Gamma(1-\alpha)}\sum_{k=0}^n h(t_n)\left(-\frac{Y^{1-\alpha}}{1-\alpha}\right)\big|_{t_{n+1}-t_{k+1}}^{t_{n+1}-t_k} \tag{13.87}$$

$$= \frac{1}{\Gamma(1-\alpha)(1-\alpha)}\sum_{k=0}^n h(t_n)\left\{(t_{n+1}-t_{k+1})^{1-\alpha} - (t_{n+1}-t_k)^{1-\alpha}\right\}. \tag{13.88}$$

Thus, replacing $n + 1$ by n, we obtain:

$$F(t_n) = \frac{1}{\Gamma(\alpha)} \sum_{k=0}^{n-1} h(t_n) \left\{ (t_n - t_{k+1})^{1-\alpha} - (t_n - t_k)^{1-\alpha} \right\}. \tag{13.89}$$

Thus,

$$\begin{aligned}
&{}^{RL}_0 D_t^\alpha h(t_n) \tag{13.90} \\
&= \frac{1}{\Delta t \Gamma(\alpha)} \sum_{k=0}^{n} h(t_n) \left\{ (t_{n+1} - t_{k+1})^{1-\alpha} - (t_{n+1} - t_k)^{1-\alpha} \right\} \\
&\quad + \sum_{k=0}^{n-1} h(t_n) \left\{ (t_n - t_{k+1})^{1-\alpha} - (t_n - t_k)^{1-\alpha} \right\}.
\end{aligned}$$

Using the same derivation for ${}^{ABR}_0 D_t^\alpha h(t)$ we obtain:

$$ {}^{ABR}_0 D_t^\alpha h(t) = \frac{AB(\alpha)}{\Delta t (1 - \alpha)} \left[\sum_{k=0}^{n} h(t_n) \Phi_n^{\alpha,1} - \sum_{k=0}^{n-1} h(t_n) \Phi_n^{\alpha,2} \right] \tag{13.91} $$

where:

$$\begin{aligned}
\Phi_n^{\alpha,1} &= (t_{n+1} - t_{k+1}) E_{\alpha,2} \left[-\frac{\alpha}{1-\alpha} (t_{n+1} - t_{k+1})^\alpha \right] \tag{13.92} \\
&\quad + (t_{n+1} - t_k) E_{\alpha,2} \left[-\frac{\alpha}{1-\alpha} (t_{n+1} - t_{k+1})^\alpha \right] \\
\Phi_n^{\alpha,2} &= (t_{n+1} - t_{k+1}) E_{\alpha,2} \left[-\frac{\alpha}{1-\alpha} (t_n - t_{k+1})^\alpha \right] \tag{13.93} \\
&\quad + (t_n - t_k) E_{\alpha,2} \left[-\frac{\alpha}{1-\alpha} (t_n - t_{k+1})^\alpha \right].
\end{aligned}$$

Using the same derivation for ${}^{CFR}_0 D_t^\alpha h(t)$ we obtain:

$$ {}^{CFR}_0 D_t^\alpha h(t) = \frac{M(\alpha)}{\Delta t (\alpha)} \left[\sum_{k=0}^{n} h(t_k) \Phi_\alpha^k - \sum_{k=0}^{n-1} h(t_k) \Phi_{\alpha,1}^k \right] \tag{13.94} $$

where:

$$ \Phi_\alpha^k = \exp \left[-\frac{\alpha}{1-\alpha} (t_{n+1} - t_{k+1}) \right] - \exp \left[-\frac{\alpha}{1-\alpha} (t_{n+1} - t_k) \right] \tag{13.95} $$

and

$$\Phi_{\alpha,1}^k = \exp\left[-\frac{\alpha}{1-\alpha}(t_n - t_{k+1})\right] - \exp\left[-\frac{\alpha}{1-\alpha}(t_n - t_k)\right]. \tag{13.96}$$

The justification of this new concept of differentiation or approach is to further present a modified groundwater recharge model, so that the parameters within the model became a function of both space and time which is more acceptable or closer to the real-world observation as that suggested in many papers in the literature where the coefficients are rather constant, as this existing model is also a linear model which cannot take into account non-constant parameters. Therefore in this part, we will stress on by suggesting or presenting a modified model for which both the recharge and the drainage resistance parameters within the equation are a function of time, this of course implying that these parameters depend on space and time within a given point of a geological formation. Moreover, by employing or applying the uncertain function and also using the Eton approach which was proposed by Atangana and Badr [150], these two aforementioned parameters representing aquifers properties obtained a new definition, whereby they could be considered non-constant. In this case, it is important to point out that the modified equation accounts for recharge percentage at a specific time and drainage resistance at a specific time [150]. In addition, this paper entailed iterative and numerical solutions to solve the recharge equation. The eventual solutions obtained after applying the Liouville, Caputo–Fabrizio, and Atangana–Baleanu derivatives, so that the local differential operator is replaced by linear differential operators, present equations which can be used to model real-world groundwater recharge scenarios where there is effect of elasticity, heterogeneity, viscoelasticity, and the memory.

CHAPTER 14

Atangana Derivative With Memory and Application

14.1 Definition and Properties

In mathematics, a dynamic scheme is a tuple (D, h, B) with D a manifold which can be a locally Banach space or Euclidean space, B the domain for time which is a set of non-negative real numbers, and h is an evolution rule $t \rightarrow f^t$ the range of which is of course a diffeomorphism of a manifold to itself [10,150].

Definition 14.1.1. *Let D be a dynamic system with domain B (time or space), let u be a positively defined function called uncertainty function of D within the domain B. Then, if $h \in D$, the Atangana derivative with memory of a function h denoted by $U^u(f)$ is defined as:*

$$U^{u(t)}(h(t)) = (1 - u(t))h(t) + u(t)h'(t). \tag{14.1}$$

Remark 1. If $u = 1$, we recover the first derivative (local derivative), if $u = 0$, we recover the initial function, this is conformable with the primary law of derivative.

14.1.1 Properties of Atangana Derivative With Memory

- Addition:

$$U^{u(t)}(ah(t) + bf(t)) = aU^{u(t)}(h(t)) + bU^{u(t)}(f(t)). \tag{14.2}$$

- Multiplication:

$$U^{u(t)}(fg(t)) = f(t)U^{u(t)}(g(t)) + u(t)g(t)f'(t). \tag{14.3}$$

Proof. Using the definition of Atangana derivative with memory, we obtain the following:

$$
\begin{aligned}
U^{u(t)}(fg(t)) &= (1 - u(t))f(t)g(t) + u(t)(f(t)g(t))', \\
U^{u(t)}(fg(t)) &= (1 - u(t))f(t)g(t) + u(t)[f(t)g'(t) + g(t)f'(t)], \\
U^{u(t)}(fg(t)) &= f(t)U^{u(t)}(g(t)) + u(t)g(t)f'(t). \tag{14.4}
\end{aligned}
$$

This completes the proof. □

Fractional Operators with Constant and Variable Order
with Application to Geo-Hydrology
DOI: 10.1016/B978-0-12-809670-3.00014-X

- Division:

$$U^{u(t)}\left(\frac{f(t)}{g(t)}\right) = \frac{g(t)U^{u(t)}f(t) - u(t)f(t)g'(t)}{g^2(t)}. \tag{14.5}$$

- Lipschitz condition: Let $f(t)$ and $g(t)$ be two functions, then

$$
\begin{aligned}
\|U^{u(t)}(g(t)) - U^{u(t)}(f(t))\| &\leq \|(1 - u(t))g(t) + u(t)g'(t) \\
&\quad - (1 - u(t))f(t) - u(t)f'(t)\| \\
&\leq |1 - u(t)|\|g(t) - f(t)\| + |u(t)|\|g'(t) - f'(t)\| \\
&\leq a\|g(t) - f(t)\| + b\|g'(t) - f'(t)\| \\
&\leq a\|g(t) - f(t)\| + b\alpha\|g(t) - f(t)\| \\
&\leq H\|g(t) - f(t)\|. \tag{14.6}
\end{aligned}
$$

This proves that the Atangana derivative with memory possesses the Lipschitz condition.

14.2 Application to the Flow Model in a Confined Aquifer

We present in this section the new concept of differentiation that will be applied to the groundwater flow model. The aim is to obtain a suitable model that takes into account the variability of the aquifer parameters including the storativity and the transmissivity. To have in this case the application of this differential operator in time, applying this differential operator on the left-hand side of Theis equation, we obtain

$$S\partial_t^{u(t)}h(r, t) = T\left(\frac{1}{r}\partial_r h(r, t) + \partial_r^2 h(r, t)\right). \tag{14.7}$$

Replacing the left-hand side with the uncertain derivative formula, we obtain the following:

$$S\left((1 - u(t)h(r, t)) + u(t)\partial_t h(r, t)\right)h(r, t) = T\left(\frac{1}{r}\partial_r h(r, t) + \partial_r^2 h(r, t)\right). \tag{14.8}$$

Notice that if the uncertain function $u(t) = 1$, we recover the Theis groundwater flow model in a confined aquifer. Thus, if we apply the uncertain function $v(r)$ in space, we obtain the following:

$$
\begin{aligned}
S\left((1 - u(t)h(r, t)) + u(t)\frac{\partial h(r, t)}{\partial t}\right) &= T\left(\frac{1}{r}((1 - v(r)))h(r, t)\right. \\
+ v(r)\partial_r h(r, t))h(r, t) + (1 - v(r))^2 - & \\
- v'(r)v(r)h(r, t) + \left(2(1 - v(r))v(r) + v(r)v'(r)\right)&\left.\frac{\partial h(r, t)}{\partial_r} + v^2(r)\frac{\partial^2 h(r, t)}{\partial_r^2}\right) \tag{14.9}
\end{aligned}
$$

From the above since the uncertain function $v(r) < 1$ then it is possible to divide the above equation by $1 - v(r)$; thus, using the Taylor series of the inverse of $1 - v(r)$, then we obtain the following:

$$S(r, t)\frac{\partial h(r, t)}{\partial t} = \frac{T(r)}{r}\frac{\partial h(r, t)}{\partial r} + T(r)\frac{\partial^2 h(r, t)}{\partial r^2} + F(r, t)h(r, t) \tag{14.10}$$

where

$$S(r, t) = S\left(u(t) + u(t)v(r)\right), \tag{14.11}$$

$$T(r, t) = Tv(r)(1 + v(r)), \tag{14.12}$$

$$F(r, t) = (1 - u(t))(1 + v(t)) + \frac{T}{r}. \tag{14.13}$$

In the above function, $T(r)$, $S(r, t)$, and $F(r, t)$ show that the transmissivity is a function of space, the storativity is a function of space and time, and finally a new force that explains the variability of the head of water changes in time and space. That new force shall be able to inform the full trajectory of a drop of water within the geological formations. Thus, a fractional differential operator is replacing the classical differential operator to obtain

$$S(r, t){}_0^{FC}D_t^\alpha h(r, t) = \frac{T(r)}{r}\frac{\partial h(r, t)}{\partial r} + T(r)\frac{\partial^2 h(r, t)}{\partial r^2} + F(r, t)h(r, t), \tag{14.14}$$

where ${}_0^{FC}D_t^\alpha h(r, t)$ is a differential fractional operators including Riemann–Liouville, Caputo, Caputo–Fabrizio, or Atangana–Baleanu types.

14.2.1 Existence and Uniqueness of Exact Solution

In this section, we ensure the existence and uniqueness of a solution of Eq. (14.14) under some mathematical conditions but using the fixed-point theorem.

Theorem 14.2.1. *Let $T(r)$, $S(r, t)$, and $F(r, t)$ be continuous and positive functions such that: $T(r) < M$, $S(r, t) < M_1$, and M_2, and also for $(x, t) \in [0, X] \times [0, t_0]$ the function head $h(r, t)$ verifies*

$$\left\|\frac{\partial h(r, t)}{\partial r}\right\| < O_1\|h(r, t)\|,$$

$$\left\|\frac{\partial h(r, t)}{\partial t}\right\| < O_2\|h(r, t)\|.$$

Then, the obtained equation has a unique solution.

Proof. Let $h(r, t)$ satisfy the above conditions, then, without lost of generality, we consider the fractional operator to be of the Riemann–Liouville type, and then, if we apply the Riemann–Liouville fractional integral on both sides of Eq. (14.14), we obtain

$$h(r, t) = \frac{1}{\Gamma[1-\alpha]} \int_0^t (t-\tau)^{-\alpha} \left(\frac{T(r)}{r} \frac{\partial h(r, \tau)}{\partial r} + T(r) \frac{\partial^2 h(r, \tau)}{\partial r^2} + F(r, t)h(r, \tau) \right) d\tau.$$

(14.15)

Let the function B be

$$B(h(r, t)) = \frac{1}{\Gamma[1-\alpha]} \int_0^t (t-\tau)^{-\alpha} \left(\frac{T(r)}{r} \frac{\partial h(r, \tau)}{\partial r} + T(r) \frac{\partial^2 h(r, \tau)}{\partial r^2} + F(r, t)h(r, \tau) \right) d\tau.$$

(14.16)

Then,

$$\| B(h(r, t)) - B(g(r, t)) \| \leq \| \\ \frac{1}{\Gamma[1-\alpha]} \int_0^t (t-\tau)^{-\alpha} \left(\frac{T(r)}{r} \frac{\partial (h(r, \tau) - g(r, \tau))}{\partial r} + T(r) \frac{\partial^2 (h(r, \tau) - g(r, \tau))}{\partial r^2} \right. \\ \left. + F(r, \tau)(h(r, \tau) - g(r, \tau)) \right) d\tau \|.$$

Then,

$$\| B(h(r, t)) - B(g(r, t)) \| \leq \frac{1}{\Gamma[1-\alpha]} \int_0^t (t-\tau)^{-\alpha} \left(\| \frac{T(r)}{r} \frac{\partial (h(r, \tau) - g(r, \tau))}{\partial r} \| \right. \\ \left. + \| T(r) \frac{\partial^2 (h(r, \tau) - g(r, \tau))}{\partial r^2} \| + \| F(r, \tau)(h(r, \tau) - g(r, \tau)) \| \right) d\tau. \quad (14.17)$$

Using the properties of h, the above inequality is reduced to:

$$\| B(h(r, t)) - B(g(r, t)) \| \leq \frac{(M_1 O_1 + M_2 O_1^2 + M_3)}{\Gamma[2-\alpha]} \| h - g \| t_0^{1-\alpha}.$$

(14.18)

This shows that the function B possesses Lipschitz condition.   □

Let us construct a sequence based on Eq. (14.15) as follows:

$$h_n(r, t) = \frac{1}{\Gamma[1-\alpha]} \int_0^t (t-\tau)^{-\alpha} \left(\frac{T(r)}{r} \frac{\partial h_{n-1}(r, \tau)}{\partial r} + T(r) \frac{\partial^2 h_{n-1}(r, \tau)}{\partial r^2} \right. \\ \left. + F(r, t)h_{n-1}(r, \tau) \right) d\tau.$$

(14.19)

Then we evaluate the following

$$\|h_n(r,t) - h_{n-1}\| \le \left(\frac{(M_1 O_1 + M_2 O_1^2 + M_3)}{\Gamma[2-\alpha]} t_0^{1-\alpha} \right) \|h_{n-1} - h_{n-2}\|, \quad (14.20)$$

$$\|h_n(r,t) - h_{n-1}\| \le \left(\frac{(M_1 O_1 + M_2 O_1^2 + M_3)}{\Gamma[2-\alpha]} t_0^{1-\alpha} \right)^2 \|h_{n-2} - h_{n-3}\|, \quad (14.21)$$

$$\|h_n(r,t) - h_{n-1}\| \le \left(\frac{(M_1 O_1 + M_2 O_1^2 + M_3)}{\Gamma[2-\alpha]} t_0^{1-\alpha} \right)^3 \|h_{n-3} - h_{n-4}\|, \quad (14.22)$$

$$\|h_n(r,t) - h_{n-1}\| \le \left(\frac{(M_1 O_1 + M_2 O_1^2 + M_3)}{\Gamma[2-\alpha]} t_0^{1-\alpha} \right)^{n-1} \|h_1 - h_0\|, \quad (14.23)$$

under the condition that

$$\frac{(M_1 O_1 + M_2 O_1^2 + M_3)}{\Gamma[2-\alpha]} < 1. \quad (14.24)$$

Then taking the limit as m tends to infinity, the constructed sequence converges towards zero. Thus the constructed sequence is Cauchy in a Banach space and converges towards $h(r,t)$ which is considered here as a fixed point. In concluding, the new model has a unique solution.

14.3 Modeling the Movement of Pollution With Uncertain Derivative

The groundwater is a very important source of drinkable water, in fact it was revealed in many studies that 80% of fresh water is found in the geological formation called aquifer. This water is in high risk of pollution and the mathematical equation describing the migration of pollution underground is the well-known advection–dispersion equation given as

$$D\frac{\partial^2 c}{\partial x^2} - v\frac{\partial c}{\partial x} - R\frac{\partial c}{\partial t} = 0. \quad (14.25)$$

In the above equation, D is the dispersion coefficient, v the advection velocity, and R the retardation factor. The above equation also tells us that all over a given aquifer, the dispersion, advection, and retardation factors are the same. This is not practically correct because the properties of soils change from one point of the aquifer to another that means the dispersion; advection and retardation factor must vary in space and time. With the variability of the dispersion coefficient, the retardation factor and advection, we will be able to trace the movement of the pollution from a certain point of time to another. To achieve this we apply the uncertain derivative in space and time as follows:

$$DU^{u(x)}\left(U^{u(x)}\left(C(x,t) \right) \right) - vU^{u(x)}\left(C(x,t) \right) - RU^{u(t)}\left(C(x,t) \right) = 0. \quad (14.26)$$

Replacing the derivative by it definition and simplifying, we obtain

$$
D\left\{[(1-u(x))^2 - u'(x)u(x)]C(x,t) + \left\{\begin{array}{c} 2(1-u(x))u(x) \\ +u(x)u'(x) \end{array}\right\} \frac{\partial C(x,t)}{\partial x}\right\} \quad (14.27)
$$
$$
+ u^2(x)\frac{\partial^2 C(x,t)}{\partial x^2} - v\left\{(1-u(x))\,C(x,t) + u(x)\frac{\partial C(x,t)}{\partial x}\right\} -
$$
$$
R\left\{(1-f(t))\,C(x,t) + f(t)\frac{\partial C(x,t)}{\partial x}\right\}
$$
$$
=\quad 0.
$$

Assuming that the uncertain order in respect to time is small, then we divide both sides by $1-f(t)$, and using some asymptotic technique, the above equation can be approximated as follows:

$$
D\left\{\begin{array}{l} [(1-u(x)+u(t))^2 - u'(x)(u(x)+f(t))]C(x,t)+ \\ \left\{\begin{array}{c} 2\,(1-u(x)+f(t))\,(u(x)+f(t)) \\ +(u(x)+f(t))\,u'(x) \end{array}\right\} \frac{\partial C(x,t)}{\partial x} \end{array}\right\} \quad (14.28)
$$
$$
+ (u(x)+f(t))^2\frac{\partial^2 C(x,t)}{\partial x^2}
$$
$$
- v\{(1-u(x)+f(t))\,C(x,t) + (u(x)+f(t))\}\frac{\partial C(x,t)}{\partial x} - RC(x,t)
$$
$$
+ f(t)R\frac{\partial C(x,t)}{\partial t}
$$
$$
=\quad 0.
$$

Rearranging the above equation, we obtain the following:

$$
\left\{\begin{array}{l} \left[D\left(\begin{array}{c} (1-u(x)+u(t))^2 \\ -u'(x)(u(x)+f(t)) \\ -v\,(1-u(x)+f(t)) \end{array}\right)\right]C(x,t) \\ +\left\{D\left(\begin{array}{c} 2\,(1-u(x)+f(t))\,(u(x)+f(t)) \\ +(u(x)+f(t))u'(x) \\ -v\,\{(u(x)+f(t))\} \end{array}\right)\right\} \end{array} \frac{\partial C(x,t)}{\partial x}\right\} \quad (14.29)
$$
$$
+ D(u(x)+f(t))^2\frac{\partial^2 C(x,t)}{\partial x^2} - RC(x,t) + f(t)R\frac{\partial C(x,t)}{\partial t}
$$
$$
=\quad 0.
$$

And finally we can reduced the above to

$$H(x,t)C(x,t) + v(x,t)\frac{\partial C(x,t)}{\partial x} + D(x,t)\frac{\partial^2 C(x,t)}{\partial x^2} = R(t)\frac{\partial C(x,t)}{\partial x}. \qquad (14.30)$$

The above equation shows that the advection and dispersion are functions of time and space, while the retardation is a function of time. However, there is a new force H that is viewed as the proportion that allows the value of that chemical concentration to remember its trajectory in the geological formation system and the time where it was retarded since its departure from the point of injection. It is also possible for a given portion of pollution to remember where it was retarded in the aquifer. In order to include into mathematical equation the filter effect in time, meaning in order to have an accurate representation of the change in time of concentration of pollution within the geological formation, we introduce the Caputo–Fabrizio derivative with fractional order into equation to obtain

$$H(x,t)C(x,t) + v(x,t)\frac{\partial C(x,t)}{\partial x} + D(x,t)\frac{\partial^2 C(x,t)}{\partial x^2} = R(t)_0^{CF}D_t^\alpha(C(x,t)). \qquad (14.31)$$

The above equation is the result of Uncertain Fractional Modeling (UFM). It is clear that the above equation is more descriptive than the fractional advection dispersion equation that was proposed by many scholars.

14.4 Numerical Analysis

In this section we will discuss the numerical solution of Eq. (14.30). To do this, we first present the numerical approximation of the Caputo–Fabrizio derivative with fractional order. For some positive integer N, the grid size in time for finite difference technique I defined by

$$k = \frac{1}{N}. \qquad (14.32)$$

The grid points in the time interval $[0, T]$ are labeled $t_n = nj$, $n = 0, 1, 2, ..., TN$. The value of the function f at the grid point is $f_i = f(t_i)$. A discrete approximation to the Caputo–Fabrizio derivative with fractional order can be obtained by simple quadrature formula as follows:

$$_0^{CF}D_t^\alpha(f(t_n)) = \frac{M(\alpha)}{1-\alpha}\int_a^{t_n} f'(x)\exp\left[-\alpha\frac{t_n - x}{1-\alpha}\right]dx. \qquad (14.33)$$

The above equation can be modified using the first-order approximation to

$$_0^{CF}D_t^\alpha(f(t_j)) = \frac{M(\alpha)}{1-\alpha}\sum_{j=1}^n\int_{(j-1)k}^{jk}\left(\frac{f^{j+1}-f^j}{\Delta t}+O(k)\right)\exp\left[-\alpha\frac{t_j-x}{1-\alpha}\right]dx. \quad (14.34)$$

Before integration we obtain the following expression:

$$\frac{M(\alpha)}{1-\alpha}\sum_{j=1}^n\left(\begin{array}{c}\frac{f^{j+1}-f^j}{\Delta t}\\+O(\Delta t)\end{array}\right)\int_{(j-1)k}^{jk}\exp\left[-\alpha\frac{t_n-x}{1-\alpha}\right]dx \quad (14.35)$$

$$_0^{CF}D_t^\alpha(f(t_j)) = \frac{M(\alpha)}{\alpha}\sum_{j=1}^n\left(\frac{f^{j+1}-f^j}{\Delta t}+O(\Delta t)\right)d_{j,k}$$

where

$$d_{j,k}=\exp\left[-\alpha\frac{k}{1-\alpha}(n-j+1)\right]-\exp\left[-\alpha\frac{k}{1-\alpha}(n-j)\right]. \quad (14.36)$$

We have finally that

$$_0^{CF}D_t^\alpha(f(t_n)) = \frac{M(\alpha)}{\alpha}\sum_{j=1}^n\left(\frac{f^{j+1}-f^j}{\Delta t}\right)d_{j,k}+\frac{M(\alpha)}{\alpha}\sum_{j=1}^n d_{j,k}O(\Delta t).$$

Derivative at a point t_n is

$$_0^{CF}D_t^\alpha(f(t_n)) = \frac{M(\alpha)}{\alpha}\sum_{j=1}^n\left(\frac{f^{j+1}-f^j}{\Delta t}\right)d_{j,k}+O\left((\Delta t)^2\right). \quad (14.37)$$

Replacing the above together with first and second approximations of local derivative, we obtain

$$H_i^j\left\{\frac{C_i^{j+1}-C_i^j}{2}\right\}-v_i^j\left\{\frac{\left(C_{i+1}^{j+1}-C_{i-1}^{j+1}\right)-\left(C_{i+1}^j-C_{i-1}^j\right)}{4\Delta x}\right\} \quad (14.38)$$

$$+D_i^j\left\{\frac{\left(C_{i+1}^{j+1}-2C_i^{j+1}+C_{i-1}^{j+1}\right)-\left(C_{i+1}^j-2C_i^j+C_{i-1}^j\right)}{2(\Delta x)^2}\right\}$$

$$= R^j\frac{M(\alpha)}{\alpha}\sum_{k=1}^j\left(\frac{C_i^{k+1}-C_i^k}{\Delta t}\right)d_{j,k}.$$

For simplicity, we let

$$a_i^j = \frac{H_i^j}{2}, \quad b_i^j = \frac{v_i^j}{4\Delta x}, \quad c_i^j = \frac{D_i^j}{2(\Delta x)^2}, \quad d_i^j = R^j \frac{M(\alpha)}{\alpha \Delta t},$$

then Eq. (14.38) becomes

$$
\begin{aligned}
& a_i^j \left(C_i^{j+1} - C_i^j \right) - b_i^j \left(C_{i+1}^{j+1} - C_{i-1}^{j+1} - C_{i+1}^j + C_{i-1}^j \right) + \\
& c_i^j \left(\left(C_{i+1}^{j+1} - 2C_i^{j+1} + C_{i-1}^{j+1} \right) + \left(C_{i+1}^j - 2C_i^j + C_{i-1}^j \right) \right) \\
= \ & d_i^j \left(C_i^{j+1} - C_i^j \right) d_{i,j} + d_i^j \sum_{k=1}^{j-1} \left(C_i^{k+1} - C_i^k \right) d_{k,j}.
\end{aligned}
\tag{14.39}
$$

Then,

$$
\begin{aligned}
& \left(a_i^j - 2c_i^j - d_i^j d_{i,j} \right) C_i^{j+1} \\
= \ & \left(a_i^j + 2c_i^j - d_i^j d_{i,j} \right) C_i^j \\
& + b_i^j \left(C_{i+1}^{j+1} - C_{i-1}^{j+1} - C_{i+1}^j + C_{i-1}^j \right) \\
& + c_i^j \left(\left(C_{i+1}^{j+1} + C_{i-1}^{j+1} \right) + \left(C_{i+1}^j + C_{i-1}^j \right) \right) \\
& + d_i^j \sum_{k=1}^{j-1} \left(C_i^{k+1} - C_i^k \right) d_{k,j}.
\end{aligned}
\tag{14.40}
$$

We shall now present the stability analysis of the numerical scheme for solving the modified model.

14.4.1 Stability Analysis of the Numerical Scheme

The aim of this section is to show the efficiency of the numerical scheme by means of the stability analysis. To achieve this, we assume that $g_i^j = C_i^j - y_i^j$ where y_i^j is the approximate solution of the modified equation at the given point in time and space (x_i, t_j), $(i = 1, 2, ..., N, \ j = 1, 2, ..., M)$; also the error for approximation is given as $g^j = \left[g_1^j, g_2^j, ..., g_N^j \right]$. The error committed while solving the new advection equation is given as:

$$
\begin{aligned}
\left(a_i^j - 2c_i^j - d_i^j d_{i,j} \right) g_i^{j+1} \ = \ & \left(a_i^j + 2c_i^j - d_i^j d_{i,j} \right) g_i^j \\
& + b_i^j \left(g_{i+1}^{j+1} - g_{i-1}^{j+1} - g_{i+1}^j + g_{i-1}^j \right)
\end{aligned}
\tag{14.41}
$$

$$- c_i^j \left(\left(g_{i+1}^{j+1} + g_{i-1}^{j+1} \right) + \left(g_{i+1}^{j} + g_{i-1}^{j} \right) \right)$$

$$+ d_i^j \sum_{k=1}^{j-1} \left(g_i^{k+1} - g_i^k \right) d_{k,j}.$$

To study the stability, we let

$$g_m(x,t) = \exp[at]\exp[ik_m x]. \tag{14.42}$$

In our study the stability characteristics can be studied using just the above form for error; with no loss in generality,

$$
\begin{aligned}
g_n^j &= \exp[at]\exp[ik_m x], &\tag{14.43}\\
g_n^{j+1} &= \exp[a(t+\Delta t)]\exp[ik_m x],\\
g_{n+1}^{j} &= \exp[at]\exp[ik_m (x+\Delta x)],\\
g_{n-1}^{j} &= \exp[at]\exp[ik_m (x-\Delta x)],\\
g_{n+1}^{j+1} &= \exp[a(t+\Delta t)]\exp[ik_m (x+\Delta x)],\\
g_{n-1}^{j+1} &= \exp[a(t+\Delta t)]\exp[ik_m (x-\Delta x)],\\
g_{n-1}^{j-1} &= \exp[a(t-\Delta t)]\exp[ik_m (x-\Delta x)]
\end{aligned}
$$

where $k_m = \frac{\pi m}{L}$, $m = 1, 2, ..., M = \frac{L}{\Delta x}$. Now, replacing the above in Eq. (14.41), we obtain

$$\left(a_i^j + 2c_i^j - d_i^j d_{i,j} \right) \exp[a(t+\Delta t)]\exp[ik_m x] \tag{14.44}$$

$$= \left(a_i^j - 2c_i^j - d_i^j d_{i,j} \right) \exp[at]\exp[ik_m x] -$$

$$b_i^j \left(\begin{array}{c} \exp[a(t+\Delta t)]\exp[ik_m (x+\Delta x)] \\ -\exp[a(t+\Delta t)]\exp[ik_m (x-\Delta x)]- \\ \exp[at]\exp[ik_m (x+\Delta x)] + \exp[at]\exp[ik_m (x-\Delta x)] \end{array} \right) +$$

$$+ c_i^j \left(\begin{array}{c} \left(\begin{array}{c} \exp[a(t+\Delta t)]\exp[ik_m (x+\Delta x)] \\ +\exp[a(t+\Delta t)]\exp[ik_m (x-\Delta x)] \end{array} \right) \\ + (\exp[at]\exp[ik_m (x+\Delta x)] + \exp[at]\exp[ik_m (x-\Delta x)]) \end{array} \right)$$

$$d_i^j \sum_{k=1}^{j-1} ((\exp[a(t+\Delta t)]\exp[ik_m x] - \exp[a(\Delta t)]\exp[ik_m x])) d_{k,j}.$$

After simplification we obtain the following

$$\left(a_i^j + 2c_i^j - d_i^j d_{i,j}\right) \exp[a\,(\Delta t)] \tag{14.45}$$

$$= \left(a_i^j - 2c_i^j - d_i^j d_{i,j}\right) -$$

$$b_i^j \left(\begin{array}{c} \exp[a\,(\Delta t)]\exp[ik_m\,(\Delta x)] \\ -\exp[a\,(\Delta t)]\exp[ik_m\,(-\Delta x)]- \\ \exp[ik_m\,(\Delta x)] + \exp[ik_m\,(-\Delta x)] \end{array} \right) +$$

$$c_i^j \left(\begin{array}{c} \left(\begin{array}{c} \exp[a\,(\Delta t)]\exp[ik_m\,(\Delta x)] \\ +\exp[a\,(\Delta t)]\exp[ik_m\,(-\Delta x)] \end{array} \right) \\ +(\exp[ik_m\,(\Delta x)] + \exp[ik_m\,(-\Delta x)]) \end{array} \right)$$

$$+ d_i^j \sum_{k=1}^{j-1} ((\exp[a\,(\Delta t)] - 1))\, d_{k,j}.$$

Rearranging the above, we obtain

$$\left\{ \begin{array}{c} \left(a_i^j + 2c_i^j - d_i^j d_{i,j}\right) + b_i^j \left(\begin{array}{c} \exp[ik_m\,(\Delta x)] \\ -\exp[ik_m\,(-\Delta x)] \end{array} \right) \\ -c_i^j \left(\left(\begin{array}{c} \exp[ik_m\,(\Delta x)] \\ +\exp[ik_m\,(-\Delta x)] \end{array} \right) \right) - j d_i^j d_{k,j} \end{array} \right\} \exp[a\,(\Delta t)] \tag{14.46}$$

$$= \left(a_i^j - 2c_i^j - d_i^j d_{i,j}\right) + b_i^j \left(\exp[ik_m\,(\Delta x)] + \exp[ik_m\,(-\Delta x)]\right) +$$

$$c_i^j \left(\exp[ik_m\,(\Delta x)] + \exp[ik_m\,(-\Delta x)]\right) - j d_i^j d_{i,j}.$$

Then

$$\left(a_i^j - 2c_i^j - d_i^j d_{i,j}\right) - b_i^j \left(\exp[ik_m\,(\Delta x)] + \exp[ik_m\,(-\Delta x)]\right) + \exp[a\,(\Delta t)] \tag{14.47}$$

$$= \frac{-c_i^j \left((\exp[ik_m\,(\Delta x)] + \exp[ik_m\,(-\Delta x)])\right) - j d_i^j d_{i,j}}{\left\{ \begin{array}{c} \left(a_i^j + 2c_i^j - d_i^j d_{i,j}\right) + b_i^j \left(\exp[ik_m\,(\Delta x)] - \exp[ik_m\,(-\Delta x)]\right) \\ -c_i^j \left((\exp[ik_m\,(\Delta x)] + \exp[ik_m\,(-\Delta x)])\right) - j d_i^j d_{k,j} \end{array} \right\}}.$$

Note that the condition for stability analysis is given by the following inequality

$$\frac{g_i^{j+1}}{g_i^j} = \exp[a\,(\Delta t)].$$

Thus if

$$\left| \frac{g_i^{j+1}}{g_i^j} \right| \leq 1,$$

from Eq. (14.46) we have the following

$$\left| \frac{g_i^{j+1}}{g_i^j} \right| = |\exp[a\,(\Delta t)]|$$

$$= \left| \frac{\left(a_i^j - 2c_i^j - d_i^j d_{i,j}\right) - b_i^j \left(\exp[ik_m\,(\Delta x)] + \exp[ik_m\,(-\Delta x)]\right) + }{\left\{ \begin{array}{c} \left(a_i^j + 2c_i^j - d_i^j d_{i,j}\right) + b_i^j \left(\exp[ik_m\,(\Delta x)] - \exp[ik_m\,(-\Delta x)]\right) \\ -c_i^j \left((\exp[ik_m\,(\Delta x)] + \exp[ik_m\,(-\Delta x)])\right) - jd_i^j d_{k,j} \end{array} \right\}} \right|$$

$$\cos[k_m \Delta x] = \frac{\exp[ik_m x] + \exp[-ik_m x]}{2}, \quad \sin^2[k_m \Delta x] = \frac{1 - \cos[2k_m \Delta x]}{2}.$$

Then the condition for stability is given as

$$b_i^j \leq c_i^j.$$

Thus from that above statement we can present the following theorem.

Theorem 14.4.1. *The Crank–Nicolson scheme for solving the uncertain fractional advection dispersion equation is stable provided that the following inequality is satisfied*

$$\frac{v_i^j}{D_i^j} \leq \frac{2}{\Delta x}.$$

14.5 Numerical Simulations

In this section, we present the numerical simulations of the resulted model from the uncertain-fractional advection dispersion equation. In this simulation, we will choose the uncertain derivative orders to be $u(x) = 2 + \sin(x + \frac{\pi}{3})$, $f(t) = 1 + \cos(t + \frac{\pi}{2})$, we consider the dispersion coefficient to be 0.96, the retardation coefficient to be 2, and the advection coefficient to be 0.74. The numerical simulation will be done for different values of the fractional order derivative; we will also alter the uncertain functions to see the effectiveness of the input. The numerical results are depicted in Figs. 14.1, 14.2, 14.3, and 14.4.

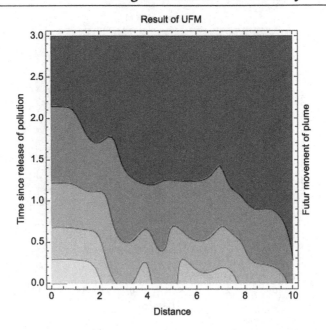

Figure 14.1: Numerical simulation for $\alpha = 0.95$.

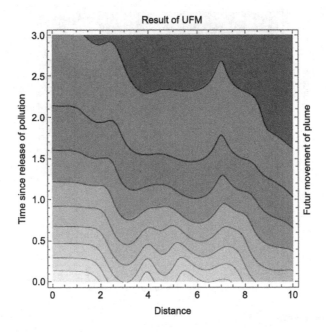

Figure 14.2: Numerical simulation for $\alpha = 0.65$.

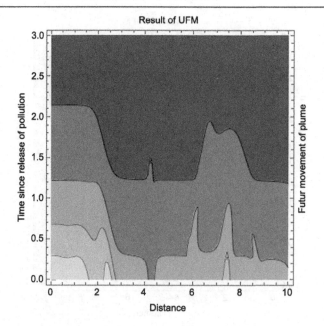

Figure 14.3: Numerical simulation for $\alpha = 0.45$.

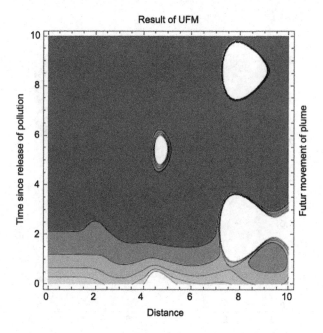

Figure 14.4: Numerical simulation for $\alpha = 0.25$.

References

[1] D. Barson, S. Bachu, P. Esslinger, Flow systems in the Mannville Group in the East-central Athabasca area and implications for steam-assisted gravity drainage (SAGD) operations for in situ bitumen production, Bulletin of Canadian Petroleum Geology 49 (3) (2001) 376–392.

[2] V.E.A. Post, J. Groen, H. Kooi, M. Person, S. Ge, W.M. Edmunds, Offshore fresh groundwater reserves as a global phenomenon, Nature 504 (7478) (2013) 71–78.

[3] Leonard F. Konikow, Groundwater Depletion in the United States (1900–2008), Scientific Investigations Report (2013-5079), U.S. Department of the Interior, U.S. Geological Survey, Reston, VA, 63 pp.

[4] A.A. Bal, Valley fills and coastal cliffs buried beneath an alluvial plain: evidence from variation of permeabilities in gravel aquifers, Canterbury Plains, New Zealand, Journal of Hydrology (NZ) 35 (1) (1996) 1–27.

[5] J.G. Begg, M.R. Johnstone, Geology of the Wellington Area. Institute of Geological and Nuclear Sciences 1:250 000 Geological Map 10, Institute of Geological and Nuclear Sciences, Lower Hutt, 2000, 1 sheet, 64 pp.

[6] M. Broadbent, P.F. Callander, A resistivity survey near Waimakariri River, Canterbury Plains, to improve understanding of local groundwater flow and of the capabilities of the survey method, New Zealand Journal of Geology and Geophysics 34 (1991) 441–453.

[7] L.J. Brown, Late Quaternary geology of the Wairau Plain, Marlborough, New Zealand, New Zealand Journal of Geology and Geophysics 24 (1981) 477–490.

[8] M.D. Razack, D. Huntley, Assessing transmissivity from specific capacity data in a large and heterogeneous alluvial aquifer, Groundwater 29 (6) (1991) 856–861.

[9] J. Logan, Estimating transmissivities from routine tests of waterwells, Groundwater 2 (1964) 35–37.

[10] P. White, Shallow Groundwater Flow Direction and Velocity at the Renwick Refuse Tip, Marlborough, Report No. WS 1109, Water and Soil Science Centre Christchurch, Ministry of Works and Development for the National Water and Soil Conservation Organisation, 1986.

[11] D.B. Boorman, J.M. Hollis, A. Lilly, Hydrology of Soil Types: a Hydrologically-Based Classification of the Soils of the United Kingdom, Report, 126, Institute of Hydrology, Wallingford, 1988.

[12] D.K. Buckley, Some case histories of geophysical downhole logging to examine borehole site and regional groundwater movement in Celtic regions, in: N.S. Robins, B.D.R. Misstear (Eds.), Groundwater in the Celtic Regions: Studies in Hard-Rock and Quaternary Hydrogeology, in: Geological Society Special Publications, vol. 182, Geological Society, London, 2000, pp. 219–238.

[13] C.R.C. Jones, A.J. Singleton, Public water supplies from alluvial and glacial deposits in northern Scotland, in: N.S. Robins, B.D.R. Misstear (Eds.), Groundwater in the Celtic Regions: Studies in Hard Rock and Quaternary Hydrogeology, in: Geological Society Special Publications, vol. 182, Geological Society, London, 2000, pp. 133–140.

[14] Y. Jin, M.-F. Uth, A.V. Kuznetsov, H. Herwig, Numerical investigation of the possibility of macroscopic turbulence in porous media: a direct numerical simulation study, Journal of Fluid Mechanics 766 (2015) 76–103.

[15] Charles V. Theis, The relation between the lowering of the piezometric surface and the rate and duration of discharge of a well using ground-water storage, Transactions, American Geophysical Union 16 (1935) 519–524.

[16] H.H. Cooper, C.E. Jacob, A generalized graphical method for evaluating formation constants and summarizing well field history, American Geophysical Union Transactions 27 (1946) 526–534.

[17] G.P. Kruseman, N.A. de Ridder, Analysis and Evaluation of Pumping Test Data, 2nd ed., Publication, vol. 47, Intern. Inst. for Land Reclamation and Improvement, Wageningen, The Netherlands, 1994, 370 pp.

[18] F.G. Driscoll, Groundwater and Wells, 2nd ed., Johnson Filtration Systems, Inc., St. Paul, Minnesota, 1986, 1089 pp.

[19] R.G. Agarwal, A New Method to Account for Producing Time Effects When Drawdown Type Curves Are Used to Analyze Pressure Buildup and Other Test Data, SPE Paper 9289, presented at the 55th SPE Annual Technical Conference and Exhibition, Dallas, TX, Sept. 1980, pp. 21–24, 1980.

[20] F. Daviau, G. Mouronval, G. Bourdarot, P. Curutchet, Pressure Analysis for Horizontal Wells, SPE Paper 14251, presented at the 60th Annual Technical Conference and Exhibition in Las Vegas, NV, Sept. 1985, pp. 22–25, 1980.

[21] N.S. Boulton, The drawdown of water table under non-steady conditions near a pumped well in an unconfined formation, Proceedings of the Institution of Civil Engineers, London 3 (1954) 564–579.

[22] N.S. Boulton, Analysis of data from nonequilibrium pumping tests allowing for delayed yield from storage, Proceedings of the Institution of Civil Engineers, London 26 (1963) 469–482.

[23] Abdon Atangana, Drawdown in prolate spheroidal–spherical coordinates obtained via Green's function and perturbation methods, Communications in Nonlinear Science and Numerical Simulation 19 (5) (2014) 1259–1269.

[24] A.R. Ingraffea, Fluid Migration Mechanisms Due to Faulty Well Design and/or Construction: An Overview and Recent Experiences in the Pennsylvania Marcellus Play, Physicians Scientists and Engineers for Healthy Energy, 2012.

[25] M.K. Boling, Model regulatory framework for hydraulic fracturing operations, Presentation, Washington, DC, USA, 2011.

[26] J.C. Helton, F.J. Davis, Latin hypercube sampling and the propagation of uncertainty in analyses of complex systems, Reliability Engineering and System Safety 81 (1) (2003) 23–69.

[27] J.C. Helton, Uncertainty and sensitivity analysis in the presence of stochastic and subjective uncertainty, Journal of Statistical Computation and Simulation 57 (1–4) (1997) 3–76.

[28] J.C. Helton, Uncertainty and sensitivity analysis techniques for use in performance assessment for radioactive waste disposal, Reliability Engineering and System Safety 42 (2–3) (1993) 327–367.

[29] R.L. Iman, W.J. Conover, Small sample sensitivity analysis techniques for computer models, with an application to risk assessment, Communications in Statistics A 9 (17) (1980) 1749–1874.

[30] R.L. Iman, J.C. Helton, An investigation of uncertainty and sensitivity analysis techniques for computer models, Risk Analysis 8 (1) (1988) 71–90.

[31] Eric Zuesse, Love canal: the truth seeps out, Reason Enterprises (1981) 1–17.

[32] Sam How Verhovek, After 10 years, the trauma of love canal continues, New York Times (1988).

[33] Lois R. Ember, Uncertain science pushes Love Canal solutions to political, legal arenas, Chemical and Engineering News 58 (August 11, 1980) 22–29.

[34] Thomas H. Maugh, 11. Health effects of exposure to toxic wastes, Science 215 (January 29, 1982) 490–493.

[35] R. Jeffrey Smith, The risks of living near Love Canal. Controversy and confusion follow a report that the Love Canal area is no more hazardous than areas elsewhere in Niagara Falls, Science 212 (August 27, 1982) 808–809, 811.

[36] National Research Council, Committee on Environmental Epidemiology, Environmental Epidemiology, vol. 1: Public Health and Hazardous Wastes, National Academy Press, Washington, 1991.

[37] Michael Jordan, Hush Hush: The Dark Secrets of Scientific Research, Firefly Books, Buffalo, 2003, p. 108.

[38] Occidental to pay $129 Million in Love Canal Settlement, U.S. Department of Justice, December 21, 1995.

[39] Jennyfer Wolf, Annette Prüss-Ustün, Oliver Cumming, Jamie Bartram, Sophie Bonjour, Sandy Cairncross, Thomas Clasen, John M. Colford, Valerie Curtis, Jennifer De France, Lorna Fewtrell, Matthew C. Freeman, Bruce Gordon, Paul R. Hunter, Aurelie Jeandron, Richard B. Johnston, Daniel Mäusezahl, Colin Mathers, Maria Neira, Julian P.T. Higgins, Systematic review: assessing the impact of drinking water and sanitation on diarrhoeal disease in low and middle-income settings: systematic review and meta-regression, Tropical Medicine and International Health 19 (8) (2014) 928–942.

[40] M. Buitenkamp, A. Richert Stintzing, Europe's Sanitation Problem 20 Million Europeans Need Access to Safe and Affordable Sanitation, Women in Europe for a Common Future (WECF), The Netherlands, 2008.

[41] Lynda Knobeloch, Barbara Salna, Adam Hogan, Jeffrey Postle, Henry Anderson, Blue Babies and Nitrate-Contaminated Well Water, 2000.

[42] M. Nasir Khan, F. Mohammad, Eutrophication: challenges and solutions, in: A.A. Ansari, S.S. Gill (Eds.), Eutrophication: Causes, Consequences and Control, Springer Science+Business Media, Dordrecht, 2014.

[43] C.J. Rosen, B.P. Horgan, Preventing pollution problems from lawn and garden fertilizers, extension.umn.edu, 2009.

[44] L.E. Jackson, M. Burger, T.R. Cavagnaro, Roots, nitrogen transformations, and ecosystem services, Annual Review of Plant Biology 59 (2008) 341–363.

[45] World Health Organization, Fluoride in drinking-water, 2004.

[46] Burden of disease and cost-effectiveness estimates, World Health Organization, 2012.

[47] John Janovy, Gerald D. Schmidt, Larry S. Roberts, Gerald D. Schmidt, Larry S. Roberts, Foundations of Parasitology, Wm. C. Brown, Dubuque, Iowa, 1996.

[48] Gary W. Brunette (Ed.), CDC Health Information for International Travel 2012. The Yellow Book, chapter 3, Oxford University Press, ISBN 978-0-19-976901-8, 2011.

[49] O. Lattimore, The Caravan Routes of Inner Asia, The Geographical Journal 72 (6) (1928) 500, http://dx.doi.org/10.2307/1783443, quoted in Wood, Frances, The Silk Road: Two Thousand Years in the Heart of Asia, Springer, Berlin, 2002, 19 pp.

[50] Bastian Schnabel, Drastic consequences of diarrhoeal disease, 2009.

[51] H.D. Sharma, K.R. Reddy, Geoenvironmental Engineering: Site Remediation, Waste Containment, and Emerging Waste Management Technologies, John Wiley and Sons, Hoboken, New Jersey, 2004.

[52] USEPA, Risk Assessment Guidance for Superfund, Washington, D.C., 1989.

[53] A. Saltelli, Sensitivity analysis for importance assessment, Risk Analysis 22 (3) (2002) 1–12.

[54] A. Saltelli, M. Ratto, T. Andres, F. Campolongo, J. Cariboni, D. Gatelli, M. Saisana, S. Tarantola, Global Sensitivity Analysis: The Primer, John Wiley and Sons, 2008.

[55] D.J. Pannell, Sensitivity analysis of normative economic models: theoretical framework and practical strategies, Agricultural Economics 16 (1997) 139–152.

[56] A. Bahremand, F. De Smedt, Distributed hydrological modeling and sensitivity analysis in Torysa watershed, Slovakia, Water Resources Management 22 (3) (2008) 293–408.

[57] A. O'Hagan, et al., Uncertain Judgements: Eliciting Experts' Probabilities, Wiley, Chichester, 2006.

[58] J. Sacks, W.J. Welch, T.J. Mitchell, H.P. Wynn, Design and analysis of computer experiments, Statistical Science 4 (1989) 409–435.

[59] J. Campbell, et al., Photosynthetic control of atmospheric carbonyl sulfide during the growing season, Science 322 (5904) (2008) 1085–1088.

[60] R. Bailis, M. Ezzati, D. Kammen, Mortality and greenhouse gas impacts of biomass and petroleum energy futures in Africa, Science 308 (2005) 98–103.

[61] J. Murphy, et al., Quantification of modelling uncertainties in a large ensemble of climate change simulations, Nature 430 (2004) 768–772.

[62] M.D. Morris, Factorial sampling plans for preliminary computational experiments, Technometrics 33 (1991) 161–174.

[63] F. Campolongo, J. Cariboni, A. Saltelli, An effective screening design for sensitivity analysis of large models, Environmental Modelling and Software 22 (2007) 1509–1518.

[64] P. Paruolo, M. Saisana, A. Saltelli, Ratings and rankings: Voodoo or science?, Journal of the Royal Statistical Society Series A 176 (3) (2013) 609–634.

[65] R. Dennis Cook, Sanford Weisberg criticism and influence analysis in regression, Sociological Methodology 13 (1982) 313–361.

[66] Willem Waegeman, Bernard De Baets, ROC analysis in ordinal regression learning, Pattern Recognition Letters 29 (2008) 1–9.

[67] N. Ravishankar, D.K. Dey, A First Course in Linear Model Theory, Chapman and Hall/CRC, Boca Raton, 2002, p. 101.

[68] R.G.D. Steel, J.H. Torrie, Principles and Procedures of Statistics with Special Reference to the Biological Sciences, McGraw Hill, 1960, p. 288.

[69] C.L. Chiang, Statistical Methods of Analysis, World Scientific, ISBN 981-238-310-7, 2003, p. 274, section 9.7.4 – interpolation vs extrapolation.

[70] B. Baeumer, D. Benson, M. Meerschaert, Advection and dispersion in time and space, Physica A 350 (2005) 245–262.

[71] D.N. Bradley, G.E. Tucker, D.A. Benson, Fractional dispersion in a sand-bed river, Journal of Geophysical Research 115 (2009) F00A09.

[72] G.W. Leibniz, Letter from Hanover, Germany to G.F.A. L'Hospital, September 30, 1695, Leibniz Mathematische Schriften, Olms-Verlag, Hildesheim, Germany, 1962, pp. 301–302.

[73] G.W. Leibniz, Letter from Hanover, Germany to Johann Bernoulli, December 28, 1695, Leibniz Mathematische Schriften, Olms-Verlag, Hildesheim, Germany, 1962, p. 226.

[74] G.W. Leibniz, Letter from Hanover, Germany to John Wallis, May 28, 1697, Leibniz Mathematische Schriften, Olms-Verlag, Hildesheim, Germany, 1962, p. 25.

[75] N. Abel, Solution de quelques problemes a l'aide d'intregrales definies, Magazin for Naturvidenskaberne 1 (2) (1823) 1–27.

[76] R. Herrmann, Towards a geometric interpretation of generalized fractional integrals – Erdelyi-Kober type integrals on r^n, as an example, Fractional Calculus and Applied Analysis 17 (2) (2014) 361–370.

[77] R. Hilfer, Applications of Fractional Calculus in Physics, World Scientific, River Edge, New Jerzey, 2000.

[78] R. Hilfer, Threefold introduction to fractional derivatives, Anomalous transport: Foundations and applications (2008) 17–73.

[79] S.G. Samko, A.A. Kilbas, O.I. Marichev, Fractional Integrals and Derivatives: Theory and Applications, Gordon and Breach, Yverdon, 1993.

[80] Abdon Atangana, Aydin Secer, A note on fractional order derivatives and table of fractional derivatives of some special functions, Abstract and Applied Analysis 2013 (2013) 279681.

[81] A.A. Kilbas, H.M. Srivastava, J.J. Trujillo, Theory and Applications of Fractional Differential Equations, Elsevier, Amsterdam, 2006.

[82] P.L. Butzer, A.A. Kilbas, J.J. Trujillo, Mellin transform analysis and integration by parts for Hadamard-type fractional integrals, Journal of Mathematical Analysis and Applications 270 (2002) 1–15.

[83] S. Pooseh, R. Almeida, D. Torres, Expansion formulas in terms of integer-order derivatives for the Hadamard fractional integral and derivative, Numerical Functional Analysis and Optimization 33 (3) (2012) 301–319.

[84] U.N. Katugampola, New approach to a genaralized fractional integral, Applied Mathematics and Computation 218 (3) (2011) 860–865.

[85] M. Caputo, M. Fabrizio, A new definition of fractional derivative without singular kernel, Progress in Fractional Differentiation and Applications 1 (2) (2015) 73–85.

[86] A. Atangana, On the new fractional derivative and application to nonlinear Fisher's reaction-diffusion equation, Applied Mathematics and Computation 273 (2016) 948–956.

[87] J. Losada, J.J. Nieto, Properties of a new fractional derivative without singular kernel, Progress in Fractional Differentiation and Applications 1 (2) (2015) 87–92.

[88] A. Atangana, B.S.T. Alkahtani, Analysis of the Keller–Segel model with a fractional derivative without singular kernel, Entropy 17 (6) (2015) 4439–4453.

[89] Abdon Atangana, Juan Jose Nieto, Numerical solution for the model of RLC circuit via the fractional derivative without singular kernel, Advances in Mechanical Engineering 7 (10) (2015) 1–7.

[90] Abdon Atangana, Emile Franc Doungmo Goufo, Extension of matched asymptotic method to fractional boundary layers problems, Mathematical Problems in Engineering 2014 (2014) 107535, http://dx.doi.org/10.1155/2014/107535.

[91] E.F.D. Goufo, A. Atangana, Analytical and numerical schemes for a derivative with filtering property and no singular kernel with applications to diffusion, The European Physical Journal Plus 131 (8) (2016) 269.

[92] A. Atangana, E.F. Doungmo Goufo, Extension of match asymptotic method to fractional boundary layers problems, Mathematical Problems in Engineering 2014 (2014) 107535.

[93] E.G. Bazhlekova, Subordination principle for fractional evolution equations, Fractional Calculus & Applied Analysis 3 (3) (2000) 213–230.

[94] E.G. Bazhlekova, Perturbation and Approximation Properties for Abstract Evolution Equations of Fractional Order, Research Report RANA 00-05, Eindhoven University of Technology, Eindhoven, 2000.

[95] M. Caputo, Linear models of dissipation whose Q is almost frequency independent II, Geophysical Journal of the Royal Astronomical Society 13 (5) (1967) 529–539.

[96] E.F. Doungmo Goufo, A. Atangana, Some generalized evolution systems using time derivative with a new parameter: stationarity, relaxation and diffusion, Advances in Mathematics (2014), under review.

[97] E.F. Doungmo Goufo, A mathematical analysis of fractional fragmentation dynamics with growth, Journal of Function Spaces 2014 (2014) 201520.

[98] E.F. Doungmo Goufo, R. Maritz, J. Munganga, Some properties of Kermack–McKendrick epidemic model with fractional derivative and nonlinear incidence, Advances in Difference Equations 2014 (2014) 278.

[99] R. Hilfer, Application of Fractional Calculus in Physics, World Scientific, Singapore, 1999.

[100] J. Prüss, Evolutionary Integral Equations and Applications, Birkhäuser, Basel–Boston–Berlin, 1993.

[101] M.H. Tavassoli, A. Tavassoli, M.R. Ostad Rahimi, The geometric and physical interpretation of fractional order derivatives of polynomial functions, Differential Geometry – Dynamical Systems 15 (2013) 93–104.

[102] I. Podlubny, Geometric and physical interpretation of fractional integration and fractional differentiation, Fractional Calculus and Applied Analysis 5 (4) (2002) 367–386.

[103] Zölzer Udo (Ed.), DAFX: Digital Audio Effects, 2002, pp. 48–49.

[104] Atangana, Koca, On uncertain-fractional modeling: the future way of modeling real world problems, 2016, paper submitted for publication.

[105] M. Caputo, M. Fabrizio, Applications of new time and spatial fractional derivatives with exponential kernels, Progress in Fractional Differentiation and Applications 2 (1) (2016) 1–11.

[106] Abdon Atangana, Derivative with a New Parameter, 1st edition, Theory, Methods and Applications, Academic Press, ISBN, ISBN 978-0-08-100644-3, 2015, 170 pp.

[107] J.H. Black, J.A. Barker, D.J. Noy, Crosshole Investigations: The Method, Theory and Analysis of Crosshole Sinusoidal Pressure Tests in Fissured Rocks, Stripa Proj., Int. Rep. 86-03, SKB, Stockholm, 1986.

[108] J.F. Botha, J.P. Verwey, I. Van der Voort, J.J.P. Viviers, W.P. Collinston, J.C. Loock, Karoo Aquifers. Their Geology, Geometry and Physical Behaviour, WRC Report 487/1/98, Water Research Commission, Pretoria, 1998.

[109] G.J. van Tonder, J.F. Botha, W.H. Chiang, H. Kunstmann, Y. Xu, Estimation of sustainable yields of boreholes in fractured rock formations, Journal of Hydrology 241 (2001) 70–90.

[110] B. Berkowitz, I. Balberg, Percolation theory and its application to groundwater hydrology, Water Resources Research 29 (4) (1993) 775–794.

[111] A. Cloot, J.F. Botha, A generalised groundwater flow equation using the concept of non-integer order derivatives, Water SA 32 (1) (2006) 1–7.

[112] C.F.M. Coimbra, Mechanics with variable-order differential operators, Annalen der Physik 12 (11–12) (2003) 692–703.

[113] T.H. Solomon, E.R. Weeks, H.L. Swinney, Observation of anomalous diffusion and Lévy flights in a two-dimensional rotating flow, Physical Review Letters 71 (24) (1993) 3975–3978.

[114] R.L. Magin, O. Abdullah, D. Baleanu, X.J. Zhou, Anomalous diffusion expressed through fractional order differential operators in the Bloch–Torrey equation, Journal of Magnetic Resonance 190 (2) (2008) 255–270.

[115] A.V. Chechkin, R. Gorenflo, I.M. Sokolov, Fractional diffusion in inhomogeneous media, Journal of Physics A 38 (42) (2005) L679–L684.

[116] F. Santamaria, S. Wils, E. de Schutter, G.J. Augustine, Anomalous diffusion in Purkinje cell dendrites caused by spines, Neuron 52 (4) (2006) 635–648.

[117] S. Umarov, S. Steinberg, Variable order differential equations with piecewise constant order-function and diffusion with changing modes, Zeitschrift für Analysis und ihre Anwendungen 28 (4) (2009) 431–450.

[118] D. Ingman, J. Suzdalnitsky, Application of differential operator with servo-order function in model of viscoelastic deformation process, Journal of Engineering Mechanics 131 (7) (2005) 763–767.

[119] A. Abdon, Newclass of boundary value problems, Information Sciences Letters 1 (2012) 67–76.

[120] A. Anatoly, J. Juan, M.S. Hari, Theory and Application of Fractional Differential Equations, Elsevier, Amsterdam, The Netherlands, 2006.

[121] A. Atangana, J.F. Botha, Analytical solution of groundwater flow equation via homotopy decomposition method, Journal of Earth Science & Climatic Change 3 (115) (2012) 2157.

[122] A. Atangana, A. Kilicman, Analytical solutions of the space–time fractional derivative of advection dispersion equation, Mathematical Problems in Engineering 2013 (2013) 853127.

[123] Y. Chen, H.-L. An, Numerical solutions of coupled Burgers equations with time- and space-fractional derivatives, Applied Mathematics and Computation 200 (2008) 87–95.

[124] V. Daftardar-Gejji, H. Jafari, Adomian decomposition: a tool for solving a system of fractional differential equations, Journal of Mathematical Analysis and Applications 301 (2005) 508–518.

[125] J.S. Duan, R. Rach, D. Bulean, A.M. Wazwaz, A review of the Adomian decomposition method and its applications to fractional differential equations, Communications in Fractional Calculus 3 (2012) 73–99.

[126] A short-distance integral-balance solution to a strong subdiffusion equation: a weak power-law profile, International Review of Chemical Engineering-Rapid Communications 2 (2010) 555–563.

[127] S. Momani, Z. Odibat, Numerical solutions of the space–time fractional advection–dispersion equation, Numerical Methods for Partial Differential Equations 24 (2008) 1416–1429.

[128] N.T. Shawagfeh, Analytical approximate solutions for nonlinear fractional differential equations, Applied Mathematics and Computation 131 (2002) 517–529.

[129] J. Singh, D. Kumar, Sushila, Homotopy perturbation Sumudu transform method for nonlinear equations, Advances in Applied Mathematics and Mechanics 4 (2011) 165–175.

[130] G.C. Wu, D. Baleanu, Variational iteration method for the Burgers' flow with fractional derivatives—new Lagrange multipliers, Applied Mathematical Modelling 37 (2013) 6183–6190.

[131] D.Q. Zeng, Y.M. Qin, The Laplace–Adomian–Pade technique for the seepage flows with the Riemann–Liouville derivatives, Communications in Fractional Calculus 3 (2012) 26–29.

[132] M. Gomo, G. Steyl, G.J. Van Tonder, Investigation of groundwater recharge and stable isotopic characteristics of an alluvial channel, Hydrology Current Research 1 (12) (2012) 7.

[133] Z. Gribovszki, J. Szilágyi, P. Kalicz, Diurnal fluctuations in shallow groundwater levels and streamflow rates and their interpretation – a review, Journal of Hydrology 385 (2010) 371–383.

[134] R.W. Healy, Estimating Groundwater Recharge, Cambridge University Press, 2010.

[135] F. Kreith, S.A. Berger, S.W. Churchill, J.P. Tullis, A.T. McDonald, A. Kumar, et al., Mechanical engineering handbook, in: F. Kreith (Ed.), Fluid Mechanic, CRC Press LLC, Boca Raton, United States, 1999, pp. 1–208.

[136] N. Kresic, Hydrogeology and Groundwater Modelling, 2nd ed., CRC Press, 2007.

[137] J. Nyende, G. van Tonder, D. Vermeulen, Application of isotopes and recharge analysis in investigating surface water and groundwater in fractured aquifer under influence of climate variability, Journal of Earth Science and Climatic Change 4 (4) (2013) 148.

[138] I. Simmers, Estimation of Natural Groundwater Recharge, D. Reidel Publishing Company, Antalya, Turkey, 1987.

[139] T. Sochi, Single-Phase Flow of Non-Newtonian Fluids in Porous Media, University College London, London, England, 2009.

[140] A. Toure, B. Diekkrüger, A. Mariko, Impact of climate change on groundwater resources in the Klela Basin, Southern Mali, Hydrology 3 (2016) 1–17.

[141] W.H. van der Model, J. Martínez Beltrán, W.J. Ochs, in: Guidelines and Computer Programs for the Planning and Design of Land Drainage Systems, 2007, 228 pp., Nature.

[142] H.F. Yeh, C.H. Lee, K.C. Hsu, P.H. Chaunge, GIS for the assessment of the groundwater recharge potential zone, Environmental Geology 58 (2008) 185–195.

[143] Badr Saad T. Alkahtani, Chua's circuit model with Atangana–Baleanu derivative with fractional order, Chaos, Solitons and Fractals 89 (2016) 547–551.

[144] A. Atangana, B. Dumitru, New fractional derivatives with non-local and non-singular kernel: theory and application to heat transfer model, Thermal Science 20 (2) (2016) 763–769.

[145] A. Atangana, I. Koca, Chaos in a simple nonlinear system with Atangana–Baleanu derivatives with fractional order, Chaos, Solitons & Fractals 89 (2016) 447–454.

[146] Abdon Atangana, Necdet Bildik, The use of fractional order derivative to predict the groundwater flow, Mathematical Problems in Engineering 2013 (2013) 1–9.

[147] D. Benson, S. Wheatcraft, M. Meerschaert, Application of a fractional advection–dispersion equation, Water Resources Research 36 (2000) 1403–1412.

[148] D. Benson, R. Schumer, S. Wheatcraft, M. Meerschaert, Fractional dispersion, Lévy motion, and the MADE tracer tests, Transport Porous Media 42 (2001) 211–240.

[149] F. Ali, M. Saqib, I. Khan, N.A. Sheikh, Application of Caputo–Fabrizio derivatives to MHD free convection flow of generalized Walters'-B fluid model, The European Physical Journal Plus 131 (377) (2016) 1–10.

[150] B.S.T. Alkahtani, A. Atangana, Modeling the potential energy field caused by mass density distribution with Eton approach, Open Physics 14 (2016) 106–113.

[151] R.T. Alqahtani, Atangana–Baleanu derivative with fractional order applied to the model of groundwater within an unconfined aquifer, Journal of Nonlinear Science and its Applications 9 (2016) 3647–3654.

[152] A. Atangana, B.S.T. Alkahtani, Analysis of the Keller–Segel model with a fractional derivative without singular kernel, Entropy 17 (2015) 4439–4453.

[153] A. Atangana, R.T. Alqahtani, Numerical approximation of the space–time Caputo–Fabrizio fractional derivative and application to groundwater pollution equation, Advances in Difference Equations 156 (2016) 1–13.

[154] A. Atangana, I. Koca, Chaos in a simple nonlinear system with Atangana–Baleanu derivatives with fractional order, Chaos, Solitons and Fractals 89 (2016) 447–454.

[155] D.S. Cherkauer, Quantifying groundwater recharge at multiple scales using PRMS and GIS, Groundwater 42 (1) (2004) 97–110.

[156] M. Ciesielski, J. Leszczynski, Numerical Simulations of Anomalous Diffusion, Institute of Mathematics and Computer Science, Technical University of Czestochowa, Czestochowa, Poland, 2003.

[157] M. Du, Z. Wang, H. Hu, Measuring memory with the order of fractional derivative, Scientific Reports 3 (2013) 3431.

[158] H.G. Gebreyohannes, Groundwater Recharge Modelling A case study in the Central Veluwe, The Netherlands, Enschede, International Institute for Geo-Information Science and Earth Observation, The Netherlands, 2008.

[159] J.F. Gómez-Aguilar, R.F. Escobar-Jiménez, M.G. López-López, V.M. Alvarado-Martínez, Atangana–Baleanu fractional derivative applied to electromagnetic waves in dielectric media, Journal of Electromagnetic Waves and Applications (2016) 1–17.

[160] H. Nasrolahpour, Time fractional formalism: classical and quantum phenomena, Prespacetime Journal 3 (1) (2015) 99–108.

[161] I. Simmers, Estimation of Natural Groundwater Recharge, D. Reidel Publishing Company, Antalya, Turkey, 1987.

[162] V.E. Tarasov, Review of some promising fractional physical models, International Journal of Modern Physics B 27 (9) (2013) 1–38.

[163] A. Toure, B. Diekkruger, A. Mariko, Impact of climate change on groundwater resources in the Klela Basin, Southern Mali, Hydrology 3 (17) (2016) 1–17.

[164] B. Wang, et al., Estimating groundwater recharge in hebei plain China under varying land use practices using tritium and bromide tracers, Journal of Hydrology 356 (2008) 209–222.

[165] A. Atangana, J.F. Botha, Generalized groundwater flow equation using the concept of variable order derivative, Boundary Value Problems 2013 (2013) 53.

[166] A. Atangana, A. Kilicman, A possible generalization of acoustic wave equation using the concept of perturbed derivative order, Mathematical Problems in Engineering 2013 (2013) 696597.

[167] C.-M. Chen, F. Liu, I. Turner, V. Anh, A Fourier method for the fractional diffusion equation describing sub-diffusion, Journal of Computational Physics 227 (2007) 886–897.

[168] Y.Q. Chen, K.L. Moore, Discretization schemes for fractional-order differentiators and integrators, IEEE Transactions on Circuits and Systems I 49 (2002) 363–367.

[169] J. Crank, P. Nicolson, A practical method for numerical evaluation of solutions of partial differential equations of the heat-conduction type, Mathematical Proceedings of the Cambridge Philosophical Society 43 (1947) 50–67.

[170] W.H. Deng, Numerical algorithm for the time fractional Fokker–Planck equation, Journal of Computational Physics 227 (2007) 1510–1522.

[171] E. Hanert, On the numerical solution of space–time fractional diffusion models, Computers and Fluids 46 (2011) 33–39.

[172] M.S. Hantush, Analysis of data from pumping tests in leaky aquifers, Transactions, American Geophysical Union 37 (1956) 702–714.

[173] M.S. Hantush, C.E. Jacob, Non-steady radial flow in an infinite leaky aquifer, Transactions, American Geophysical Union 36 (1955) 95–100.

[174] G.P. Kruseman, N.A. de Ridder, Analysis and Evaluation of Pumping Test Data, 2nd edition, International Institute for Land Reclamation and Improvement, Wageningen, The Netherlands, 1990.

[175] C.P. Li, A. Chen, J.J. Ye, Numerical approaches to fractional calculus and fractional ordinary differential equation, Journal of Computational Physics 230 (2011) 3352–3368.

[176] C.P. Li, C.X. Tao, On the fractional Adams method, Computers and Mathematics with Applications 58 (2009) 1573–1588.

[177] R. Lin, F. Liu, V. Anh, I. Turner, Stability and convergence of a new explicit finite-difference approximation for the variable-order nonlinear fractional diffusion equation, Applied Mathematics and Computation 212 (2009) 435–445.

[178] M.M. Meerschaert, C. Tadjeran, Finite difference approximations for fractional advection–dispersion flow equations, Journal of Computational and Applied Mathematics 172 (2004) 65–77.

[179] I. Podlubny, A. Chechkin, T. Skovranek, Y.Q. Chen, B.M. Vinagre Jara, Matrix approach to discrete fractional calculus II. Partial fractional differential equations, Journal of Computational Physics 228 (2009) 3137–3153.

[180] B. Ross, S. Samko, Fractional integration operator of variable order in the Hölder spaces $H^{\lambda(x)}$, International Journal of Mathematics and Mathematical Sciences 18 (1995) 777–788.

[181] H. Sun, W. Chen, C. Li, Y. Chen, Finite difference schemes for variable-order time fractional diffusion equation, International Journal of Bifurcation and Chaos in Applied Sciences and Engineering 22 (2012) 1250085.

[182] C. Tadjeran, M.M. Meerschaert, H.-P. Scheffler, A second-order accurate numerical approximation for the fractional diffusion equation, Journal of Computational Physics 213 (2006) 205–213.

[183] C.V. Theis, The relation between the lowering of the piezometric surface and the rate and duration of discharge of a well using ground-water storage, Transactions, American Geophysical Union 16 (1935) 519–524.

[184] S. Umarov, S. Steinberg, Variable order differential equations with piecewise constant order-function and diffusion with changing modes, Journal of Analysis and Its Applications 28 (2009) 431–450.

[185] S.B. Yuste, L. Acedo, An explicit finite difference method and a new von Neumann-type stability analysis for fractional diffusion equations, SIAM Journal on Numerical Analysis 42 (2005) 1862–1874.

[186] Y. Zhang, A finite difference method for fractional partial differential equation, Applied Mathematics and Computation 215 (2009) 524–529.

[187] A. Atangana, A. Kilicman, On the generalized mass transport equation to the concept of variable fractional derivative, Mathematical Problems in Engineering 2014 (2014) 542809.

[188] M.M. Meerschaert, H.P. Scheffler, Limit Distributions for Sums of Independent Random Vectors: Heavy Tails in Theory and Practice, John Wiley and Sons, New York, 2001, pp. 45–46.

[189] S.P. Neuman, D.M. Tartakovsky, Perspective on theories of non-Fickian transport in heterogeneous media, Advances in Water Resources 32 (5) (2009) 670–680.

[190] R. Herrmann, Fractional Calculus, 2nd edition, World Scientific Publishing Co. Pte. Ltd., Hackensack, NJ, 2014.

[191] R. Hilfer, Threefold introduction to fractional derivatives, in: R. Klages, et al. (Eds.), Anomalous Transport, Wiley-VCH Verlag GmbH & Co. KGaA, 2008, pp. 17–77.

[192] K.B. Oldham, J. Spanier, The Fractional Calculus, Academic Press, New York–London, 1974.

[193] S. Samko, A.A. Kilbas, O. Marichev, Fractional Integrals and Derivatives, Taylor & Francis, 1993.

[194] I. Area, D.J. Djida, J. Losada, J.J. Nieto, On fractional orthonormal polynomials of a discrete variable, Discrete Dynamics in Nature and Society 2015 (2015) 141325.

[195] Luis Caffarelli, Juan Luis Vazquez, Nonlinear porous medium flow with fractional potential pressure, arXiv:1001.0410 [math.AP], 2010.

[196] A. Atangana, B. Dumitru, New fractional derivatives with non-local and non-singular kernel: theory and application to heat transfer model, Thermal Science 20 (2) (2016) 763–769.

[197] Mark Allen, A non-divergence parabolic problem with a fractional time derivative, preprint available on arxiv.org, 2015.

[198] Luis Silvestre, Hölder estimates for advection fractional-diffusion equations, arXiv:1009.5723 [math.AP], 2011.

[199] Luis Silvestre, On the differentiability of the solution to the Hamilton–Jacobi equation with critical fractional diffusion, Advances in Mathematics 226 (2) (2011) 2020–2039, MR 2737806 (2011m:35408).

[200] Héctor Chang Lara, Gonzalo Dávila, Regularity for solutions of non local parabolic equations, Calculus of Variations and Partial Differential Equations 49 (1–2) (2014) 139–172, MR 3148110.

[201] L. Caffarelli, C.H. Chan, A. Vasseur, Regularity theory for parabolic nonlinear integral operators, Journal of the American Mathematical Society 24 (3) (2011) 849–869.

[202] R. Zacher, Weak solutions of abstract evolutionary integro-differential equations in Hilbert spaces, Funkcialaj Ekvacioj 52 (1) (2009) 1–18.

[203] M. Kassmann, M. Rang, R.W. Schwab, Integro-differential equations with nonlinear directional dependence, Indiana University Mathematics Journal 63 (5) (2014) 1467–1498, http://dx.doi.org/10.1512/iumj.2014.63.5394.

[204] Mark Allen, A nondivergence parabolic problem with a fractional time derivative, preprint on arXiv, 2016.

[205] Mark Allen, Luis Caffarelli, Alexis Vasseur, A parabolic problem with a fractional time derivative, Archive for Rational Mechanics and Analysis 221 (2) (2016) 603–630, MR 3488533.

[206] Mark Allen, Luis Caffarelli, Alexis Vasseur, Porous medium flow with both a fractional potential pressure and fractional time derivative, Chinese Annals of Mathematics 38 (1) (January 2017) 45–82.

[207] J. Losada, J.J. Nieto, Properties of a new fractional derivative without singular kernel, Progress in Fractional Differentiation and Applications 1 (2) (2015) 87–92.

[208] Luis A. Caffarelli, Xavier Cabré, Fully Nonlinear Elliptic Equations, American Mathematical Society Colloquium Publications, vol. 43, American Mathematical Society, Providence, RI, 1995, MR 1351007 (96h:35046).

[209] D. del Castillo-Negrete, B.A. Carreras, V.E. Lynch, Fractional diffusion in plasma turbulence, Physics of Plasmas 11 (8) (2004) 3854–3864.

[210] Nondiffusive transport in plasma turbulene: a fractional diffusion approach, Physical Review Letters (2005) 1693–1696, arXiv:physics/0403039.

[211] Kai Diethelm, An application-oriented exposition using differential operators of Caputo type, in: The Analysis of Fractional Differential Equations, in: Lecture Notes in Mathematics, vol. 2004, Springer-Verlag, Berlin, 2010, MR 2680847 (2011j:34005).

[212] David Gilbarg, Neil S. Trudinger, Elliptic Partial Differential Equations of Second Order, Classics in Mathematics, Springer-Verlag, Berlin, 2001, reprint of the 1998 edition, MR 1814364.

[213] N.V. Krylov, M.V. Safonov, A property of the solutions of parabolic equations with measurable coefficients, Izvestiâ Akademii Nauk SSSR. Seriâ Matematičeskaâ 44 (1) (1980) 161–175, 239, MR 563790.

[214] Héctor Chang Lara, Gonzalo Dávila, Regularity for solutions of non local parabolic equations, Calculus of Variations and Partial Differential Equations 49 (1–2) (2014) 139–172, MR 3148110.

[215] Ralf Metzler, Joseph Klafter, The random walk's guide to anomalous diffusion: a fractional dynamics approach, Physics Reports 339 (1) (2000) 77, MR 1809268 (2001k:82082).

[216] Luis Silvestre, On the differentiability of the solution to the Hamilton–Jacobi equation with critical fractional diffusion, Advances in Mathematics 226 (2) (2011) 2020–2039, MR 2737806 (2011m:35408).

[217] Lihe Wang, On the regularity theory of fully nonlinear parabolic equations. I, Communications on Pure and Applied Mathematics 45 (1) (1992) 27–76, MR 1135923.

[218] G.M. Zaslavsky, Chaos, fractional kinetics, and anomalous transport, Physics Reports 371 (6) (2002) 461–580, MR 1937584, 2003i:70030.

[219] P.D. Pacome, A. Atangana, P.D. Vermeulen, Modelling groundwater fractal flow with fractional differentiation via Mittag–Leffler law, The European Physical Journal Plus 132 (2017) 165, http://dx.doi.org/10.1140/epjp/i2017-11434-8.

[220] M. Jessica Spannenbrug, Develoment of a Groundwater Recharge Model, Master thesis, 2017.

[221] Abdon Atangana, Juan Jose Nieto, Numerical solution for the model of RLC circuit via the fractional derivative without singular kernel, Advances in Mechanical Engineering 7 (10) (2015) 1–7, http://dx.doi.org/10.1177/1687814015613758.

[222] J.A. Barker, A generalized radial flow model for hydraulic tests in fractured rock, Water Resources Research 24 (10) (1988) 1796–1804.

[223] Ben O'Shaughnessy, Itamar Procaccia, Analytical solutions for diffusion on fractal objects, Physical Review Letters 54 (1985) 455.

[224] Shlomo Havlin, Daniel Ben-Avraham, Diffusion in disordered media, Advances in Physics 51 (1) (2002) 187–292.

[225] J. Chang, Y.C. Yortsos, Pressure transient analysis of fractal reservoir, SPE Formation Evaluation 5 (1) (1990) 31–38.

[226] J. Chang, Y.C. Yortsos, A note on pressure-transient analysis of fractal reservoirs, SPE Advanced Technology Series 1 (2) (1993) 170–171, http://dx.doi.org/10.2118/25296-PA.

[227] Ning Hu, Xi-Qiao Feng, Shao-Yun Fu, Cheng Yan, Guang-Ping Zhang, Jihua Gou, Mechanical behavior of nanostructured materials, Journal of Nanomaterials 2015 (2015) 829367, http://dx.doi.org/10.1155/2015/829367.

Index

Printed in the United States
By Bookmasters